Manufacturing Scheduling Systems

Jose M. Framinan · Rainer Leisten
Rubén Ruiz García

Manufacturing Scheduling Systems

An Integrated View on Models, Methods and Tools

Springer

Jose M. Framinan
Departamento de Organización
Industrial y Gestión de Empresas
Universidad de Sevilla Escuela
Superior de Ingenieros
Isla de la Cartuja
Seville
Spain

Rubén Ruiz García
Grupo de Sistemas de Optimización
Aplicada, Instituto Tecnológico
de Informática
Universitat Politècnica de València
Valencia
Spain

Rainer Leisten
Fakultät für Ingenieurwissenschaften
Allgemeine Betriebswirtschaftslehre und
Operations Management
Universität Duisburg-Essen
Duisburg
Germany

ISBN 978-1-4471-6271-1 ISBN 978-1-4471-6272-8 (eBook)
DOI 10.1007/978-1-4471-6272-8
Springer London Heidelberg New York Dordrecht

Library of Congress Control Number: 2013956332

Printed on acid-free paper

Springer is part of Springer Science+Business Media (www.springer.com)

To Carmen, Julio, and Daniel

Jose M. Framinan

To Karin, Steven, and Annemarie

Rainer Leisten

To Gema and Diego

Rubén Ruiz García

Preface

Obtaining economic and reliable schedules constitutes the core of excellence in customer service and of efficiency in manufacturing operations. While few areas have been more productive in decision sciences than scheduling, only a fraction of the vast amount of scheduling research has been translated into practice. On one hand, the inherent complexity of most scheduling problems has led to an excessively fragmented and specialised field in which different facets and issues are treated in an independent manner as goals themselves, lacking a unifying view of the scheduling problem. Aside, scientific brilliance and mathematical elegance has been sometimes prevalent to practical, hands-on approaches, producing specific procedures of narrow application. Finally, the nearly non-existence of research on implementation issues in scheduling as well as the scarcity of case studies has not helped in translating valuable research findings into industry.

In this book we attempt to overcome these limitations by presenting an integrated framework for formulating, developing and implementing systems that can be effectively used for supporting manufacturing scheduling decisions. To do so, we have intentionally avoided the traditional taxonomy style of many scheduling approaches and focused on the explanation and integration of the different issues into a (hopefully) coherent framework for modeling, solving and implementing scheduling decisions. Therefore, the book does not contain a detailed description of specific models and/or procedures, but rather will present a general overview of the scheduling field in order to provide the tools to conduct an up-to-date scheduling research and practice. To compensate this, we include a generous compilation and discussion of related literature for further reading, so the reader can easily track the main contributions of particular interest.

The book is intended for an ample audience who wish to get an overview of actual scheduling topics. It may constitute a starting point for researchers in the scheduling field. Also, post-graduate students with a focus on operations management can be attracted by the content of the book. Practitioners in the field would feel comfortable with many of the models and systems presented in the book as well, as they can easily see in them real-life cases. As a result, the book is not intended to be read sequentially, but is designed to support different itineraries, providing comprehensive introductory chapters/sections before detailed explanation of specific topics.

There are few but important prerequisites for this book. Knowledge of the most basic concepts of production management is required, although an effort is done in Chap. 2 to place scheduling into context. Maths will appear profusely in some parts of the book, mostly in the chapters devoted to scheduling models.

Organisation of the book

The book is structured into five parts. In the first part, we introduce the main definitions and notation, and present the framework that we will use throughout the book. In this framework, a scheduling system is defined as a collection of models (representations of scheduling problems), methods (procedures to obtain efficient solutions out of scheduling models), and tools (software devices to embed models and procedures in order to support the scheduling decision problem), together with the human elements operating the system. Models, procedures, and tools will constitute the next three parts of the book. A final part on scheduling systems is devoted to assemble all these elements together in a roadmap to guide the development of a scheduling system. The structure of the book is summarised in Fig. 1.

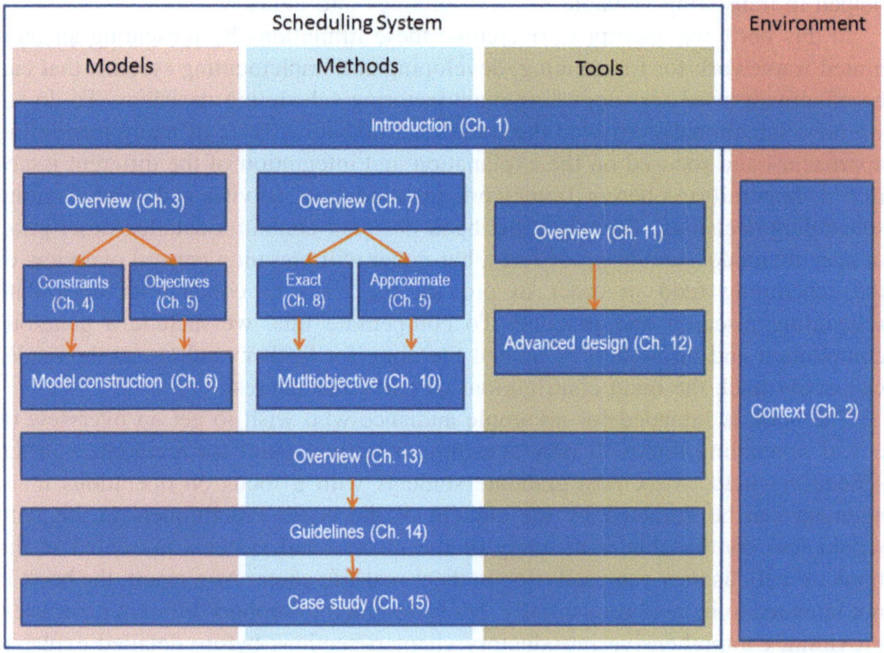

Fig. 1 Structure of the book

As mentioned before, the book does not have to be read comprehensively and sequentially. Manufacturing scheduling constitutes a topic of interest for different professional (including managers and top/intermediate staff in charge of operations management) and academic profiles (including production/manufacturing engineers, computer scientists, and different bachelors in operations research and management science). When used as a textbook, the instructors would have a clear idea about the content, sequence, and depth in which the different topics contained in the book should be learnt. Therefore, the subsequent lines are mainly intended as a rough advice whenever such supervision is not present.

We believe that a rather basic knowledge of the scheduling field can be gained by reading the introductory chapters of each part, i.e. Chaps. 1, 3, 7, 11, and 13. A more classical (although, in our opinion, also more theoretical) itinerary for the scheduling field would exclude Chap. 11 to Chap. 15, and would stress issues related to modeling and solution procedures. Modeling would be reinforced by adding Chaps. 4, 5, and 6. These three chapters are heavily interdependent and therefore should be read together, although readers with some scheduling background may skip some parts of Chaps. 4 and 5, and focus instead on the discussions in Chap. 6. A rather comprehensive understanding of solution procedures is provided in Chaps. 8 and 9, whereas multiobjective scheduling issues are treated in Chap. 10. Since multiobjective scheduling is a hot topic with practical and theoretical interest, we would favor its inclusion even at a basic level course. Nevertheless, most of it could be skipped if the reader adopts a more classical/basic view of the topic.

We think that Chaps. 11 and 12 would be of particular value for an audience interested in the design and implementation of scheduling tools. Chapter 11 would serve to grasp a general view on the topic, whereas Chap. 12 is intended for readers with some background and experience on business information systems, and –in our opinion– should be spared for basic scheduling courses, unless they are heavily geared toward information systems/computer science.

We hope that users and consultants of manufacturing scheduling systems would find interesting Chaps. 13 to 14, as the practical issues treated there are not usually subject of classical scheduling books. Despite the verbosity and apparent simplicity of some of the ideas contained in these chapters, they are mainly intended for an experienced reader who may better grasp the inherent complexity of the deployment of manufacturing scheduling systems and their integration with the human schedulers. We are not sure that many of the ideas contained there could be fully appreciated at a more basic level, although we think that at least a glimpse of those in Chap. 13 should be given in order to avoid an excessively technical approach to the field, which in our opinion is and has been a rather common flaw in scheduling teaching and research.

Finally, Chap. 15 may be used in many different itineraries, ranging from a basic level in which modeling and solution procedures issues are seen into practice, to a more advanced discussion case in which success factors, shortcomings, lost opportunities, and learnt lessons can be debated.

Acknowledgements

In the process of writing this book, we are indebted to many people. The authors benefited from the careful reading of earlier versions of the book done by Manu Dios, Paz Perez-Gonzalez and Natalia Prischepov. The authors are also indebted to the many members and colleagues of their institutions at Seville, Duisburg, and Valencia. The collaborations, help and working atmosphere have proven to be key when writing this book. They also had to endure many of our absences when writing "the book" and we are really thankful for their support and patience.

January 2014

Jose M. Framinan
Rainer Leisten
Rubén Ruiz García

Contents

Part II Scheduling Models

Part I
Introduction to Manufacturing Scheduling

This part of the book consists of Chaps. 1 and 2. In the first chapter, we move from a general definition of scheduling as the allocation of resources to tasks along time to a definition of manufacturing scheduling, where we identify its main features and context in which it takes place. Then we define a scheduling system as a collection of methods, models and tools that support scheduling-related decisions in a company. These three elements are heavily influenced by the organisational context in which the scheduling-related decisions have to be taken. All these three parts: models, methods and tools together with the organisation (humans) involved in the process constitute a framework that we will use throughout the book.

The second chapter is devoted to describing the context in which manufacturing scheduling decisions are taken. Quite often, specialized scheduling literature fails to acknowledge that scheduling decisions are usually integrated into a more general (explicit or implicit) decision system, which influences (and is influenced by) scheduling decisions. In addition, the aforementioned decision system is not restricted to an individual company, but extended along a whole supply network composed of independent enterprises. Therefore, we review the concept of operations management and its place in obtaining competitive advantages for a firm. After exemplifying the complexity and interdependency of different decisions that constitute the management of operations, we introduce the different approaches to handle this complexity, including hierarchical planning and decentralised approaches (agent-based systems). In the centralised approach, we identify the main decision blocks, place scheduling decisions and discuss its relation with other decision blocks. The difference between scheduling and rescheduling (which can be interpreted as the frontier among planning and control) is discussed. Next, we discuss the interrelation of scheduling decisions with other manufacturing-related decisions, such as simultaneous lot-sizing and scheduling, and emphasize the need of alignment of these decisions with the goals in the company. Finally, we also present the context of the supply chain, discuss the concept of supply chain management and present how this affects scheduling decisions.

Chapter 1
Overview of Manufacturing Scheduling

1.1 Introduction

Broadly speaking, scheduling deals with the allocation of resources to tasks along time. In general, such definition may encompass a huge number of real-life applications, such as allocating nurses to the shifts of their hospitals, aircraft scheduling in airports, processing units in a computing environment, etc. In this book, we will focus onto *production scheduling* or *manufacturing scheduling*, i.e. assigning the various resources in the company to the manufacturing of a range of products that are requested by the customers. Since, as we will see later, certain level of abstraction is required to manage these resources, many of the models and methods that will be shown here may be applied outside the manufacturing scope. Nevertheless, manufacturing scheduling presents a number of features that cannot be, in general, extrapolated to other decision sciences. Most notably, scheduling is not carried out in an isolated manner, as on one hand it uses as input the results of a previous plan that determined the set of products to be manufactured by the company along with the real (or expected) demand of these products, among other issues. On the other hand, the usually highly variable shop floor conditions may complicate or impede the fulfilment of a schedule, no matter how detailed it is, so there are further decisions to be taken, possibly implying the review or modification of the existing schedule. As a consequence, manufacturing scheduling cannot be considered an isolated process, but integrated into a set of managerial decisions collectively known as production management (or operations management).

More specifically, in this chapter we:

- briefly summarize the context in which manufacturing decisions take place (Sect. 1.2),
- review a number of features in scheduling common to most manufacturing companies (Sect. 1.3),
- use the previous understanding of the nature of scheduling decisions to elaborate the concept of a *manufacturing scheduling system* (Sect. 1.4),

J. M. Framinan et al., *Manufacturing Scheduling Systems*,
DOI: 10.1007/978-1-4471-6272-8_1, © Springer-Verlag London 2014

- briefly introduce the basic concepts regarding manufacturing scheduling (Sect. 1.5), and
- provide some hints for further analysis on the issues discussed in this chapter (Sect. 1.6)

1.2 Manufacturing Scheduling into Context

Production management is a managerial process in which a large number of decisions are taken over time in order to ensure the delivery of goods with the maximum quality, minimum cost, and minimum lead time. These decisions differ—among other issues—on their impact in the company, their scope, and on the time period to take them. They range from strategic, high-impact, long-range decisions such as deciding whether certain plant will manufacture or not a new product, to short-term, low-level, small-impact decisions such as which product is next to be manufactured in certain machine in the shop floor.

Given the different nature and timing of these decisions, and the different decision makers involved, production management has been traditionally decomposed by adopting a hierarchical structure in which the different decision problems are successively solved. At the top of this structure, there are decisions related to the *strategic design of the manufacturing network*, and include the selection of providers, the location of the plant(s) to manufacture the products, the design of the plant layout and the structure of the distribution network, among others. Broadly speaking, these decisions—called *strategic decisions*—serve to establish, on an aggregate level, the capacity of the manufacturing and distribution resources, including the co-operation with the providers. Obviously, these decisions are strategic the sense that they have an enormous impact on the bottom line of the companies and that the economic and physical resources involved make them unlikely or even impossible to be reviewed within less than a horizon usually measured in years.

Once the manufacturing and distribution structure has been designed, there are a number of decisions related to how to make the best usage of this structure. These decisions are again usually decomposed into medium-term decisions (*tactical decisions*), and short-term decisions (*operating decisions*). This distinction may seem a bit arbitrary—and indeed it may blur for some manufacturing scenarios—but it is usually related to the complexity of the manufacturing process and to the granularity and quality of the data required to make the decisions. When planning the usage of the manufacturing resources for the next year (something that must be done to decide on the workforce required, to plan the contracts with the providers, etc.), the key information driving the whole manufacturing process (i.e. the orders from the customers) is not known and it can only be guessed or estimated (even for the best of the cases) via historical or casual data. However, it is clear that the quality of this estimation decreases with the level of detail, as it is easier to estimate the yearly sales of a model of a car than to estimate the weekly sales of the cars of a certain model including left-side steering wheel. Therefore, it is usual to make an *aggregated production plan* in which the input for this decision is not the estimation of the actual products to be sold each day, but an estimation of the families or groups of products with similar usage of resources and/or similar behaviour with respect to the demand that are to be sold over a period of time ranging from weeks to months. As a consequence, this aggregated production plan will serve to estimate the (aggregated) capacity to be used, the amount of raw materials to be purchased, etc., but not as a detailed indication of the production orders that have to be released to the shop floor. This exhaustive plan is usually left for a later stage in which a detailed status of the shop floor and on the firm orders to be processed is known, and this is precisely the mission of *manufacturing scheduling*: matching the jobs (tasks) to be executed against the resources in the company. The result of this process is, ideally, a *schedule* where it is specified which job should enter each resource, and when.

As plans rarely occur as estimated (this may be an overstatement for many environments, but certainly not for manufacturing scheduling), modifications to clear the way for unforeseen events (machine breakdown, urgent orders, cancelation of existing orders,...) have to be carried out continuously. Therefore production plans may have to be *rescheduled*. Note that this rescheduling process may actually consist of making a full detailed new plan (schedule) capturing the new situation on the shop floor, but for us, the distinction between scheduling and rescheduling would lie basically in the aims sought in the decision: While scheduling decisions would aim at improving certain performance measure in the shop floor, rescheduling aims to minimize some measure of the disturbance of the current status on the shop floor, usually upon the arrival of unexpected events. An implicit assumption behind is that—at least under certain circumstances—it may be more interesting to modify an existing schedule to accommodate new events occurring in the shop rather than generating a new schedule (almost) from scratch in order to fully take into account these new events. While under ideal circumstances the later would always render a better solution, there are costs associated to this so-called *nervousness* resulting from the scheduling process.

Note that sometimes different definitions are employed to distinguish between scheduling and rescheduling, as some authors use the terms predictive scheduling and reactive scheduling instead. A predictive schedule is an instruction to the shop floor, causing the shop to execute events in the sequence and time indicated in the schedule, while the process of modifying the predictive schedule in the face of executional disruptions is denoted as reactive scheduling. Note that the definition of reactive scheduling not only encompasses our definition of rescheduling, but also generating a completely new schedule that is followed until the next disruption occurs. These aspects will be discussed in detail in Sect. 2.3.1.

It is clear that the scope and the time period to take these decisions greatly influences the quality of the data used to support the decision process. Broadly speaking, long-term decisions usually involve speculations on the evolution of the market and on the internal resources in the company, and thus are by nature less amenable to be formalised using quantitative techniques that require a high quality in the input data to make a sensible decision. Besides, there is little sense in formalising the decision process itself, since the changing nature of the business and the long intervals among two consecutive decisions of this nature makes them to be unique in practice. Here, human judgement based on expertise and on a few key estimations is usually what is required to make this type of decisions. On the contrary, the repetitive nature of short-term decisions make them suitable to be formalised and possibly encapsulated into decision models, which can produce remarkable results as one can reasonably trust on data related to the current state of the shop floor, the average processing times of the machines, and on the (relative) stability of the list of orders to be completed to be able to capture the decision process into a quantitative model that can possibly be solved as an optimisation problem. Indeed, these two aspects (repetitive nature of the decision process, and the high volume of data required to support it) make quantitative, procedural techniques to be competitive as compared to the human subjective expertise.

It is also interesting to note that scheduling does not only serve to generate detailed plans to be executed, but can be also used to provide a source of information for different business functions in the company. For instance, it may serve to provide a capacity-check for the Materials Requirement Planning (MRP) system, or to help quoting an incoming order from customers, or to decide about its acceptance/rejection. Again, we can see the close interrelation of scheduling to other business functions, and quite often the distinction between scheduling and these functions would blur. For purely practical purposes, in this book we would exclude from scheduling those decisions for which the level of detail considered with respect to time, resources, products, or customers is not enough to provide a specific plan to be immediately translated into actions on the shop floor.

Given the high level of interrelation among different production management decisions (including scheduling) shown in the previous paragraphs, one may be tempted to try establishing a prioritisation of these decisions, in order to more efficiently allocate the usually scarce financial and human resources to those decisions with higher impact in the bottom line of the company. However, we believe that discussing whether the scheduling function is more important than quality management, supply chain coordination, or product design (to name a few examples of operations management decisions) is a useless discussion for any practical purpose. It is obvious that without an overall good operations management, better scheduling will not help to improve the results of a company no matter how well the scheduling function is performed, and the same can be said about the different decisions. Nevertheless, it should be clear that scheduling is a core business decision for industrial companies. As such, its importance greatly depends on the strategic importance of manufacturing decisions to gain competitive advantage. For instance, it may not be as critical for companies trying to sell a small range of new products avidly accepted by the market with little opposition from competitors, but it would help companies manufacturing a large range of products in obtaining a competitive advantage as compared to their competitors. Indeed, the quality of scheduling is often cited as a key factor if the products and the factory are well-designed.

Since scheduling is closely related to the allocation of resources, it is also obvious that its importance would increase with the possibility of having a conflict for the usage of these resources. Along the history of manufacturing, this conflict for the usage of resources (which is not desirable) has been traditionally avoided by designing shop floors manufacturing a limited range of products with similar technological requirements. Unfortunately, the trend towards mass customisation, shortening of product life-cycles, time-based competition, and the globalisation (with the corresponding increase of competition) has made this solution unacceptable for most companies. Another option has been what is described as 'management by buzzword' (Hopp and Spearman 2008) meaning the vast stream of three letter acronyms encompassing the ultimate solutions for production management. While these apparent business panaceas (one every few years, just another proof that they are not panaceas at all) may contain valuable insights, mostly translate into approaches neglecting or ignoring the realities of manufacturing scheduling. Given the wrong answer provided by the two previous options, we believe that there is no other way than addressing scheduling in full detail, with its complexities and their enormous opportunities for operational improvement.

1.3 Features of Scheduling

Scheduling decisions may be very different from one company to another, but all they share a number of common features. In this section, we will discuss these features in order to obtain certain degree of abstraction that may serve us to formulate a generic framework useful for different companies. These features are:

- They are complex decisions, as they involve developing detailed plans for assigning tasks to resources over time. Although this may vary greatly from one company to another, there is a universal trend on increasing the sophistication of the products and on their customisation, which in turn affects to the complexity of the manufacturing process.
- Scheduling decisions are short time interval decisions to be taken over and over. The average lifetime of a schedule is very short, and indeed many authors refer to a continuous scheduling decision process. Here we mean that what is repeated is the decision process, which is different from stating that the outcome of a single decision is put into practice rhythmically again (cyclic scheduling).
- Despite being a short-time decision, scheduling is relevant for companies' bottom line, as it determines the lead times and the cost of the products, which on the long run affects the service level of the company as well as its ability to compete both in manufacturing costs and on delivery times.
- As a decision process at the core of the operations of a manufacturing company, the constraints and objectives affecting scheduling are extremely company-specific. The nature and usage of the resources in a plant producing chemical commodities has little in common with manufacturing ball bearers, or assembly of highly customised electronic devices.
- Finally, scheduling decisions are—as we already discussed—relatively structured decisions, at least as compared to other decision problems within the company. Its

operational nature makes scheduling to require fairly well-defined data, constraints and objectives.

In any case, we should recall that there is no way to escape from making (implicit or explicit) scheduling decisions in manufacturing companies. Perhaps ironically, this fact has led sometimes to the underestimation of the potential advantages of efficient scheduling, since, as some plant managers put, 'we already do scheduling'. The truth is that, most often, such scheduling is accomplished in a non-structured, expertise-dependent manner, with partial information both about the inputs of the process and with respect to the desired output. The consequences of this approach are obvious: underutilisation of resources, longer lead times than required, higher inventories, ... to name a few. This book is about how to conduct this process in a systematic, efficient manner.

1.4 A Framework for Manufacturing Scheduling System

As we have seen in the previous sections, scheduling consists of a dynamic and adaptive process of iterative decision making related to assigning resources in the plant to manufacture a range of products over time. This process is carried out by humans (schedulers or decision makers) which may be assisted by a software tool that helps them to model some decision problems arising during this process, and to find a solution for these by employing methods (or solution procedures) on these models. Four remarks are inherent to this definition:

- First, scheduling is a *process*. It is not about formulating and solving a specific problem once, or even repeatedly solving the same problem again and again. This dynamic view contrasts with the static approach found in most scheduling literature, and emphasizes the fact that manufacturing scheduling is a business process that should be managed over time. As such business process, it is a dynamic, evolving situation for which 'cooked', monolithically designed procedures seldom work.
- Second, the scheduling process is directed by a human, a team, or an organisation. The vision of fully automated systems that perform and execute scheduling decisions without human guidance does not reflect the current reality in most shop floors. Therefore, there is a need to understand and incorporate the human element into this decision process.
- Third, the relatively structured nature of the scheduling process *and* its inherent complexity leads naturally to the convenience of formulating the decision problem in terms of mathematical models to better assess the decisions to be taken by the schedulers, as well as the corresponding methods to solve these models. However, the formulation of such models and the implementation of the solutions provided by the corresponding methods is far from being trivial/immediate and should be driven by the human element mentioned in the previous item.

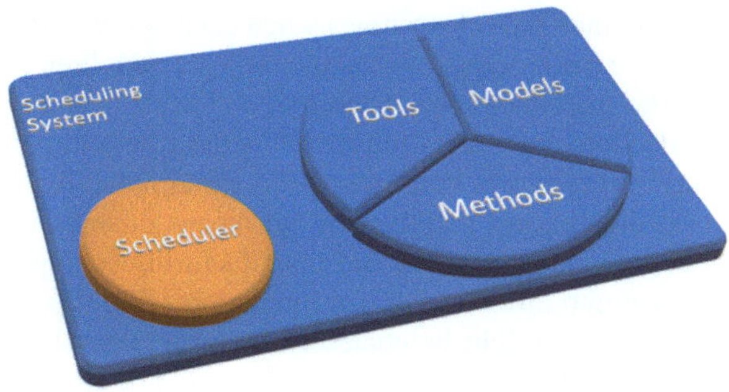

Fig. 1.1 Framework for manufacturing scheduling systems

- Fourth, although strictly speaking there is no need of a tool to support the schedulers in their decisions, in most cases it is completely unpractical to conduct them without any help from a software tool. Simply the sheer volume of data referring to resources and tasks found in most shop floors would be sufficient to justify the need of such tool. Whether this software is a spreadsheet, an 'off-the-shelf' software, or an advanced planning and scheduling tool is a question that will be discussed later in this book, but from now on we will assume that some type of support exists.

From the above paragraph, it should be clear that, in order to effectively perform scheduling decisions, a system—i.e., a collection of pieces—is needed. The combination of *models*, *methods* and *tools*—together with the human schedulers involved in the process—is what we call a *scheduling system*. Scheduling models, solution procedures and software tools all play an important part alone, but they have to be put together to make up scheduling decisions. Therefore, a holistic view—in contrast to the more extended technical or optimization-related view—is needed to have a complete picture of scheduling.

This book is thus adapted to the above view, which is summarised in Fig. 1.1. In the next chapters, the different elements of the scheduling system will be described one by one. To start this discussion, a number of basic concepts are defined in the next section.

1.5 Basic Concepts

By formalising the ideas given in the previous sections, we can define manufacturing scheduling as the decision-making process consisting of assigning the set of operations/tasks required to manufacture a set of products to the existing resources in the shop floor, as well as the time periods to initiate these operations/tasks. A *schedule* is defined as a specific assignment of these operations/tasks to the resources on a time-scale. A schedule thus fully determines when each operation/task should start and it is regraded as the main output of manufacturing scheduling.

The products to be manufactured, or the units of work that the manufacturing activity can be divided into are usually referred to in the scheduling literature simply as *jobs*. Although in most of our book we will assume naturally the existence of such units of work, the concept of job is by no means unambiguous as it might be dependent on the company, context, and manufacturing setting. Clients' orders might be broken down into different Fabrication Orders (FO). Several FOs coming from different client orders might be coalesced to form larger FOs that might be more economical to produce. Later, these large FOs could be again broken down, at a shop level, into smaller, but not necessarily and immediately consecutive, sublots into what is called a lot streaming process. We will discuss these aspects in Sect. 3.4.

The operations/tasks required by the jobs are provided by different productive resources available in the shop floor. These could range from cheap tooling and fixtures to human resources and expensive machinery. Again, we will assume that it is possible to identify units of resources that can perform these operations/tasks, and, accordingly to the scheduling literature, we will refer to them simply as *machines*. As with the jobs, the concept of 'machine' is an abstraction (or mental model) of real-world operations and is therefore subject to the purposes of the modeling activity, i.e.: for estimating the throughput of a factory, the whole factory with all its resources can be considered as a 'machine', while for providing a daily timetabling of the work that the operator of a drilling machine in this factory must perform, the previous assumption of machine is clearly insufficient.

In addition, there are many technical/production/economic characteristics of the jobs, of the machines, and of the operation of a job on a machine that must be taken into account in the scheduling process. Although these will be formalised and extended in Sect. 3.1, let us state perhaps the most obvious:

- There exists machine specialisation: Not all machines can perform every operation required by each job, as manufacturing resources are usually capable to perform just one or at most few operations. Even in the case that such a thing is possible (e.g. by polyvalent workers or general-purpose machines), it is extremely doubtful that their efficiency is operation-independent, so at least preferences/exclusions would have to be defined.
- The order in which the different operations have to be performed on a job is usually fixed given the specific technology of the product to be manufactured, i.e. there exists a predefined order (or orders) of operations for each job that must be strictly followed. This order is labelled *job processing route* and will be further formalised in Sect. 3.2.2.
- In general, different jobs have different requirements with respect to the technical, production management, and economic aspects. From a technical viewpoint, even if two jobs have the same processing route, some characteristics of the operation (such as e.g. processing time, or set-up times) may be different. From a production management/economic viewpoint, technically identical jobs may have different

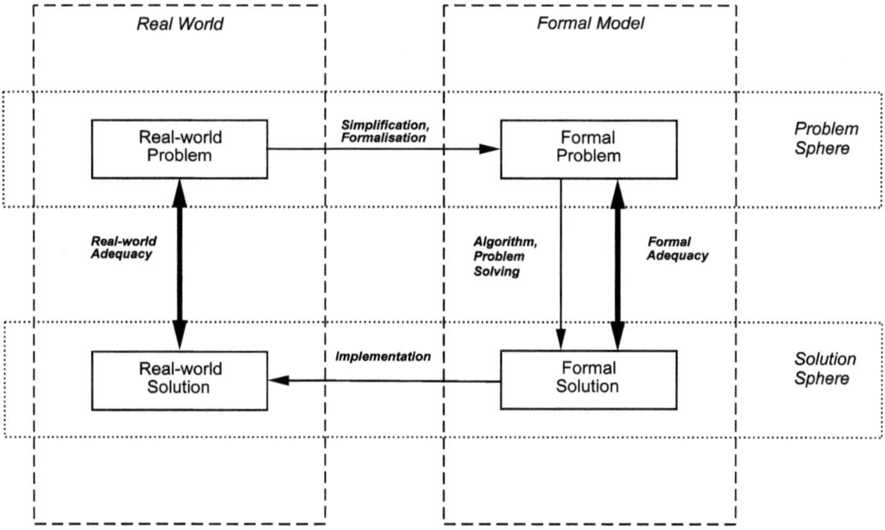

Fig. 1.2 Real world and formal model, problem sphere and solution sphere for decision-making

due dates committed to the customer(s), or are to be sold with different benefits, or may require raw materials which are not served by the provider at the same time.

- All/some of operations/tasks may allow the so-called *preemption*, which refers to the possibility (for all/some jobs) of interrupting the operation once it has started. If such a possibility exists, the execution of the operation can be stopped and resumed later with/without any/some penalty (these cases are denoted as resumable or semi-resumable operations, respectively). There are manufacturing examples of both preemptive and non preemptive operations. Heat/Chemical treatments are basically non preemptive operations, as they correspond to gradually changing the physical or chemical properties of the material under treatment, a process that, in general, cannot be reversed to the original properties of the material. On the other hand, some assembly operations are preemptive, at least from a technological viewpoint, although there may be managerial/economic considerations deterring or impeding the preemption of such operations.

As stated, these are just a fraction of the jobs/machines/operations characteristics that should be taking into accoung in the scheduling decision process. As a consequence, not every schedule that we may think of can take into account all these characteristics. In such case, this schedule is denoted as *unfeasible schedule*, as it cannot be directly translated into commands for the shop floor. Obviously, the primary goal in a scheduling process is to find at least one feasible schedule. Provided that more than one feasible schedule exists, the goal is to select (at least) one feasible schedule among the set of feasible schedules available.

Therefore, manufacturing scheduling can be viewed as a *decision-making problem*, i.e. a process to select a course of action (i.e. a schedule) among several alternative options. To do so, one or more metrics (or *criteria*) have to be defined to characterise the desired features of the schedules. Once the criterion is established, it is (at least technically) possible to compare different schedules, and to select one yielding the best value (maximum or minimum) for this criterion among all feasible schedules (*optimal schedule*).

In manufacturing, metrics are usually classified into the categories of costs (sometimes profit), time, quality and flexibility. As it will be discussed in Chap. 5, in scheduling decision-making these categories are associated with criteria related to the *completion time* of the jobs. Loosely speaking, the completion time of a job in a specific schedule is the time spent by the job in the shop floor—according to the schedule—before all its corresponding operations have been performed. Note that the completion times of the jobs in a schedule serve to know whether executing this schedule would result in delays on the delivery of the jobs with respect their committed due date, or the average time required, according to this schedule, to process the whole set of jobs.

If the criterion employed is a non-decreasing function of the completion times of the jobs, then this criterion is named *regular*. One obvious example of a regular criterion is the average completion time of the jobs, while the average delay of the jobs with respect to their due dates is an example of a non-regular criterion. Finally, note that more than one criteria can be of interest, and that there might be conflicts between them. An exhaustive discussion and classification of the different criteria employed in scheduling is given in Chap. 5, while multicriteria scheduling is discussed in Chap. 10.

In order to highlight the main issues related to scheduling decision-making, we map in Fig. 1.2 the well-known flow of a decision-making process, at least for quantifiable decision problems: Starting at the real-world problem, for which a decision is required, a formal model is derived by means of simplification and formalisation. By using some formal procedure or *algorithm* (see Sect. 7.2.1 for a more precise definition of algorithm), this formal model is solved, i.e. a formal solution for this problem is derived. Afterwards, within implementation this formal solution is transferred to a real-world solution which is implemented for solving the real-world decision problem.

The scheme presented in Fig. 1.2 is rather basic and superficial, and it can be complexified to almost every level. We just refer to this simplification because it depicts the basic issues related to manufacturing scheduling decision-making. We will go back to this figure in Chap. 6 (when discussing the transfer of real-world relations into a formal, mathematical decision model and the techniques to set up a respective formal model), and in Sect. 7.1 (when addressing the adequacy of the formal solutions found by the algorithms to the real-world problem).

Since it has been established that obtaining (one or several) schedules is the main goal of the manufacturing scheduling decision problem, in the next sections we will present ways to represent schedules (Sect. 1.5.1), their main types (Sect. 1.5.2), and the number of existing schedules (Sect. 1.5.3).

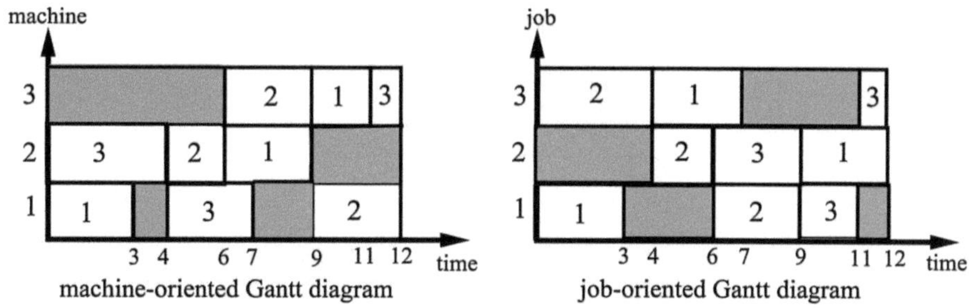

Fig. 1.3 Machine-oriented and job-oriented Gantt diagrams (Domschke et al. 1997)

1.5.1 Representation of Schedules

A schedule is traditionally represented by the so-called Gantt diagrams. While always horizontally mapping the time scale, with respect to the vertical axis Gantt diagrams can be machine-oriented or job-oriented. Figure 1.3 shows these two types of diagrams for the same schedule. The length of each box represents the length of the operation in the job or machine, depending on the diagram.

Note that the first of the two diagrams indicates the sequence in which jobs are processed on each machine, i.e. the first machine processes job 1, then job 3, and finally job 2. In the following, we will denote this sequence of jobs using a vector, i.e. the previous job sequence is denoted as $(1, 3, 2)$. This is an important notation since, as we will discuss later, there are occasions in which a full specific schedule can be obtained just by giving these sequences of jobs (or sequences for short) on each machine.

A modern and perhaps more adequate way of representing schedules is the disjunctive graph presentation which will be shortly described below. A disjunctive graph is a graph $G = (V, C \cup D)$ where V denotes a set of vertices corresponding to the operations of jobs. C is a set of conjunctive arcs connecting every two consecutive operations of the same job. The undirected edges of set D connect mutually unordered operations which require the same machine for their execution. Each vertex is weighted with the processing time of the operation at its start. The edges have weight 0. In the disjunctive graph representation, obtaining one schedule is equivalent to select one arc in each disjunction, i.e. to turn each undirected disjunctive edge into a directed conjunctive arc. By fixing directions of all disjunctive edges, the execution order of all conflicting operations requiring the same machine is determined and a complete schedule is obtained. Moreover, the resulting graph has to be acyclic.

An example of a disjunctive graph is given in Fig. 1.4 In this example there are three machines $M = \{M_1, M_2, M_3\}$ and a set of three jobs $J = \{J_1, J_2, J_3\}$ which are described by the following sequences of operations: J_1: $T_1 \rightarrow T_2 \rightarrow T_3$, J_2 : $T_4 \rightarrow T_5$, $J_3 : T_6 \rightarrow T_7 \rightarrow T_8$. For any operation T_i the required machine $M(T_i)$ and the processing time p_i are assumed to be known.

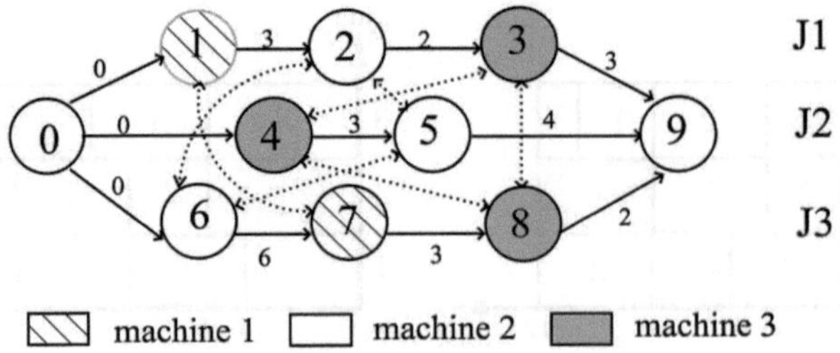

Fig. 1.4 An example for a disjunctive graph (Błażewicz et al. 2000)

1.5.2 Classes of Schedules

For non-preemptive scheduling problems, several classes (or types) of schedules can be identified. As we will see, these classes are used

1. to reduce the number of possible schedules which reduces the search effort for good or optimal solutions as compared with a continuum of solutions, and
2. to identify properties of the solution space and/or good or optimal solutions to adjust solution procedures accordingly.

The classes of schedules that can be considered are:

- *Non-delay schedules:* A schedule having no unforced idle time on any machine is called a non-delay schedule, i.e. a feasible schedule belongs to the class of non-delay schedules if and only if no operation is kept waiting while the machine assigned to this operation is available to process it.
- *Active schedules:* A schedule is said to be active if no operation can start earlier without delaying at least one other operation in this schedule. Obviously, non-delay schedules are active whereas the opposite does not hold.
  It can be shown that for scheduling decision problems with a regular criterion there is an active schedule which is optimal. Note that this does not hold for non-regular criteria.
- *Semi-active schedule:* A feasible schedule is called semi-active if no operation can be completed earlier without changing the order of processing on any one of the machines. Obviously, an active schedule has to be semi-active. However, the reverse is not necessarily true. Intuitively, semi-active schedules are characterised by the fact that every operation is left-shifted as much as possible, i.e. given the current sequence of operations on every machine, every operation is started as soon as possible.

The classes of schedules are further detailed in Fig. 1.5. Note that, in order to take advantage of restricting the search of the optimal solution to a certain class of schedules, it has to be checked in advance whether all optimal solutions (or at least one of them) fulfil the respective property. In this case, the effort to find an adequate

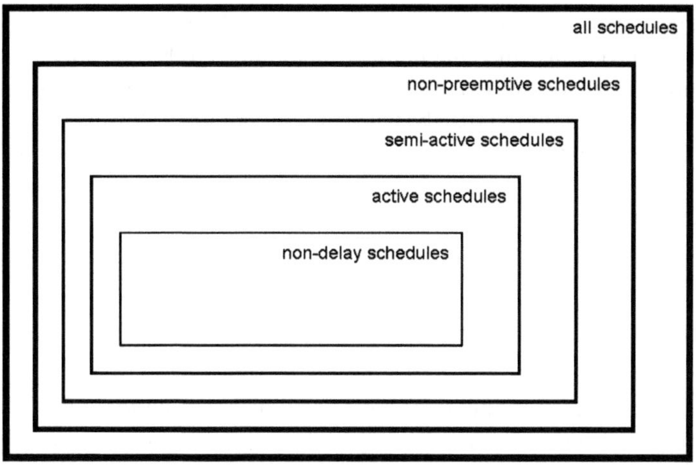

Fig. 1.5 Classes of non-preemptive schedules for job shop problems (similar to Pinedo 2012; T'Kindt and Billaut 2006)

(possibly optimal) solution might be reduced significantly as compared with the evaluation of all possible solutions.

1.5.3 Combinatorial Nature of Manufacturing Scheduling

Although the expressions schedule and sequence are often used synonymously in the literature, one schedule uniquely defines a sequence but not vice versa. Often, especially in a multi-stage context, (at least non-critical) operations will have some time buffer within a given sequence. These time buffers will open ranges for the assignment of these operations to the time-scale. Operations might be assigned 'left-shifted' to their earliest possible processing period (as is often supposed in scheduling procedures), 'right-shifted' to their latest possible processing period (as might make sense from a profitability point of view, i.e. to minimise tied-up capital), or somewhere in between (e.g. if externally specified due dates should be reached and neither early nor late completion of operations or jobs, respectively, is intended).

In general, the number of feasible schedules for most manufacturing scheduling problems will be infinite while the number of sequences will be countable and even finite, however possibly rather large. I.e. although the number of sequences on every single machine and in combination over all machines will be finite, a single operation might often be shifted continuously on the time scale, at least within some time limits. However, if we can prove, as holds in many scheduling settings, that an optimal solution will always be active (at least in some non-preemptive scheduling problems, see Sect. 1.5.2), no matter whether left-shifted or right-shifted, the number of solutions to be considered for the determination of an optimal solution will be finite.

As a consequence, manufacturing scheduling problems are usually classified as combinatorial problems since the number of solutions to be considered in the optimisation process can be limited to a finite number. Therefore, procedures to generate feasible and optimal solutions can be basically enumerative. However, as we shall see in Sect. 2.3.2.1, (complete) enumeration will usually require too much of time and of other computational resources. Consequently, other maybe approximate methods will be more appropriate.

1.6 Conclusions and Further Readings

During the last decades, manufacturing scheduling has been identified to be one of the most important decisions in planning and control of industrial plant operations, both in science and in practice. Scheduling is seen as a decision-making process that is used in manufacturing industries as well as in service industries.

In this chapter we have introduced the conceptual framework which we will use throughout this book. We have given a definition of manufacturing scheduling and (even if only in a coarse manner) delineated which aspects are within the scope of the book, and which are not. A framework for studying and connecting the different topics involved in manufacturing scheduling (models, methods and tools under human guidance) has been presented, and the basic scheduling concepts have been introduced.

The remainder of the book follows the structure summarised in Fig. 1.1. First, in Chap. 2 we finalize Part I, which has been devoted to analysing the manufacturing context in which scheduling takes place. We then discuss the issues related to the different parts in a scheduling system as depicted in Fig. 1.1 one by one: Part II of the book (Chaps. 3, 4, 5 and 6) is devoted to scheduling models, Part III (7, 8, 9, and 10) to scheduling methods, and Part IV (Chaps. 11 and 12) to scheduling tools. Finally, Part V (Chaps. 13, 14, and 15) relinks all these elements into the concept of a scheduling system by discussing its relations with its organisational (i.e. human) environment, and the process of developing such systems. This part concludes by presenting a real implementation in which the main concepts and issues discussed in the book are presented.

As scheduling is an area of utmost importance in Operations Research and Management Science, there is a wealth of books dealing with specific aspects of scheduling. We omit those books of wider scope (e.g. in Operations Management) that partially cover scheduling decisions in one or two chapters. Among these, we would like to cite the book by Hopp and Spearman (Hopp and Spearman 2008), although there are other excellent books on Operations Management. Interestingly, the reader would detect that, in some of these books, the authors do not seem to feel very comfortable placing scheduling in the context of production management and the tone and content of the chapter(s) devoted to scheduling are quite different than that of the rest of the chapters.

Regarding the scheduling topic itself, the book acknowledged to be the first book devoted to scheduling is Conway et al. (1967). This book, and most pre-dated to 1995, seems today a bit outdated as the scheduling field has almost unrecognisably changed since, and the contents would be entirely different even if the topics addressed were identical. Regarding introductory readings on the topic of manufacturing scheduling, there are many excellent books, although most of them lean towards one or two of the components of our framework. Among the principal books, here we mention those by Baker (1974); French (1982); Błazewicz et al. (2002); Brucker (2007); Pinedo (2012); Baker and Trietsch (2009) or Pinedo (2009). Less known but interesting references are those of Domschke et al. (1997).

The terms predictive scheduling and reactive scheduling are discussed—although in a slightly different manner than here—in Aytug et al. (2005). The last reference is also a key source regarding the use of scheduling for providing information for different business functions in the company. Specifically, the uses of scheduling for due date quoting and acceptance/rejection decisions are discussed in Framinan and Leisten (2010). An excellent book describing human performance in scheduling is MacCarthy and Wilson (2001). Brucker (2007) is devoted mostly to models and methods for scheduling problems, not confined to the manufacturing field.

References

Aytug, H., Lawley, M. A., McKay, K., Mohan, S., and Uzsoy, R. (2005). Executing production schedules in the face of uncertainties: A review and some future directions. European Journal of Operational Research, 161(1):86–110.

Baker, K. R. (1974). Introduction to Sequencing and Scheduling. John Wiley & Sons, New York.

Baker, K. R. and Trietsch, D. (2009). Principles of Sequencing and Scheduling. Wiley, New York.

Błazewicz, J., Ecker, K. H., Pesch, E., Schmidt, G., and Węglarz, J. (2002). Scheduling Computer and Manufacturing Processes. Springer-Verlag, Berlin, second edition.

Błażewicz, J., Pesch, E., and Sterna, M. (2000). The disjunctive graph machine representation of the job shop scheduling problem. European Journal of Operational Research, 127(2):317–331.

Brucker, P. (2007). Scheduling Algorithms. Springer, New York, fifth edition.

Conway, R. W., Maxwell, W. L., and Miller, L. W. (1967). Theory of Scheduling. Dover Publications, New York. Unabridged publication from the 1967 original edition published by Addison-Wesley.

Corsten, H. (2009). Produktionswirtschaft - Einfuehrung in das industrielle Produktionsmanagement. Oldenbourg, Muenchen. 12th, revised and upgraded edition.

Domschke, W., Scholl, A., and Voss, S. (1997). Produktionsplanung: Ablauforganisatorische Aspekte. Springer, Berlin. 2th, revised and upgraded edition.

Framinan, J. and Leisten, R. (2010). Available-to-promise (ATP) systems: A classification and framework for analysis. International Journal of Production Research, 48(11):3079–3103.

French, S. (1982). Sequencing and Scheduling: An Introduction to the Mathematics of the Job-Shop. Ellis Horwood Limited, Chichester.

Hopp, W. J. and Spearman, M. L. (2008). Factory Physics. McGraw-Hill, New York.

MacCarthy, B. L. and Wilson, J. R., editors (2001). Human performance in Planning and Scheduling. Taylor & Francis.

Pinedo, M. (2009). Planning and Scheduling in Manufacturing and Services. Springer, New York, second edition.

Pinedo, M. L. (2012). Scheduling: Theory, Algorithms, and Systems. Springer, New York, fourth edition.

T'Kindt, V. and Billaut, J.-C. (2006). Multicriteria Scheduling: Theory, Models and Algorithms. Springer, New York, second edition.

Chapter 2
The Context of Manufacturing Scheduling

2.1 Introduction

In the previous chapter, a unified view of manufacturing scheduling has been given. Furthermore, we have also outlined that manufacturing scheduling is not carried out in an isolated manner, but as part of a set of interrelated decisions—collectively known as production management—dealing with efficiently ensuring the delivery of goods provided by the company. Therefore, before analysing manufacturing scheduling decisions in detail, it is worth looking at the context in which manufacturing scheduling is embedded: i.e the company's production management and in the supply network to which the company belongs.

More specifically, in this chapter we

- present a framework for scheduling decisions (Sect. 2.2),
- analyse manufacturing scheduling as an isolated decision process and study the main aspects influencing these decisions (Sect. 2.3) and
- investigate the relationship of manufacturing scheduling with other decisions in the company and its supply chain network (Sect. 2.4).

2.2 A Framework of Manufacturing Scheduling

The basic definitions for manufacturing scheduling have been given in the previous chapter. In this section, we now discuss the decision framework of manufacturing scheduling, i.e. the main aspects influencing the scheduling decision process. We adopt a systems analysis approach: A system can be characterised by its objects and their relations and whether it is closed or open (i.e. whether it does not include relations to the off-system environment or it does). Manufacturing scheduling can be interpreted as a closed system when it is studied as an isolated decision process, and also as an open system when the focus is set on its relationship to other decision processes.

J. M. Framinan et al., *Manufacturing Scheduling Systems*,

DOI: 10.1007/978-1-4471-6272-8_2,

This dual (open and closed) view will be extended in the next sections. First (Sect. 2.3), we study manufacturing scheduling as an isolated decision process and will discuss the main aspects influencing these decisions. Next (Sect. 2.4), the relationship of scheduling with other decisions is investigated. These relationships can be classified into three types:

- Relationship between manufacturing scheduling and the rest of decisions in production management (Sect. 2.4.1).
- Relationship between manufacturing scheduling and the rest of decisions (apart from those in production management) within the company (Sect. 2.4.2).
- Relationship between manufacturing scheduling and the rest of decisions in the supply network in which the company is integrated (Sect. 2.4.3).

2.3 The Closed View of Scheduling Problems and Decisions

Manufacturing scheduling problems and decisions can be classified along different dimensions or features. Perhaps the most employed dimensions are the following:

1. Deterministic vs. stochastic. In the case where all the characteristics of the decision problem (processing time of each operation, release date, due date, etc.) are well known and single-valued, the decision problem is of deterministic type. In contrast, it is considered of stochastic type if at least one of these characteristics is not known deterministically but only with its probability distribution.
2. Static vs. dynamic. If all the relevant data of the decision problem are known in advance, i.e. at the point in time where the planning procedure starts, then the problem is classified as static. If it is taken into account that intermediately (i.e. before the realisation of the schedule is finished) appearing new information is included, the scheduling problem is called dynamic.
3. Organisation of the decision process (centralised vs. decentralised). In a centralised manufacturing environment all jobs/operations are scheduled to all machines by a central planning authority. A main task to make such systems run effectively is the provision of this institution with all relevant information. In decentralised systems, in its extreme version, one planning institution exists separately on each stage. These authorities, of course, have to interact with each other, given the reason for agent-based scheduling approaches both in practice as well as item of the scientific consideration. Carefully setting up the respective information system, therefore, is one main task when setting up decentralised scheduling systems.
4. Scope of the solution to be generated (complete vs. partial schedule). If the outcome resulting from the scheduling process is not complete, it is called a partial schedule. The incompleteness of a partial schedule can refer to many items. For example, not every machine and/or type of operation might be considered, the time horizon regarded might be limited deliberately, etc. In a well-organised

planning and control system, also for manufacturing scheduling, the effects of such a limited scope have to be anticipated and examined carefully.

We can use the above structure to classify scheduling decisions. This classification will be employed in the next subsections when we analyse the main aspects influencing manufacturing scheduling, namely time, complexity, variability and flexibility.

2.3.1 Time

As already discussed, time is the main reference point of scheduling and scheduling decisions. Apart from duration, starting and ending specification of operations and their assignment to the timescale by scheduling decisions, there are several further time-related issues in scheduling, which are graphically described in Fig. 2.1. These are:

- Point in time of scheduling. The point in time of a scheduling decision itself, its level of information with respect to the real-world scheduling problem (e.g. the degree of determination of operations duration or even operations' determination in total) or, more precisely, its time lag to the first operation to be physically executed may play an important role concerning the quality of the scheduling decision.
- Scheduling horizon. Scheduling horizon refers to the time horizon of operations or jobs considered in the scheduling process. The scheduling horizon is strongly related to the selection of jobs included in the scheduling decision problem.
- Scheduling frequency refers to the frequency and the trigger of updating of a schedule, by including new information arrived meanwhile into the new scheduling decision can be controlled. This updating process might be rhythmic or event-based, e.g. if a new job arrives or a machine becomes (un-)available.
- Realisation horizon. Closely connected to scheduling horizon and scheduling rhythm is the determination of the realisation horizon, i.e. those part of the scheduling horizon which is implemented, usually at least until updating by the next (possibly re-)scheduling decision takes place. This happens because scheduling is often performed on the well-known *rolling horizon* basis, i.e. a schedule is constructed for the scheduling horizon, but it is executed only for the first few periods of this horizon (i.e. its realisation horizon) before it is updated by a new plan using updated input data.

Additionally, it should be mentioned that the grid of the time scale might influence the quality of the scheduling decision as well. However, usually, the time grid assumed for the process, may it be minutes, hours, shifts or days, will be 'adequately' detailed, i.e. will not contribute to significant infeasibilities and/or suboptimalities.

Closely related to time aspects in manufacturing scheduling is the consideration whether a new or updated schedule is generated from scratch as a greenfield solution or whether at least parts of it are predetermined by remaining operations from the

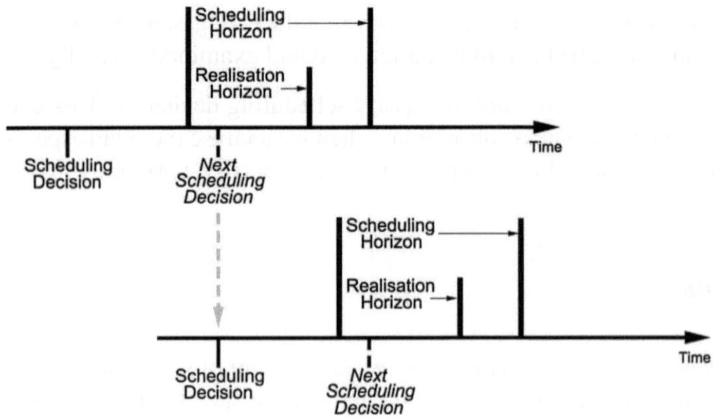

Fig. 2.1 Time aspects of manufacturing scheduling

former schedule. The latter case is, obviously, more realistic and sometimes expressed by something like 'the tails of the former schedule are the heads of the new, updated one'. However, many scheduling approaches in science as well as their software implementation in practice disregard this predetermination of the beginning of a schedule (at least on some machines) and generate the new schedule from scratch (usually including the remaining operations of the former schedule) into the planning process of the new iteration, not as fixed in the beginning of the new schedule.

In order to avoid nervousness consequences already mentioned in Sect. 1.2, plan revisions in this update might be intended to be avoided or at least be minimised from a planning point of view, and only new jobs should be appended to the old plan. However, the relevance of this update has to be interpreted at least two-fold: On one hand, jobs (or at least operations) already started should be more or less fixed and their schedule might be, therefore, fixed as input data for the new schedule. The time period covering such jobs/operations is denoted as *frozen period*. On the other hand, if previously scheduled jobs have not yet been started or physically prepared in the shop floor, respectively, their preliminary schedule might be known only to the planner while shop floor people possibly got no information on these scheduled but not yet released jobs/operations up to the replanning point in time. Therefore, these latter jobs/operations may be very well considered anew in the replanning procedure of the rolling horizon approach—without causing much additional nervousness to the shop floor level.

A further approach to reduce nervousness in the shop floor is to expand the frozen periods to more than the plan revision point in time, so plan revisions enter the shop floor more 'smoothly'. However, this advantage with respect to reduced nervousness is balanced by a reduced reaction speed of the shop floor.

Another way of updating schedules is *event-oriented scheduling*. Predefined types of events, such as newly arriving and/or cancelled jobs, machine failures or unavailability of material, tools and/or staff, processing times different from the planned ones (a very common situation), etc. give reason for updating the schedule. Sometimes, not a single modification but a certain amount (number) of modifications is awaited before a plan update is initiated. The modifications might be just added to the old schedule or a more or less complete replanning is executed while avoiding nervousness, if possible, as mentioned above with respect to the rolling horizon approach. It should be pointed out that event-oriented scheduling requires more or less permanent observation (and availability) of all relevant actual data.

As already mentioned, the scheduling process itself might be *static* or *dynamic*. Static in this context does not necessarily mean the absence of intermediate updating, e.g. in a rolling horizon context. And dynamic does not include only simply more than one planning period. Recall that static scheduling is separated from dynamic scheduling by the fact that static scheduling derive a plan/schedule always from scratch without taking explicitly into account that till the end of the execution of the plan changes might and will appear. The explicit provision of resources (time, capacity, etc.) for these changes and or scheduling with the explicit inclusion of these possible changes (e.g. by applying scenario techniques and/or techniques of flexible planning) characterise dynamic scheduling.

Within the dynamic context above mentioned, at least four different types of scheduling approaches can be identified:

- Completely reactive scheduling. This type of scheduling is defined more or less as online scheduling. Assuming the poor quality of centrally determined schedules as a drawback, it is supposed that quick and local reaction yields better scheduling results. Therefore, simple scheduling approaches (known as scheduling policies and described in Sect. 7.4), are locally applied on the shop floor level, possibly in real time to determine the next job to be processed on a specific machine. This approach is flexible and is able to include most recent information from the very detailed shop floor level. However, because of its myopic perspective, the overall performance of this approach is questionable since it usually includes an inherent lack of coordination.
- Predictive-reactive scheduling. The most common dynamic scheduling approach in manufacturing systems is called predictive-reactive scheduling. It combines static-deterministic scheduling as a first step with event-based reactive, possibly real-time updating of schedules in the meantime as second step until a new static-deterministic schedule is generated. Apart from its time coordination effect between the two steps, this approach is also intended to coordinate the broader view of the overall system to be scheduled in the first step with the myopic perspective of rescheduling in the second. However, as often confirmed empirically, large tolerances in the second step often significantly reduce the overall performance of the system since the finally implemented schedule may deviate significantly from the original schedule—and the deviations are a result of local, myopic considerations exclusively.
- Robust predictive-reactive scheduling. Taking into account the problems of predictive-reactive scheduling, additional robustness considerations should be included, both with respect to the robustness of the first step's overall schedule

and the limitation of the second step's tolerances. The focus of robust predictive-reactive scheduling is building predictive-reactive schedules which minimise or at least limit the effects of disruption on the performance measure values of the realised schedule.

- Robust proactive scheduling. Robust proactive scheduling intends to immunise the predictive schedule in advance against possible stochastic disruptions. Usually, specific time buffers are included to cope with this kind of uncertainty and to make the predictive schedule robust. Determination of the predictability measures is the main difficulty of this approach.

Predictive scheduling is an integral part of manufacturing systems planning. Predictive schedules are often produced in advance in order to direct production operations and to support other planning activities. Since most manufacturing systems operate in dynamic environments subject to various real-time events, this may render the predictive optimal schedule neither feasible nor optimal. Therefore, dynamic scheduling is of great importance for the successful implementation of approaches to real-world scheduling systems.

Rolling horizon approaches and reactive scheduling can be interpreted as manifestations of *rescheduling* approaches. These rescheduling approaches may be triggered externally, e.g. by newly arriving orders or by reaching the next planning point in time. They might be induced, however, also internally, i.e. by every significant deviation from the planned, predictive schedule derived earlier. The reasons for these deviations might be multifaceted as is well known. They can refer to resource availability, stochasticity of processing times, etc. The decision whether a deviation is significant (i.e. it initiates an update of the schedule) or not, depends on the assessment of the decision maker(s). In case of large deviations and subsequent adjustment requirements, the existing schedule can be updated, repaired or rejected/stopped. In the latter case, a completely new schedule will be determined.

The 'core' rescheduling approaches include updating and repairing of a schedule. Rescheduling in this perception is said to be 'the process of updating an existing production schedule in response to disruptions or other changes. This includes the arrival of new jobs, machine failures, and machine repairs.' (Vieira et al. 2003). Reasons for rescheduling may also be due date changes, job cancellations, delay in the arrival or shortage of materials, change in job priorities, rework or quality problems, overestimation or underestimation of processing times, operator absenteeism, etc. Reaction to rescheduling requirements does not only include the modification of the schedule itself but may also refer to the modification of its preconditions, including the alternatives of overtime, assignment of utility persons, in-process subcontracting, process change or re-routing, machine substitution, etc. These arrangements represent means to augment the manufacturing capacity as basis of a (re-)scheduling decision while opposite arrangements will reduce it.

Vieira et al. (2003) present a framework for rescheduling which includes rescheduling environments, rescheduling strategies (including rescheduling policies) and rescheduling methods as categories of classification. This framework is outlined in Fig. 2.2.

Rescheduling environments				
Static (finite set of jobs)		Dynamic (infinite set of jobs)		
Deterministic (all information given)	Stochastic (some information uncertain)	No arrival variability (cyclic production)	Arrival variability (flow shop)	Process flow variability (job shop)

Rescheduling strategies				
Dynamic (no schedule)		Predictive-reactive (generate and update)		
Dispatching rules	Control-theoretic	Rescheduling policies		
		Periodic	Event-driven	Hybrid

Rescheduling methods				
Schedule generation		Schedule repair		
Nominal schedules	Robust schedules	Right-shift rescheduling	Partial rescheduling	Complete regeneration

Fig. 2.2 Rescheduling framework (Vieira et al. 2003)

Since the rescheduling environment can be either static or dynamic (see classification above), in the static environment, a given set of jobs/orders/operations have to be scheduled while in a dynamic setting, jobs may arrive after the scheduling decision has been taken. Therefore, in principle, the set of jobs might be interpreted as infinite.

The static situation can be further separated into a deterministic and a stochastic setting. Note that a deterministic setting, either needs no rescheduling because it represents the 'true' situation, or an update is performed from scratch. In contrast, a dynamic environment might be (a) deterministic with respect to job arrivals (i.e. production cycles which are repeated again and again), (b) stochastic with respect to jobs' arrival times but with same flow of every job through the system (flow shop setting), (c) or dynamic with respect to arrival times and varying with respect to jobs' routes through the system (job shop setting).

Rescheduling strategies for simply adapting the current schedule or modifying it myopically then may use (mostly myopic) dispatching rules (see Sect. 7.4) or some control-theoretic approach which basically tries to keep the system in balance and initiates adaptations if the system under consideration runs the risk of getting imbalanced. Otherwise, the combination of predictive and reactive schemes which have been described above can be applied, periodically, event-driven or in some way hybrid. Finally, the possible rescheduling methods are obvious from Fig. 2.2, namely separating schedule generation from schedule repair approaches.

2.3.2 Complexity

Complexity, in many contexts, often is used as a rather nebulous expression or concept. References dealing with the definition and the discussion of the expression 'complexity' are as numerous as confusing. In systems theory, complexity of a system is given by the number and diversity of objects in the system as well as by the number and the diversity of the relations between the objects.

In parallel to Fig. 1.2 in Sect 1.5, complexity in manufacturing scheduling can be restricted either to the formal sphere on one hand, or can also be addressed more generally and then be extended to the real-world sphere. These two types of complexity are discussed in the next subsections.

2.3.2.1 Computational Complexity

Complexity on the formal sphere refers to the complexity of the formal problem and the corresponding algorithms for solving the formal problem. We refer to this type of complexity to as *computational complexity* and it will be treated in detail in Chap. 7. Roughly speaking, this concept of complexity serves to separate algorithms whose computational effort can be bounded by some polynomial function of some characteristics of the formal problem from those where such a polynomial limit has not been derived yet and where it will probably never be found. This separation is justified by the fact that polynomial approaches can be usually accomplished within reasonable computational effort while non-polynomial algorithms cannot.

In particular, algorithms performing an explicit or implicit enumeration of each feasible solution are of this non-polynomial type. Since, as we have discussed in Sect. 1.5.3, procedures to generate feasible and optimal schedules are basically enumerative, we can conclude that manufacturing scheduling is complex from this formal perspective.

A final remark is that the polynomial/non-polynomial behaviour of the algorithms refers to its worst case performance and does not give any insight in its average behaviour. In contrast, from a real-world application point of view, a non-polynomial algorithm might reach a good or even the optimal solution in reasonable time. However, the proof of this optimality might be the additional and very time-consuming step. This, from a practical point of view, gives reason to prematurely stop an optimising algorithm, especially if an available bound indicates that the maximum deviation of the current solution from the optimal solution is acceptable.

2.3.2.2 Real-World Complexity in Manufacturing Scheduling

It is a commonplace that real-world problems are complex and their respective formal problems are complex as well—no matter how complexity is defined. Here we just intend to structure real-world complexity and thereby to provide starting points for handling (including reducing) complexity.

Real-world decision problems usually include complexity imposed by a complex decision framework. This may include among others

- different, maybe even not clearly specified objectives,
- a large variety of sometimes not even clearly pre-specified constraints,
- a large, maybe even not clearly pre-specified number of possible actions,
- a possibly hierarchical or even not clearly specified system of planning and decision-making and decision-makers,
- complexity which is induced by dynamics and uncertainty,
- the interaction of all aspects mentioned above, within their category and between the categories.

As already indicated earlier, there is no clear definition of complexity. However, there is a shirt sleeve classification of complexity by Reiss (1993a, b) which can easily serve at least as a first approach to clarify complexity issues of a decision problem, and particularly manufacturing scheduling. Reiss classifies complexity aspects into

- *Mass aspects*, further divided into
  - Multiplicity, i.e. number of elements and interactions in the system, and
  - Variance, i.e. number of *different* elements and interactions in the system, and
- *Chaos aspects*, further divided into
  - Ambiguity, i.e. degree of uncertainty about the characteristics of the elements and interactions in the system, and
  - Changeability, i.e. the change of the characteristics of the elements and interactions over time (thus closely related to dynamics).

We point on the fact that classifying (manufacturing scheduling) decision problems by means of this scheme not only provides a systematisation but also gives hints on how to treat the complexity of a decision problem, may it be by simplification, by identifying the relevant issues, by indicating basic solution strategies, etc.

By applying this scheme of complexity classification to manufacturing scheduling, the multiplicity of a manufacturing system is determined, e.g. by the number of jobs, the number of stages etc. Diversity might, e.g. refer to the (non-)homogeneity of jobs processing times, the diversity of job routes in the system, etc. Replacing small jobs by a larger one, replacing single products by product types, ignoring possibly sequence-dependent setup times or replacing several machines on one (hopefully non-bottleneck) stage by one 'formal' machine with added up capacity on this stage are examples for strategies regarding the mass aspect of complexity in manufacturing scheduling. As is well known, these aspects are intensively dealt with in scheduling science and practice, however mostly, if at all, they are only implicitly seen within a complexity context.

In contrast, chaos aspects of Reiß' complexity classification, i.e. ambiguity/ uncertainty (deterministic vs. stochastic scheduling) and/or changeability (static vs.

Complexity aspects		Some drivers of complexity in manufacturing scheduling
Mass aspects	Multiplicity	Number of jobs, stages, etc.
	Variety	Diversity of jobs' processing times, number/type of operations required, etc.
Chaos aspects	Ambiguity	Uncertainty about processing times, release times, etc.
	Changeability	Arrival of new jobs, machine breakdowns, etc.

Fig. 2.3 Summary of the classification of complexity

dynamic scheduling), are much more explicitly addressed in the scheduling literature and in applications as discussed already earlier. Replacing a dynamic problem by a 'semi-dynamic' problem with large time grid or even by a static model (without intermediate update of information) or replacing probability distributions of processing times by their expected value are simplification strategies concerning the chaotic aspects of complexity.

Figure 2.3 summarises the discussion above. It should be mentioned that complexity of the decision-making problem (in our case, manufacturing scheduling) not only refers to the structural characteristics of the planning object(s) and their relations but also to the planning procedure, including the time needed for executing the procedure. On one hand, with respect to the formal complexity discussed in the previous section, this represents at least one interface to the approach to complexity sketched out there. On the other hand, the complexity of the solution procedure is closely connected to the speed of this procedure. Taking into account that real-time solutions have come up and are required for many scheduling problems during the last years, this aspect of complexity of manufacturing scheduling becomes increasingly important. It refers not only to the 'core' algorithms applied but also to related aspects as information and/or database management both, from the input as well as from the output perspective.

Using this concept of structuring complexity and applying it to manufacturing scheduling problems, can provide valuable hints of how to make the real-world scheduling problem manageable, e.g. by well-defined simplification strategies in model building and/or model solution and removing these simplifications when deriving and implementing a solution for the real-world problem.

2.3.3 Variability

Variability, in general, is a term that roughly spoken indicates a system's behaviour with respect to its deviation from uniformity. If variability is quantifiable, it is usually expressed by the coefficient of variation of the respective context. In this section, we address the influence of variability aspects depending on the strategic, tactical, or operating decision levels already discussed in Sect. 1.2.

On the strategic (long-term) level, this includes aspects of frequency of repetition of processes. In mass production, ideally only one process is manufactured permanently. Therefore, in such a manufacturing environment, a 'true' scheduling problem will not occur. In high volume but not mass production, often a production calendar will be generated which makes further scheduling considerations superfluous. Small and medium sized lots of products with medium to large heterogeneity are the main object of shop scheduling approaches. Many approaches and publications refer to this case. Individual design of products and process sequences lead to (capacitated) project planning in (job) shop environments being also a main subject of manufacturing scheduling approaches. Therefore, with respect to the long-term level, the more product and especially process variability occurs, the more relevant become manufacturing scheduling approaches.

Looking at the tactical (mid-term) level, i.e. referring to the allocation of resources/capacities, availability and flexibility are the main drivers of manufacturing scheduling with respect to the variability perspective. Flexible machines with small and possibly sequence-independent changeover times but presumably longer processing times per part are the counterpart to inflexible machines with long and/or sequence-dependent setup times but smaller processing times per part. Availability of machines also refers to the maintenance strategy. Small, more frequent interruptions because of (preventive) maintenance make the system usually better perform as compared to long, infrequent interruptions of (probably curative) maintenance after a machine breakdown. Similar considerations take place for workforce capacity, possibly accompanied by labour regulations.

On the operating (short-term) level, where among others scheduling decisions take place, we will refer to the scheme of Figs. 2.6 and 2.7 (see Sect. 2.4.1) for considering variability aspects. Input to scheduling from the upper levels of production management (see Sect. 2.4.1) usually comprises extended demand data (including type and size of demand, processing times, due dates, release dates, etc.). In most cases/companies, these data are taken as unalterable input to the problem. However, on one hand usually these data are often not as fixed as customers, the sales department or somebody else outside the manufacturing system claims. In a make to stock environment the targets derived from the demand prognosis might be discussed just as well as customer-driven due dates in a make to order environment. This is, e.g. reflected by the discussion on due date setting which is closely related to scheduling analysis in the recent past. By introducing flexibility to due dates (e.g. by introducing due windows or re-negotiating due dates after they have been preliminary fixed, see Sect. 2.3.4) variability of the system can be reduced, uniformity can be approached and the performance of the system may be improved. Further down, additional remarks on due date setting can be found. On the other hand, if demand data shows high and/or even increasing variability (which is not unusual in a world where demand lots decrease and the degree of customer-individual requirements for specific product features increase), this variability might not be reducible. In such cases, variability usually can only be handled by introducing buffers (time, capacity and/or inventory) or by allowing worse performance (lower service level).

In addition, at all levels, with respect to variability attention has to be paid on the bottleneck's performance of the system. On the long-term level, long-term capacity considerations will take place to harmonise the long-term capacities of the system. With respect to a mid-term horizon, availability of the bottleneck and/or guaranteeing its performance, e.g. by adequate maintenance strategies and/or allocating qualitatively and quantitatively sufficient manpower, will be main approaches to increase or to maximise the performance of the system. On the short run, and immediately related to manufacturing scheduling decisions, apart from guaranteeing bottlenecks' availability, scheduling and especially shop floor control have to assure that the bottleneck(s) neither starve (because no job is available to be processed on the bottleneck) nor is blocked (because a job finished at the bottleneck is not able to leave it, may it be because of lacking transportation facilities or limited buffer space behind the bottleneck). Avoiding these drawbacks might be significantly influenced by both, adequate scheduling and applying adequate shop floor control mechanisms such as appropriate dispatching rules. With respect to transportation between different operations of a job, it can be stated that these transportation processes are often completely excluded from consideration in manufacturing scheduling approaches. However, on one hand, the variability of these transportation processes can significantly influence the overall performance of the system. On the other hand, splitting production lots into smaller transportation lots might allow overlapping processing of a job/order on consecutive machines and reduce idle time of machines as well as waiting time of jobs.

A final remark on variability in manufacturing scheduling refers to data availability and data quality and has basically been addressed already earlier. If, especially in deterministic models and respective solutions, fixed processing times, fixed release dates and fixed due dates are given, these dates, to a large extent, often will not be as deterministic in the respective real-world settings. Therefore, given degree of uncertainty (variability) of these data, in deterministic planning procedures provisions (time buffers) for this uncertainty have to be included. Another approach would be to stabilise/to standardise the system as far as possible, just to reduce the variability itself. This could be reached, e.g. by automation, standardisation (of processes as well as of labour skills), pull-based material supply etc.

2.3.4 Flexibility

Clearly, one of the aspects affecting the decision process is the degree of flexibility of the data at hand. While virtually all data under consideration are candidates to be flexibilised, here we will refer to one type of flexibility with great impact in practice, namely *due date setting/quoting*.

Although due dates are often seen as fixed input into manufacturing scheduling problems, these due dates might be flexible in one way or another which is not reflected in standard deterministic scheduling settings. Rather often, customers are flexible with respect to due dates (at least within some limits) or the manufacturing company is simply asked to propose due dates. So due date setting and/or due date quoting represent additional decision problems closely related to manufacturing scheduling which balance customer/sales needs on one hand and manufacturing

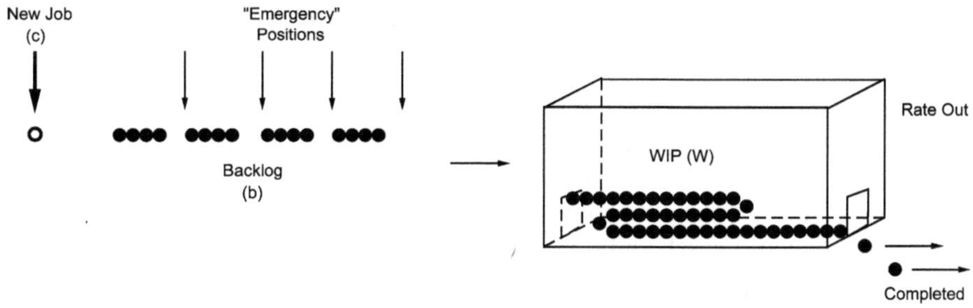

Fig. 2.4 Method for quoting lead times (Hopp and Spearman 2008)

capabilities on the other (Duenyas 1995). Thus people from the sales department and from the manufacturing department will have to agree on a due date (Fig. 2.4).

Apart from this human resources aspect, formally a due date can be determined rather easy on a coarse level. The determination of a reasonable due date has to include three time components into the calculation, i.e.

- the time required to process the current work in process, w,
- the time required to process the currently waiting jobs in front of the system and having higher priority than the job for which the due date is to be quoted, b, and
- the processing time of the job itself for which the due date is to be quoted, c.

Then the due date to be quoted is simply $d = w + b + c$. However, setting due dates in this way may turn out not to be as easy as it might look at a first glance. All 3 components include lead times which might be difficult to be calculated the more difficult/variable the manufacturing system is. Its variability might refer to jobs' flow through the system as well as to their processing times.

2.4 The Open View of Scheduling Problems and Decisions

As mentioned in Sect. 2.2, the open view of manufacturing scheduling corresponds to analysing its relationships with other (related) problems and decision-making processes. First we discuss the connection between manufacturing scheduling and the rest of decisions in production management (Sect. 2.4.1). Next, the interface between manufacturing scheduling and the rest of decisions (excluding those in production planning and control) within the company are presented in Sect. 2.4.2. Finally, we introduce the relationship between manufacturing scheduling and the rest of decisions in the supply network in which the company is integrated (Sect. 2.4.3).

2.4.1 Manufacturing Scheduling Within Production Management

In this section we investigate the relationship between manufacturing scheduling and production management. Production management decision problems are usually decomposed along two dimensions, i.e.: the scope of the decisions to be taken (briefly discussed in Sect. 1.2), and the logistic flow of the goods to be manufactured.

Once manufacturing scheduling is placed within these dimensions, we can discuss two of the most popular systems employed to conceptually describe the decisions in manufacturing, i.e. the Production Planning and Control (PPC) system, and the Advanced Planning Systems (APS) view, to study their connections with manufacturing scheduling. The two dimensions of production management decision problems are explained in Sect. 2.4.1.1, while the systems and their connections with manufacturing scheduling are discussed in Sects. 2.4.1.2 and 2.4.1.3, respectively.

2.4.1.1 The Two Dimensions of Production Management Decision Problems

The first dimension refers to the scope of the decisions in product management. In every company many different decisions have to be taken on different levels (of hierarchy and/or detail) with different importance. Usually, these decisions are not of stand-alone type but have to be coordinated with many other decisions. Depending on their relevance for the overall company, the time horizon of decisions' impact and their degree of detail of information and decision, these decisions are usually characterised as strategic, tactical and operational decisions (see Chap. 1). Strategic planning refers to long-term and conceptual decision problems to create and maintain a dynamic setting for long-term success of the company (e.g. decisions referring to locations, setting up a manufacturing hierarchy, (general and long-term) choice of process types, organisation of information flows, etc.). Tactical planning refers to mid-term and resource allocation tasks, i.e. setting up the infrastructure for a successful company (according to the strategic implications), while operational planning and control deals with short-term activities within a given infrastructure for externally specified requirements such as fulfilling orders/jobs etc.

According to this dimension, scheduling is part of planning and controlling the execution of manufacturing tasks within a given manufacturing infrastructure—and not planning of the infrastructure itself. Accordingly, many decision problems related to scheduling include a rather short-term and operational planning horizon.

The second dimension refers to the logistics of the manufacturing activity, which is the generation of marketable or intermediate goods (or, for the sake of completeness, also the elimination of bads, such as in waste combustion; we will not address this aspect further here) that constitute the products/services and are denoted as the *outputs* of the company. Manufacturing these goods requires *inputs* either procured externally or provided by own preceding production processes. This process of transforming inputs into outputs is the manufacturing or production process and, as it is the logistics stage between input and output, is sometimes called *throughput*.

Consequently, planning and control in a manufacturing system can be separated into output, input, and throughput planning and control. Throughput planning and control refers to the determination of general types of processes chosen for manufacturing down to planning and control of a physical process setting for given jobs which also includes manufacturing scheduling tasks. A main part of throughput planning and control deals with (material) flow planning and control while input planning and control refers to purchasing/procurement strategies and lot-sizing down to staging of material. Output planning and control refers to the derivation of (strategic) market strategies down to (operational) distribution.

Referring to the standard definitions and/or descriptions of logistics as an integrating, flow-oriented function which intends to optimally bridge spatial, quantitative and temporal distances between 'providers' and 'consumers', it is nearby to interpret manufacturing systems also in a logistics context: 'Providers' can be suppliers, inventories (of raw material, intermediates and/or finished goods), production sites, preceding machines, etc. Accordingly, 'consumers' can be inventories, production sites or (internal or external, referring to the system under consideration) customers. From the point of view of manufacturing, besides of the core manufacturing processes/operations, the complementary logistics operations of storing, packaging and/or transportation are linking manufacturing operations with each other and with their 'boundary' functions in procurement and distribution. These complementary operations are often not explicitly included into manufacturing scheduling. However, their influence on the overall performance of a manufacturing system should not be neglected and estimated carefully. Their influence with respect to the performance of the system as a whole might be significant.

According to the two dimensions discussed above, it is clear that most decisions in manufacturing scheduling are assigned to operational throughput planning and control (see Fig. 2.5). However, at least setting up the scheduling system itself will be part of the tactical and sometimes even of the strategic level: e.g. general choice of process type as a main framework for every scheduling approach and the decision whether centralised or decentralised scheduling systems are set up can be assigned to the strategic level. Fixing the general scheduling procedures within the strategic settings and allocation of human as well as software resources for the scheduling tasks can be interpreted as tactical decisions. Nevertheless, concrete application of decision support tools for scheduling to real-life jobs and operations will always be operational.

Figure 2.5 emphasises the necessity of *coordinating* the decisions within and across levels, both on the vertical as well as on the horizontal level: With respect to the vertical axis, strategic concepts determine the infrastructure acquired on the tactical level, and this infrastructure significantly determines the realisable alternatives on the operational level. In turn, lower levels' information and/or problems are fed back to the upper levels to possibly adjust the higher level decisions. Concerning the horizontal axis, input, throughput and output are obviously closely connected to each other, no matter whether its primal perspective is push from input to output or pull from output to input. Feedback procedures take place here as well. For the sake of brevity, we will not address diagonal influences implied by Fig. 2.5 here, although

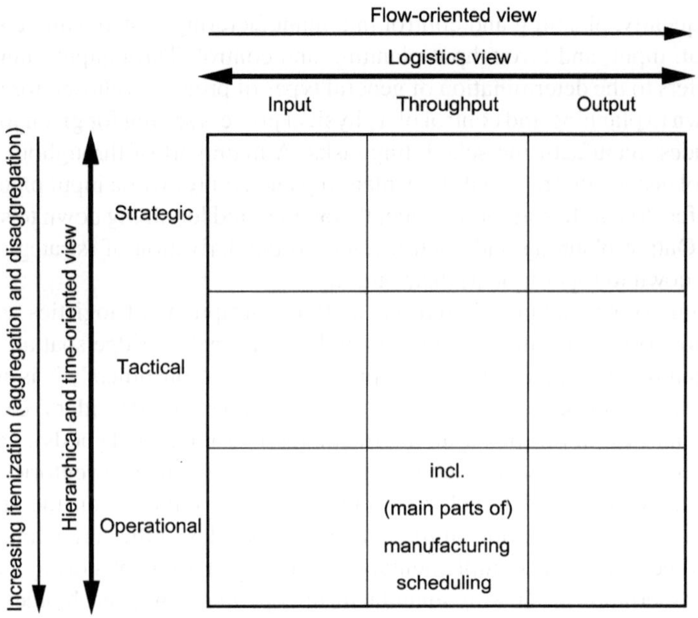

Fig. 2.5 Manufacturing scheduling in its planning and control context (see, e.g. Corsten 2009, p. 26)

these influences exist obviously. In the case of manufacturing scheduling, it has to be coordinated vertically within the multi-level structure of this field as well as with the respective horizontal 'neighbours' in input (purchasing, provision of material, etc.) and output (distribution, packaging, etc.).

Using the two dimensions, we can also classify the main decisional systems that have been employed for handling manufacturing decisions. The first one is the so-called PPC system, which basically corresponds to organising the related decisions along the vertical axis (scope of the decisions). The second view (APS) refines/extends the PPC system by introducing the logistic dimension. Both systems will be discussed in the next subsections.

2.4.1.2 Production Planning and Control (PPC) system

As discussed in Fig. 2.5, manufacturing scheduling is part of operational throughput planning and control. Within this cell, manufacturing scheduling is embedded usually into a multi-level production planning and control system which includes program planning, material planning, capacity and flow planning as well as scheduling. Figure 2.6 represents this classical scheme. Basically, Fig. 2.6 can be interpreted as a hierarchical itemisation of the operational/throughput cell in Fig. 2.5. This is denoted as the PPC system and, with slight differences, it has been presented in

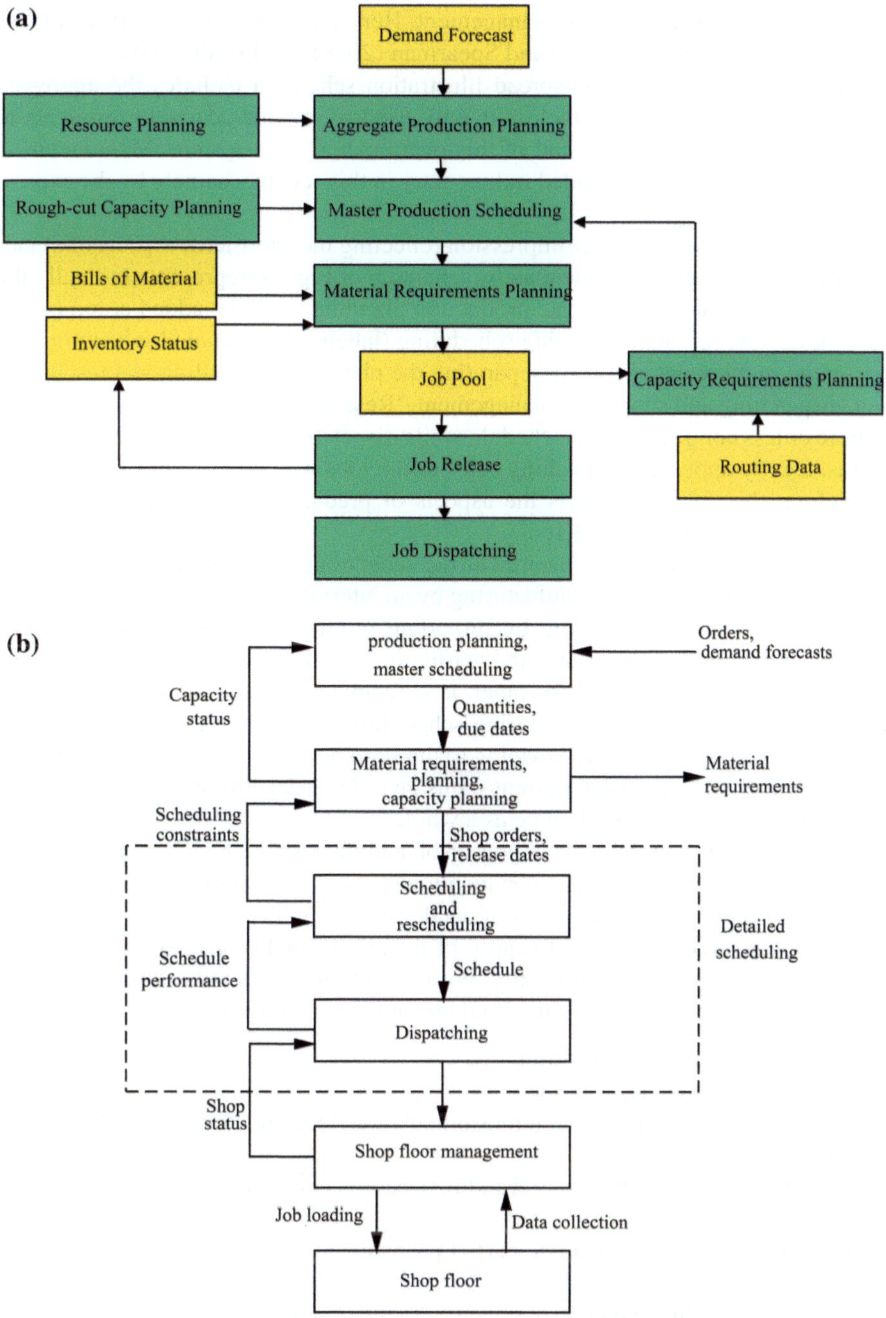

Fig. 2.6 Multi-level scheme of PPC. **a** Hopp and Spearman (2008), **b** Pinedo (2012)

many textbooks on production management. Here we show in Fig. 2.6 two of these presentations, taken from Hopp and Spearman (2008) and Pinedo (2012).

As can be seen, the wide-spread illustration scheme a includes the aggregate production planning level which is mainly used to provide adequate resources to the system on the long-term end of this overall short-term system. Manufacturing scheduling itself is only included rudimentary in this scheme, namely by the expression job dispatching.

Not as common, but to our impression reflecting the short-term aspects of manufacturing scheduling more adequately, scheme b of Fig. 2.6 represents basically the same system as in (a) (aggregate production planning might be added at top of this scheme). However, manufacturing scheduling (labelled as 'Detailed scheduling') is presented in more detail, i.e. by separating the phases of scheduling and rescheduling, dispatching and shop floor management. 'Rescheduling' here is meant to set-up a more or less completely new schedule while dispatching means to react to the shop status, e.g. by applying dispatching rules to the jobs waiting in front of a machine. Therefore, this scheme reflects the aspects of predictive and reactive scheduling addressed already in Sect. 2.3.1.

Figure 2.6b indicates that manufacturing scheduling is embedded into the classical hierarchical scheme of manufacturing by an interface with material requirements planning and capacity planning by providing implications from scheduling constraints to this level on one hand and getting information on jobs (shop orders) and their earliest release dates on the other. If scheduling/rescheduling and dispatching are seen as the level of manufacturing scheduling, then its data provides the main information with respect to job loading to the shop floor and its management as well as data collection on the jobs' current status gives feedback information for updating dispatching as well as the schedule as a whole.

Apart from this hierarchical integration of manufacturing scheduling into the planning and control context it has to be mentioned also here that the different levels of this scheme include different levels of detail (at least) with respect to products/jobs, capacities/resources and time. Because of this, horizontal (logistical) and diagonal coordination tasks occur additionally. Without getting into detail of these coordination tasks, their main components should be at least mentioned:

- decomposition and composition,
- aggregation and disaggregation,
- hierarchical coordination (including anticipation, feed-forward and feedback components),
- model building (including central and/or decentral components and their respective decision structures),
- problem solving (for the single partial problems as well as for their integration), and
- fitting the planning and control system into the organisational structure.

2.4.1.3 APS Systems and Manufacturing Scheduling

Numerous drawbacks of the classical PPC system which have been widely discussed for decades in the literature and the increasing ability to integrate different planning systems and planning levels from an IT perspective have led to the development of so-called APS systems. Additionally, the Supply Chain Management (SCM) 'movement' of the last years, may it be *intra*-organisational SCM (within one system and tending to control via hierarchical structures) or *inter*organisational (maybe crossing systems, including coordination tending to be controlled via market-based structures), promoted modified structures of PPC systems within the planning and control system of the logistics system. We only refer to some aspects of these systems which are relevant for manufacturing scheduling. More specifically, an APS system:

- gives a framework for manufacturing planning and control which includes both, the hierarchical as well as the logistics perspective,
- comprises advanced planning approaches, including 'true' optimisation approaches,
- includes state-of-the-art IT technology, i.e. among many others availability of quick (possibly real time) updated data within a network structure.

The modular structure of APS systems can be seen from Fig. 2.7. In this figure the information flows are indicated by the arrows. Manufacturing scheduling is located as short-term module within the production hierarchy which is coordinated by information flows related to production planning via lot sizes, and to procurement and distribution via due dates. It gives the main information to the shop floor control level.

Figure 2.8 presents the planning tasks in APS systems as assigned to every module. As can be seen, the scheduling module includes machine scheduling and scheduling is referring to data from the mid-term procurement, production and distribution levels using data on material availability, release dates as well as due dates/deadlines.

Finally, it should be mentioned that both, classical PPC systems as well as APS systems are integrated into standard ERP software packages. Therefore, they are applicable to real-world problems and, if they contain respective modules, they are able to perform automated manufacturing scheduling or to support respective decision problems and decision makers. However, to perform these tasks, adequate data must be available at the right time at the right place. Therefore, advanced database structures and database management are most important features of these systems. During the last years, data and model integration as well as real-time availability of data have contributed significantly to the performance of PPC and APS systems within ERP systems, including modules for manufacturing scheduling.

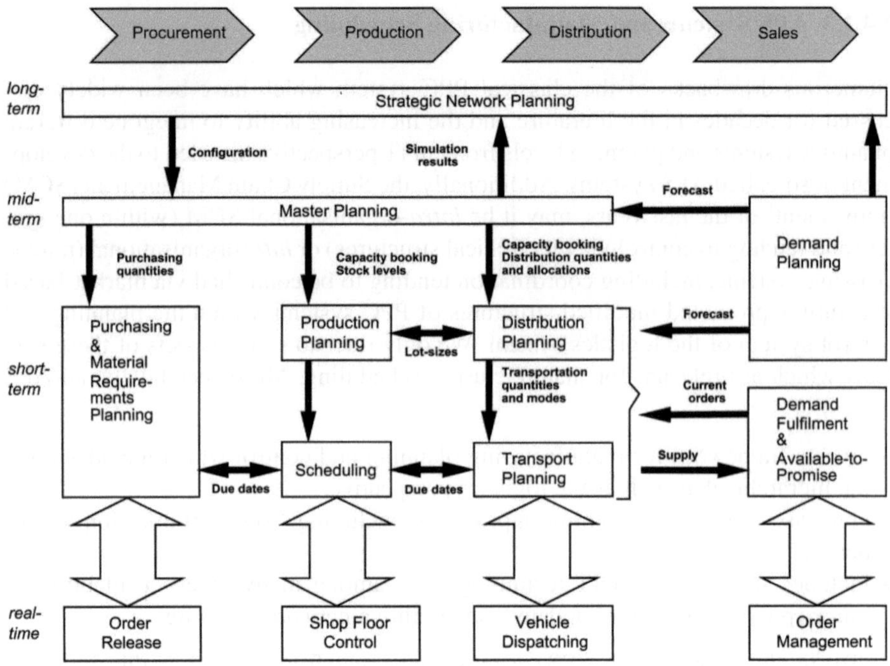

Fig. 2.7 APS modules from a logistics (supply chain) perspective (Reuter and Rohde 2008, p. 249)

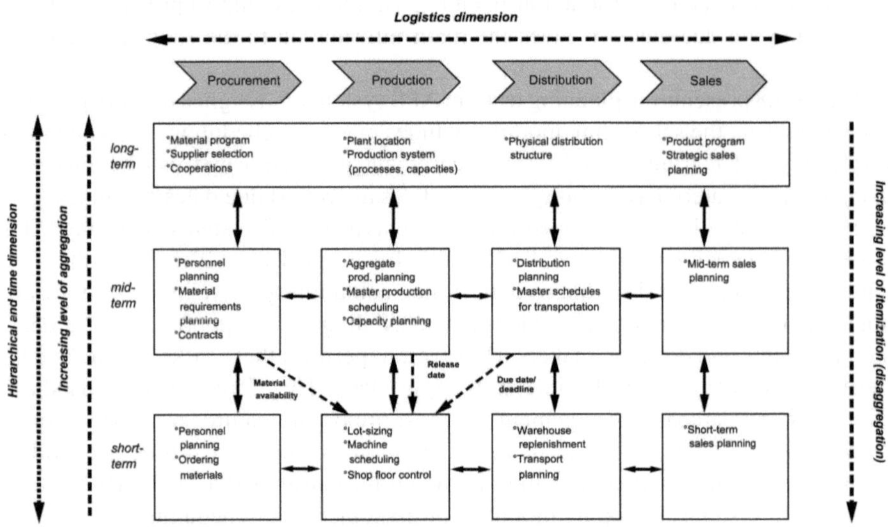

Fig. 2.8 (Supply chain) planning tasks in APS systems (modified from Fleischmann et al. 2008)

2.4.2 Interfaces of Manufacturing Scheduling with Other Decisions in Production Management

Many problems and their respective models and decisions in manufacturing combine machine scheduling aspects with other aspects in production management, such as inventory control, workforce scheduling, maintenance scheduling, capacity control or pricing. These approaches are manifold and will not be described here further. Aspects of model decomposition and solution integration are relevant here as well, however even more complicate since the objects to be planned and coordinated are even more complex than in the settings addressed earlier.

2.4.3 Manufacturing Scheduling and Intra- and Inter-Organisational Supply Networks

As already mentioned earlier, SCM (induced by movements as focus on core competences and related outsourcing), in addition to the classical and 'selfish' intra-organisational view on production, opened the view on logistics also from an inter-organisational perspective and on the combination of both, intra- and inter-organisational aspects of manufacturing.

Up to now, it can be observed that many real-world realisations of inter-organisational supply chain coordination schemes, including their IT structure, more or less simply copy the former intra-organisational information structures and transfer them to inter-organisational settings (and therewith transferring basically hierarchy-based information and coordination systems to more market-based systems–without regarding their different concepts and constraints). However, depending on the constellation of the cooperation between systems (companies) in the supply chain, every intermediate form between pure intra-organisational, hierarchically controlled coordination on one hand and pure market-based coordination on the other can appear. Although this aspect has been dealt with in science, meanwhile more or less intensively, applicable solutions to real-world systems are still missing to a large extent.

This (non-)development also alludes manufacturing scheduling. Here as well, the traditional focus on intra-organisational settings has to be supplemented by the consideration of manufacturing settings which cross borders of subsystems (e.g. plants, companies) and which are not exclusively characterised by basically hierarchical structures but also include more independent subsystems, possibly with different, maybe even conflicting objective functions. This aspect of inter-organisational manufacturing scheduling has not been dealt with very intensively so far. Instead, these inter-organisational interfaces are often seen to be determined by mid-term coordination and not by short-term scheduling. However, the closer cross-system manufacturing processes are linked, the more relevant inter-organisational approaches to manufacturing scheduling are. For example in systems including principles like

Just-in-Time or Vendor Managed Inventory, cross-system manufacturing scheduling aspects are relevant also from the short-term perspective.

As a keyword, *supply chain scheduling* occurred during the last years. For example, a hierarchy of scheduling plans can be proposed, a more aggregate master or global scheduling procedure might be linked to detailed scheduling in every subsystem, both considering predictive as well as reactive aspects of scheduling. In addition, since the manufacturing facilities in supply networks are often not located close to each other, transportation scheduling, between and within the subsystems, might become a severe issue which has to be analysed and integrated into the overall scheduling concept carefully. Further future research topics will appear with respect to these and other related topics.

Finally, it should be mentioned that, on a more abstract level, there are systematic approaches for structuring inter- and intra-organisational supply chains or supply networks. The well-known SCOR (Supply Chain Operations Reference Model) model is the reference point for these approaches. However, also in the SCOR model manufacturing scheduling aspects are not addressed explicitly and/or intensively, at least not from a model building and solution procedure point of view.

2.5 Conclusions and Further Readings

As stated in the introduction, this chapter serves to contextualise manufacturing scheduling. A framework for scheduling decisions has been presented, employing a dual view (open/closed) of manufacturing scheduling coming from systems analysis. The closed view analyses manufacturing scheduling as an isolated problem area and related decision process and the main aspects influencing these decisions (time, complexity, flexibility and variability) are discussed. In contrast, the open view emphasises the relationship of manufacturing scheduling with other decisions in the company and its supply chain network.

This chapter is conceptually demanding, and many of the ideas sketched here would deserve greater space. At the expense of rigour, these have been reduced to the minimum to give an integrated framework for understanding the context in which scheduling takes place. We believe that, in some cases, this lack of understanding is one of the causes leading to the poor design of scheduling systems and/or to under/over-estimate the importance of scheduling decisions in some manufacturing scenarios.

Regarding the presentation of the framework, most of the discussion of complexity is along the classification by Reiss (1993a, b), which is further elaborated in Gepp et al. (2013). For a comprehensive discussion of variability aspects in manufacturing, the book of Hopp and Spearman (2008) is an excellent reference. A discussion between centralised and decentralised systems can be found in Hamscher et al. (2000). Excellent places to start with the topics of rescheduling are Ouelhadj and Petrovic (2009) and Vieira et al. (2003). Holistic books about the manufacturing framework are the excellent Factory Physics books (Hopp and

Spearman 1996, 2008). This is, in any case, a very small extract of the related vast literature on scheduling topics. For a detailed discussion on planning and the PPC system, the reader is referred to Domschke et al. (1997). Aspects of horizontal (logistical) and diagonal coordination tasks within the PPC system are discussed in detail in Stadtler (2000), while a comprehensive presentation of and an overview on APS systems is given in the book chapter of Reuter and Rohde (2008). Finally, a nice presentation of the main concepts related to supply chain management is Stadtler and Kilger (2002).

References

Corsten, H. (2009). *Produktionswirtschaft - Einfuehrung in das industrielle Produktionsmanagement.* Oldenbourg, Muenchen. 12th, revised and upgraded edition.

Domschke, W., Scholl, A., and Voss, S. (1997). *Produktionsplanung: Ablauforganisatorische Aspekte.* Springer, Berlin. 2th, revised and upgraded edition.

Duenyas, I. (1995). Single facility due date setting with multiple customer classes. *Management Science,* 41(4):608–619.

Fleischmann, B., Meyr, H., and Wagner, M. (2008). Advanced planning. In Stadtler, H. and Kilger, C., editors, *Supply chain management and advanced planning,* pages 81–106, Berlin. Springer.

Gepp, M., Amberg, M., Vollmar, J., and Schaeffler, T. (2013). Structured review of complexity in engineering projects. State-of-research and solution concepts for the plant manufacturing business. *International Journal of Business and Management Studies,* 5 (1):318–327.

Hamscher, V., Schwiegelshohn, U., Streit, A., and Yahyapour, R. (2000). Evaluation of job-scheduling strategies for grid computing. In Buyya, R. and Baker, M., editors, *Grid Computing - GRID 2000; First IEEE/ACM International Workshop Bangalore, India, December 17, 2000 Proceedings,* pages 191–202, Berlin / Heidelberg. Springer.

Hopp, W. J. and Spearman, M. L. (1996). Factory physics. *Foundations of manufacturing management.* Irwin, New York, USA.

Hopp, W. J. and Spearman, M. L. (2008). *Factory Physics.* McGraw-Hill, New York.

Ouelhadj, D. and Petrovic, S. (2009). A survey of dynamic scheduling in manufacturing systems. *Journal of Scheduling,* 12(4):417–431.

Pinedo, M. L. (2012). *Scheduling: Theory, Algorithms, and Systems.* Springer, New York, fourth edition.

Reiss, M. (1993a). Komplexitaetsmanagement i. *WISU - Das Wirtschaftsstudium,* 1:54–59.

Reiss, M. (1993b). Komplexitaetsmanagement ii. *WISU - Das Wirtschaftsstudium,* 2:132–137.

Reuter, B. and Rohde, J. (2008). Coordination and integration. In Stadtler, H. and Kilger, C., editors, *Supply chain management and advanced planning,* pages 247–261, Berlin / Heidelberg. Springer.

Stadtler, H. (2000). Production planning and scheduling. In Stadtler, H. and Kilger, C., editors, *Supply chain management and advanced planning,* pages 197–213, Berlin / Heidelberg / New York. Springer.

Stadtler, H. and Kilger, C. (2002). *Supply chain management and advanced planning.* Springer, Heildelberg.

Vieira, G., Herrmann, J., and Lin, E. (2003). Rescheduling manufacturing systems: A framework of strategies, policies, and methods. *Journal of Scheduling,* 6(1):39–62.

Part II
Scheduling Models

The Part II of the book consists of Chaps. 3–6 and it is devoted to scheduling models. In Chap. 3, we carry out an introduction to scheduling models. Next, in Chap. 4, the different scheduling constraints found in practice (process-related, operations-related, storage-related, transportation-related, ...) are reviewed, and classified, as well as their interactions. Chapter 5 deals with the different objectives that may be considered, and reviews the main performance measures found in scheduling research. Finally, we discuss in Chap. 6 the integration/alignment of classical scheduling objectives within the rest of a company's objectives.

Chapter 3
Overview of Scheduling Models

3.1 Introduction

As it has been discussed in a previous chapter (see Sect. 1.5), scheduling refers to a decision-making process in which, in order to solve a real-world problem, a formal *model* is obtained. This part of the book is devoted to (formal) scheduling models and address the elements composing a scheduling model (jobs/machines/operations, constraints, criteria) as well as the main issues related to building scheduling models. In particular, this chapter overviews the basics of modelling scheduling decision problems. As we will point out, the modelling process in scheduling goes first through a rather detailed classification of scheduling models.

More specifically, in this chapter, we

- give an outline of the nature of modeling for problem solving (Sect. 3.1.1), together with the main characteristics of models, types of models, common errors and shortcomings (Sect. 3.1.2)
- introduce the basic definitions for building manufacturing scheduling models (Sect. 3.2.1), with special emphasis in the resources (Sect. 3.2.2) and processing layouts (Sect. 3.2.3), discuss the main processing constraints (Sect. 3.2.4) and outline the types of scheduling criteria (Sect. 3.2.5)
- present a classification for scheduling models (Sect. 3.3),
- discuss the identification of jobs as the processing unit in scheduling models (Sect. 3.4).

3.1.1 Modelling as an Approach to Scheduling Problem Solving

Modelling lies within the very core of the scientific approach to problem solving. Reality is utterly complex and varied. Models give some rough classification of problems and settings in an attempt to guide the researcher in a fruitful direction.

J. M. Framinan et al., *Manufacturing Scheduling Systems*,
DOI: 10.1007/978-1-4471-6272-8_3, © Springer-Verlag London 2014

In a nutshell, a model is a stripped down, simplified and abstract representation or view of an otherwise complex, detailed and broad reality. Scientific modelling is therefore the process of generating the abstract and conceptual models that will be of use to researchers and practitioners. Modelling is, in itself, a science that stems from the philosophy of science, systems theory, data and knowledge visualisation, statistics and many other branches of science.

Models represent objects or entities that have an empirical existence. In manufacturing scheduling, we will model all relevant productive resources such as machines, workforce, tooling and storage areas. How a model is constructed is closely tied to the final intended use of the model itself. A map, for example, is a geographical model. A very simplified map is useful to understand, in a few seconds, a weather forecast. However, we need a more detailed topological map in order to construct a bridge, for example.

One preliminary question at this stage is 'Why modelling?'. Models have a countless series of characteristics and benefits that can be summarised as follows:

- Models focus on certain aspects of reality. Human thought can then be applied to a more focused part and later amplified.
- A model is always related to the reality or 'target' that it modelises. As mentioned previously, the model nature and the model building process will vary as a function of the intended use (Gilbert 1991). Furthermore, the model should be as close as possible to the target so as to achieve its intended goal. A too broad or general model might fail to serve its purpose, so might be a very detailed model.
- A model is different from reality. It has to be a simplification. If a model matches exactly the target, it is no longer a model but a copy.
- As a result of the two previous items, a model is always a compromise between verboseness and simplification.
- Models, or better, the results given by the experimentation of the model, can be often easily measured. This is specially important when measurements over the modelled reality are impossible or impractical to obtain (Mayer 1992).
- Models can be used to study altered realities. Once a model is thought to be valid for the present modelled reality, altered models can be used to study 'What if?' scenarios on different realities, with some degree of confidence, that would be otherwise impossible to study.
- Models, once validated, allow to explain past events, to predict future observations and/or to refute or approve hypotheses over the modelled reality.
- Models have to be easy to work with. The reality is modelled since it is too complex. A model that is hard to handle and to interpret is not a valid model.
- Models are full of limitations. The interpretation of the results of a model are only valid for the model and extrapolation to reality has to be done in a sound and scientific way. The model has to be validated and the results are closely examined.

Once the definition of a model has been introduced, we discuss the different types of models, along with common mistakes in modelling in the next section.

3.1.2 Types of Models and Common Errors in Modelling

Models range from simple pictograms (e.g. traffic signs) to complex distributed and evolutionary agent systems. The following list is just a small summary of the types of models and methodologies that result in models and that are of use in scheduling problem solving:

- Mathematical models: mathematical programming, linear programming, non-linear programming, deterministic, probabilistic, etc.,
- Graphical model: pictograms, drawings, Venn diagrams, flow charts, etc.,
- Statistical models,
- Simulation models, and
- Algorithmic models.

Mathematical and algorithmic modelling are the bulk of the types of models used in deterministic scheduling. This will be later explained in Chap. 6.

Models are not the holy grail of the scientific method. Actually, they are a necessary evil. This is so since no matter how perfect the analysis of the outcome of a model might be, the conclusions of the whole research could be completely flawed due to a countless series of problems that abound in model building. In the following, we provide just a summarised list:

1. The model could be incomplete to the point of rendering the results impractical for the studied reality.
2. The model might be unrelated with the studied reality. This is often referred to as the 'engineering problem' which is, 'perfectly solving the wrong problem.'
3. The model might be fed with inaccurate or incomplete data. Hence, the conclusions from its study might be off from the modelled reality.
4. The model works only under severe assumptions and limitations that hinder its generality and inferential power.
5. The model might be too complex to solve realistically sized instance problems. Again, the generalisations might be hard to achieve under such circumstances.
6. Improper procedure when testing the results of the model. Lack of sound statistical testing of the sampled results.
7. Related with the above, small and unrepresentative samples used in the tests.
8. Perceptual, interested bias or even fraud in the interpretation of the results of the model.
9. Difficulty in reproducing the results. The lack of the reproducibility is behind many scientific faux pas. Many of the above items result in irreproducibility.
10. Reasoning errors. Confusing causality with casuality. Mistaking correlation for causation, etc.

All the previous problems also occur in building manufacturing scheduling models. As we will closely study in the following chapters, specially in Chap. 6, the most important problems in scheduling model construction are the ones listed in the previous list as 1, 2, 3, 4 and 5. More specifically, if a scheduling model is overly simple, the resulting production scheduling will be orientative at best for the modelled production scheduling problem. In many extreme cases, the scheduling model studied is such an overly simplification that the results could be completely disconnected with practice. Furthermore, many scheduling models work under so many severe assumptions that result also in disconnections with reality. Similarly, complex models are many times hard to solve with current technologies and solutions to all but the smallest instance problems are unattainable. As we will mention in later chapters, the correct formulation of sufficiently precise manufacturing scheduling models is one of the main issues in the scheduling field.

3.2 Formal Definitions

A rough sketch of some basic definitions related to manufacturing scheduling (such as, e.g. jobs, machines, routing process) has been already given in Sect. 1. In this section, we delve into all these items by providing a formal definition, along with a mathematical notation that we will use to model manufacturing scheduling decision problems.

3.2.1 Scheduling Model

A *scheduling model* is a formal abstraction of a (manufacturing) scheduling decision-making problem which is described by considering the system of tasks/operations, the processing constraints and the criteria.

Note that the above definition is more or less equivalent to that of *scheduling problem* in most (classical-oriented) texts. However, we want here to emphasize that we solve (real-world) scheduling problems first by modeling them (hence the need of scheduling models), second by finding methods that provide solutions for these models—and not for the real-world problem—, and third by transferring these solutions to the real world (see Chap. 1).

A scheduling model can be further concretised by selecting specific (i.e. numerical) values for the resources, tasks and rest of elements that make the model. This specific selection is termed an *instance* of the scheduling model.

In the following sections, we will provide formal definitions for the different elements constituting the scheduling model: The system of taks/operations is discussed in Sects. 3.2.2 and 3.2.3, while the processing constraints are presented both in the previous sections as well as in Sect. 3.2.4. Finally, a hint on scheduling criteria is given in Sect. 3.2.5.

3.2.2 Jobs and Machines

For the time being, we will assume that a scheduling model is determined by a known, finite and deterministic number of jobs that have to be processed on an equally known number of machines. Although we have already discussed in Chap. 1 that all these assumptions are not easily and readily acceptable in all situations, we will start with these simplest cases in order to gain the knowledge required for addressing more complex settings.

More specifically, we will assume that there is a set N of jobs that are consecutively indexed $N = \{1, 2, \ldots, n\}$. Therefore, there is a total of n jobs to be processed. Subindices j and k will be used to refer to jobs in the set N. Similarly, there is a set M of m machines or productive resources, that are indexed as $M = \{1, 2, \ldots, m\}$. We will be using mainly the subindex i to refer to any machine in the set M. If not stated otherwise, machines and jobs are independent and all the following data is deterministic, and known a priori:

- Task or Operation (O_{ij}). Each job $j \in N$ has a predefined number of operations or tasks. These tasks are at least carried out in 1 machine each. Therefore, O_{ij} denotes the task on machine i of job j. It is also common to refer to tasks simply as (i, j).[1] Furthermore, each operation has to be scheduled, this is, each operation will end up with a start and finish (end) time as a result of the production scheduling. We will refer to those times as SO_{ij} and EO_{ij}, respectively.
- Processing route. Each job has a predefined route throughout the production floor. In our terms, we will denote as R_j to the ordered list of machine visits. For example, this list could be (2, 4, 5, 1, 3) for a given job, meaning that the job has five different operations and that it has to visit first machine 2, then machine 4 and so on until the last machine 3. A discussion on the reasons for the assumption of a processing route is given in Chap. 1).
- Processing time (p_{ij}). This is the known span of time that is needed at machine i to process job j. Machine i will be busy during this period processing job j. This processing time is associated with the operation or task O_{ij}. Usually, it has to be satisfied that $p_{ij} \leq EO_{ij} - SO_{ij}$. Notice that the previous expression does not need to be necessarily equal as the processing of a job could be interrupted several times as per example, long jobs spanning through many days being stopped during the night shifts.
- Release or ready dates (r_j). Those are instants in time from which jobs might start. They also denote the first point in time where the job is available. They are helpful when modelling earliest starting times due to availability of raw materials or order placement time, just to name two possible examples. Jobs enter the production shop not earlier than r_j. This implies that $\forall j \in N, i \in M, SO_{ij} \geq r_j$. In other words, no

[1] Note that, in order to keep the complexity at minimum, this notation excludes the situation where a job has to visit the same machine more than once.

task of job j might start before r_j. In some scenarios, each task might have a release date r_{ij}, in such cases then it has to be satisfied that $\forall j \in N, i \in M, SO_{ij} \geq r_{ij}$.

- Follow-up time (fu_j). While release dates refer to the beginning of processing of a job (or even a single operation), a job or an operation may also have, after being finished on the respective machine, a *follow-up time* where it does not require any additional resources or machines but is not able to be processed further. (Such settings occur, e.g. in cooling or aging processes). We will see later that these times are many times modeled as waiting times or lags.
- Delivery or due dates (d_j). These are instants in time at which jobs have to be ideally completed. They model the delivery dates agreed with customers and constitute a compromise. Therefore, the last operation of each job should be completed before d_j. Usually, due dates are not fully mandatory and can be violated, subject to a penalty or cost. Exceeding the due date, i.e. finishing tardy, usually is connected to some penalty, e.g. by reduction of earnings.
- Mandatory due dates or deadlines ($\overline{d}_j$). Contrary to due dates, deadlines cannot be violated and it is mandatory, due to many possible reasons, to finish the job before $\overline{d}_j$. Therefore, $\forall j \in N, i \in M, EO_{ij} \leq \overline{d}_j$. As we will see in later chapters, the presence of deadlines may cause infeasibilities in scheduling problems as it might be physically impossible to satisfy all deadlines of a set of jobs in a given production environment.
- Due windows. An extension of the concept of due dates are the so-called *due windows*, i.e. a time interval where a job should be delivered. This time interval might be different per job or all or some jobs have a common due window. These windows can be modelled as $[d_j^-, d_j^+]$ where d_j^- and d_j^+ denote the start and finish of the delivery window, respectively.
- Costs, priorities, importance or weights (w_j). This figure, measured in any desired scale, models the importance or priority (as well as many other possible indicators) of each job. With the weights it is easy to establish a relative hierarchy among the set of n independent jobs. This weight could be established as a function of the total production cost of the job, importance of the client that placed the order, magnitude of penalty for finishing the job beyond its due date or many others.
- Release dates for machines (rm_i). Similar to the job release dates, machines might not be available before a point in time. This implies that $\forall i \in M, j \in N, SO_{ij} \geq rm_i$ Usually, this is much more complex as machines are seldom considered to be continuously available from rm_i. As a matter of fact, machines might have unavailability windows due to, for example, preventive maintenance or even unexpected breakdowns.

As we can easily see, the previous list is just a meagre extract of the sheer amount of data that could be associated with jobs and/or machines (for a more exhaustive list, see e.g. Pinedo 2012). More information would be added in subsequent sections and chapters whenever needed.

3.2.3 Processing Layouts

With the previous data, and among all possible classifications, by far, the scheduling literature is sectorised according to how the different machines in the set M are laid out and also according on how the different job processing routes R_j are specified. The most typical layouts are discussed next, together with some examples of schedules. Unless stated otherwise, we assume in the examples that the jobs are readily available for processing and that preemption is not allowed, for reasons of simplicity.

3.2.3.1 Single Machine Layout

As the same implies, the plant layout is formed by a single machine. Therefore, each job has to be processed exactly once on that machine. So there are really no processing routes.

Single machine models are relatively simple and are often viewed as too theoretical. However, they can be seen as special cases of other more complex layouts and therefore their study is of significant interest. In addition, we have already discussed in Chap. 1 that a single machine can adequately represent, under certain circumstances, larger real-world entities within the shop floor, or even the whole shop floor.

The flow layout schematics of a single machine model is clear, jobs enter the machine, each one of them keeping it busy for at least p_j processing time units (note that the processing times p_{ij} are simplified to just p_j in this case) and leave the machine once finished. The schematics are shown in Fig. 3.1.

Picture an example with six jobs, whose processing times are given in Table 3.1.

In a single machine model, each possible semi-active schedule (see Sect. 1.5.2) can be represented by simply specifying the sequence in which the jobs are to be processed in the machine. For instance, if we assume the following sequence $\pi = (1, 2, 3, 4, 5, 6)$, the resulting Gantt chart (see Sect. 1.5.1) of this single machine example is shown in Fig. 3.2.

Since in semi-active schedules, all jobs are processed without interruptions on the single machine, $SO_1 = 0$ and $EO_1 = SO_1 + p_1 = 0 + 8 = 8$ for this sequence. Similarly, $SO_2 = EO_1 = 8$ and $EO_2 = SO_2 + 4 = 12$. After all, it is easy to see that the machine is going to be busy during $\sum_{j=1}^{6} p_j = 42$ units of time. As a matter of fact, and as will be discussed later, the order of the jobs in the single machine setting is not going to affect the total busy time of the machine.

3.2.3.2 Parallel Machine Layout

After single machines, the straightforward extension to increase production capacity, is to replicate the single machine into several parallel machines. The processing routes for the jobs are equally simple: each job has to be processed in one out of the m available parallel machines. Usually, all m parallel machines are eligible for the jobs.

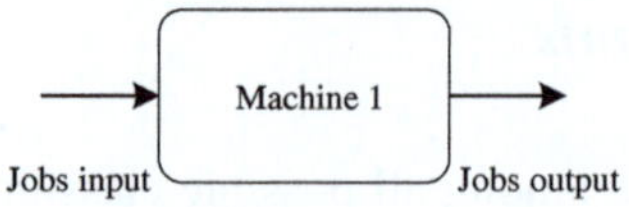

Fig. 3.1 Flow layout schematics of a single machine

Table 3.1 Processing times (p_j) of the six jobs for the single machine example

Machine (i)	Job (j)					
	1	2	3	4	5	6
1	8	4	10	5	3	12

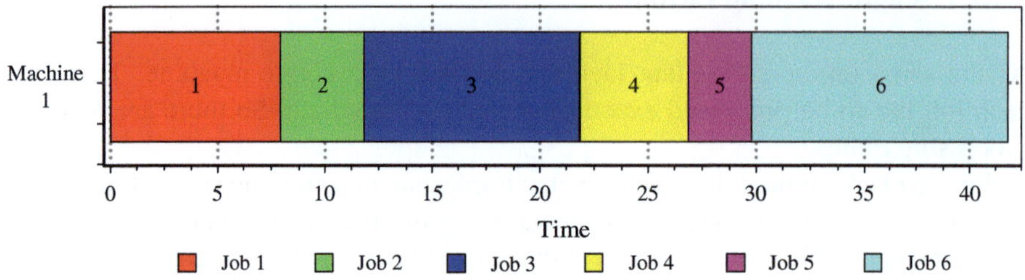

Fig. 3.2 Gantt chart with the result of the sequence $\pi = (1, 2, 3, 4, 5, 6)$ for a single machine scheduling example with the processing times of Table 3.1

Fig. 3.3 Flow layout schematics of a parallel machine layout

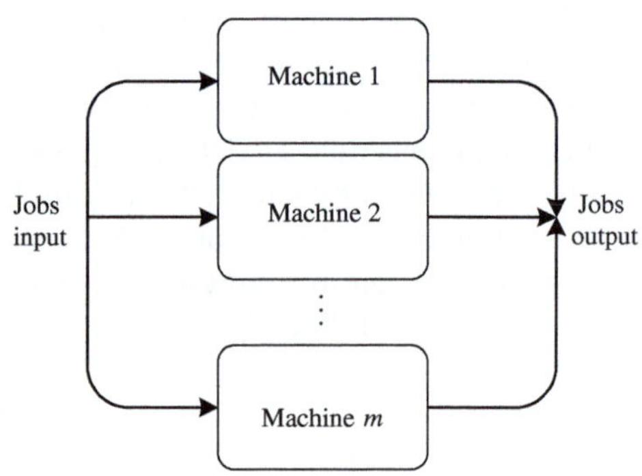

Parallel machine models are interesting since a second dimension arises in the scheduling model. Again, if we assume semi-active schedules, a schedule can be represented by the assignment of each job to each machine plus the specific sequence of processing the jobs for each machine.

The flow layout schematics of a parallel machine is also simple. Jobs enter the shop, are assigned to one out of the m machines, processed and then leave the shop. The schematics are shown in Fig. 3.3.

Table 3.2 Processing times (p_{ij}) of the eight jobs and three unrelated parallel machines example

Machine (i)	Job (j)							
	1	2	3	4	5	6	7	8
1	37	25	25	11	48	68	28	28
2	44	42	20	27	56	59	12	39
3	44	66	58	87	53	41	47	76

Parallel machine models are further divided into three categories:

1. Identical parallel machines. Each machine is identical as regards the processing of the jobs. This means that the processing time of each job (p_j) does not depend on the machine to which it is assigned to. These models are easier since, much of the time, the assignment problem can be reduced to assigning each job to the first available machine.
2. Uniform parallel machines. This setting is also called machines with different speeds. Under this scenario, some machines are slower or faster by a constant ratio for all the jobs. Therefore, a fast machine is actually faster for all the jobs. In order to calculate the processing times, each machine is assigned a speed ratio v_i. Then, the processing time of job j for that machine is calculated as $p_{ij} = p_j/v_i$. Therefore, v_i values greater than one represent faster machines whereas v_i values lower than one are given to slower machines. This uniform parallel machine model can be reduced to the identical parallel machine case if $\forall i \in M$, $v_i = c$, where c is any constant value.
3. Unrelated parallel machines. This is the most general case of the three. Machines are assumed to be different from one another. This might be helpful when modelling production shops where machines that serve the same purpose have been purchased and installed at different dates and therefore, different machine versions coexist in the shop. It is expected that each machine might perform differently in relation to the job. Under this scenario, the processing time of a given job depends on the machine to which is assigned to, and, as a result, the input data is a processing time matrix p_{ij}.

Let us exemplify one unrelated parallel machine model. We have eight jobs to be processed on three unrelated parallel machines. The processing time matrix is given in Table 3.2.

Now let us picture a processing sequence. Jobs 4, 2 and 8 have been assigned, in this sequence, to machine 1. Similarly, jobs 7, 3 and 1 have been assigned to machine 2 and jobs 6 and 5 to machine 3. The resulting Gantt chart can be observed in Fig. 3.4.

3.2.3.3 Flow Shop Layout

Flow shops (often, this term is spelled together as flowshop, we will use both forms interchangeably) are the first of a series of the so-called 'shop' layouts, which include

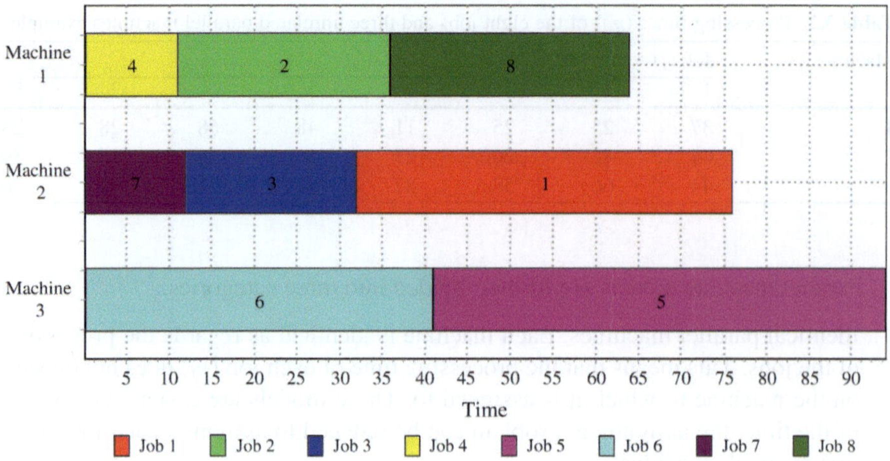

Fig. 3.4 Gantt chart with the result of the assignment and sequencing of eight jobs on three parallel machines

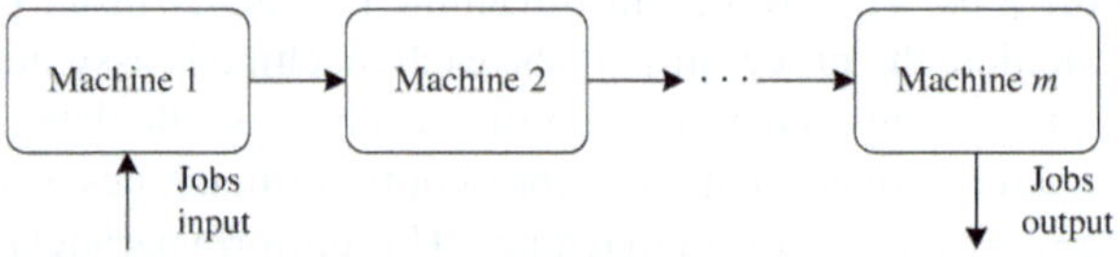

Fig. 3.5 Flow layout schematics of a flow shop

flow shops, job shops and open shops (these last two will be object of study in the following sections).

In a flow shop, there are m machines, but instead of being arranged in parallel, they are organised in series. It is assumed that each machine serves a different purpose. Every job has to visit all machines, in the same specified order. This order, without loss of generality, can be assumed to be $1, 2, \ldots, m$. As the order is the same for all jobs, we have that $\forall j \in N, R_j = (1, 2, \ldots, m)$. Furthermore, p_{ij} denotes the time job j needs at machine i. The term 'flow' refers to the fact that all jobs flow from one machine to the other in the same order, this is clearly depicted in the flow layout schematics of Fig. 3.5.

Note that the order of processing of each job on each machine is, in general, different. Therefore, each semi-active schedule can be represented by giving, for each machine, the sequence to process the jobs. In many cases, and mostly for reasons of simplicity in scheduling and shop floor control, it is assumed that the processing sequence is the same for all machines. This special case is referred to as the permutation flow shop model. Clearly, in a permutation flow shop model, each semi-active schedule can be represented by a single sequence of the jobs, which means that the maximum number of possible (semi-active) schedules in a permutation flow shop is reduced to $n!$ as compared to the $(n!)^m$ in the (general) flow shop model.

Table 3.3 Processing times (p_{ij}) of the five jobs and four machines permutation flow shop model example

Machine (i)	Job (j)				
	1	2	3	4	5
1	31	19	23	13	33
2	41	55	42	22	5
3	25	3	27	14	57
4	30	34	6	13	19

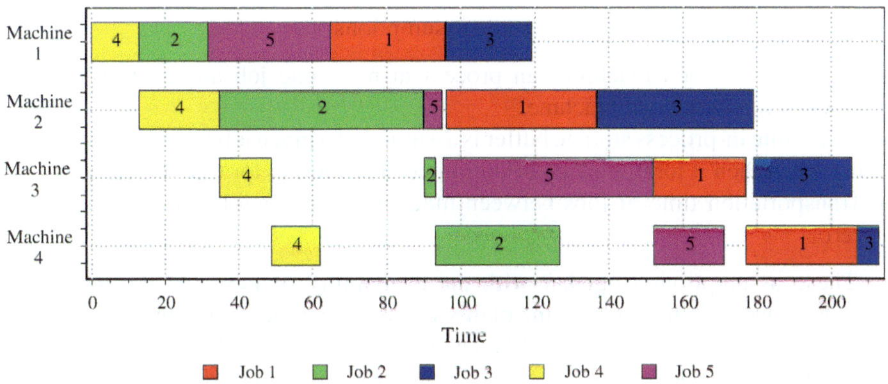

Fig. 3.6 Gantt chart with the result of the sequence $\pi = \{4, 2, 5, 1, 3\}$ for a permutation flow shop model with the processing times of Table 3.3

Due to its higher simplicity, let us start with an example of a permutation flow shop with four machines disposed in series and five jobs to schedule. The processing times are given in Table 3.3. Assuming the sequence $\pi = (4, 2, 5, 1, 3)$, the resulting Gantt chart is provided in Fig. 3.6.

Notice how some idle times start to appear on machines. This is different from the two previous single and parallel machine problems that have been depicted in Figs. 3.2 and 3.4 where machines were continuously busy. In Fig. 3.6, we see that the first machine is never idle from start of its first operation until the finishing time of its last operation, as all the jobs are being 'launched' to the shop one after another. However, and at least while the first job in the sequence is being processed on the first machine, all other machines are left waiting. Other notable idle times occur at machines three and four after processing the first job (4) of the sequence. Those large idle times obey to the fact that job (2), the second in the sequence, has a very large processing time on the second machine (55) and machine 3 has to wait until this job is completed in the previous machine.

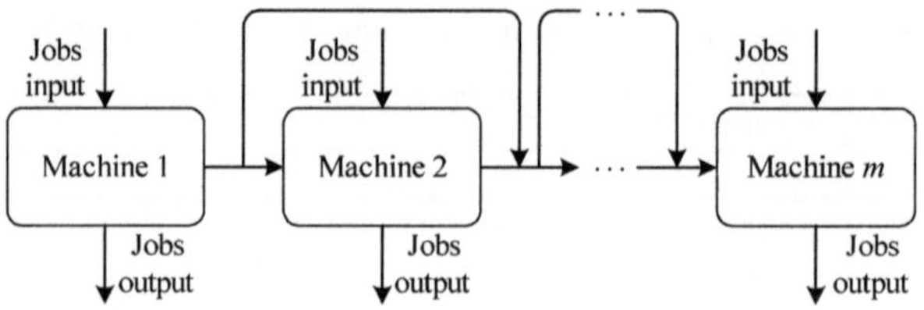

Fig. 3.7 Flow layout schematics of flexible flow shop

In the example, we have done some assumptions:

- At any time, each machine can process at most one job and each job can be processed only on one machine.
- An infinite in-process storage buffer is assumed. If a given job needs an unavailable machine, then it joins a queue of unlimited size waiting for that machine.
- Transportation time of jobs between machines is negligible and assumed to be zero.

These assumptions (together with the availability of jobs and no preemption already mentioned in the beginning of this section) are rather common in the scheduling literature, also for other shop-like layouts (see Baker and Trietsch 2009, among many others). These aspects as well as other constraints will be discussed in Chap. 4. The resulting model, although more realistic in the sense that different production facilities disposed in series are modelised, is rather theoretical. As a matter of fact, it has received a number of criticisms, most notably the ones of Dudek et al. (1992) or MacCarthy and Liu (1993). Nevertheless, the permutation flow shop model has been thoroughly studied in the past and it is still a very active and fruitful field of study nowadays.

One interesting variant of the permutation flow shop model is the so-called *flow line* in which all jobs follow the same relative route visiting the machines but some machines might be skipped by some jobs. The skipped machines can be seen as optional processing of treatments that are not necessary for all jobs. A schematic layout is given in Fig. 3.7.

As mentioned before, the (general) flow shop does not preclude the possibility of changing the processing sequence from machine to machine. Figures 3.8 and 3.9 show a simple example of a three machine, four job Gantt chart example of the same schedule on permutation and a (general or non-permutation) flow shop, respectively.

Notice how in the non-permutation version of Fig. 3.9, the sequence of the first machine $\pi_1 = (2, 3, 1, 4)$ changes in the second and third machines to $\pi_2 = \pi_3 = (2, 1, 3, 4)$. In this example, we would say that job 1 has passed job 3 on machines 2 and 3. As already mentioned, the (non-permutation) flow shop problem is more complex than the permutation version. Clearly, all permutation flow shop solutions are included in the set of solutions of the general flowshop. Therefore, the optimal solution of the general flow shop cannot be worse than the one from the permutation

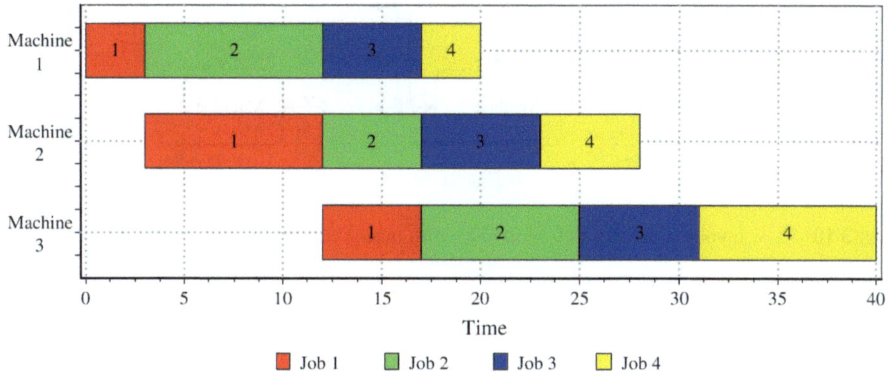

Fig. 3.8 Gantt chart of a 3 machine, four job permutation flow shop

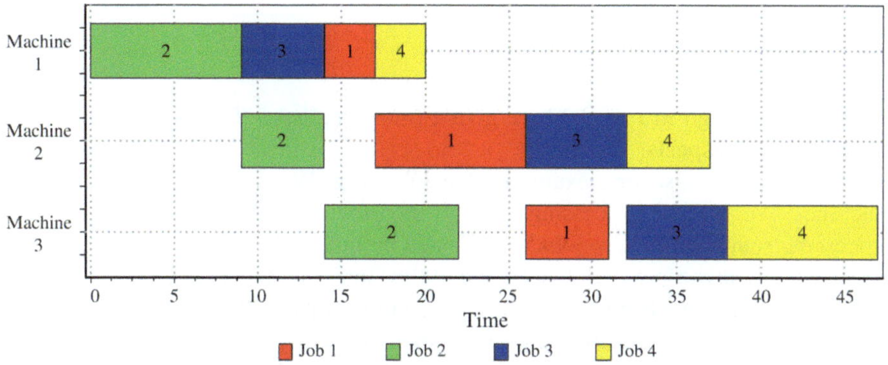

Fig. 3.9 Gantt chart of a 3 machine, four job (non-permutation) flow shop

problem. However, it might very well be that some of the non-permutation solutions are worse than some of the permutation solutions.

3.2.3.4 Job Shop Layout

The job shop is the second type of the so-called shop environments. Similar to flow shops, in a job shop there are m machines disposed in series. However, there is a big difference as compared to the flow shop: Every job has a potentially different route. For example, given 3 machines and four jobs, we could have $R_1 = (1, 2, 3)$, $R_2 = (2, 1, 3)$, $R_3 = (3, 2, 1)$ and $R_4 = (2, 3, 1)$. Job shops are very frequent in practice when modelling shops where each order is different from the others and requires a different machine routing. In most theoretical job shop research, the jobs visit all machines exactly once, with no missing machines and no re-visitation of

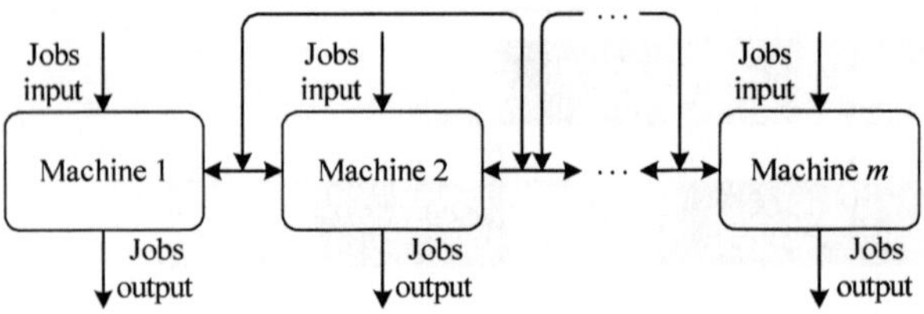

Fig. 3.10 Flow layout schematics of a job shop with machine skipping and recirculation

Table 3.4 Processing routes (R_j) for the five jobs in the job shop example

Job (*j*)	R_j			
1	1	2	3	4
2	4	2	1	3
3	3	1	2	4
4	2	4	3	1
5	4	1	3	2

machines, although in Fig. 3.10 we give a (more realistic) flow layout schematics including these aspects.

Let us picture also one example of a job shop. We have five jobs and three machines. We take the processing times of the previous Table 3.3 and consider the processing routes of the five jobs as indicated in Table 3.4. With these data, we can construct one possible schedule as given in the Gantt chart of Fig. 3.11.

Job shops are much more complex than flow shops as these last problems are special cases when all R_j are identical. Notice how the possibilities for improving a sequence are endless. For example, in the last Fig. 3.11, there is a large idle time on machine 3 between jobs 1 and 2. It would be better to start task O_{35} around time 110, just when task O_{15} finishes, delaying a bit O_{32} but in turn also advancing O_{25}.

3.2.3.5 Open Shop Layout

The open shop is the more general and most complex of the shop layouts. Again, as in the flow and job shops, there are m machines disposed in series and n jobs that have to visit all m machines. However, the big difference is that the processing route for each job is not fixed as is to be determined in the scheduling process. From another perspective, it can be assumed that all R_j for the jobs are arbitrary and that operations of a job can be processed in any order on the machines. Figure 3.12 shows a layout schematics of an open shop.

It is easy to see that flow shops and job shops are special cases of the open shop. One way to see this is to consider the enormous number of possible solutions that could be obtained following the example of Fig. 3.11 knowing that the only thing that has to be satisfied is that operations of the same job cannot be overlapped and that

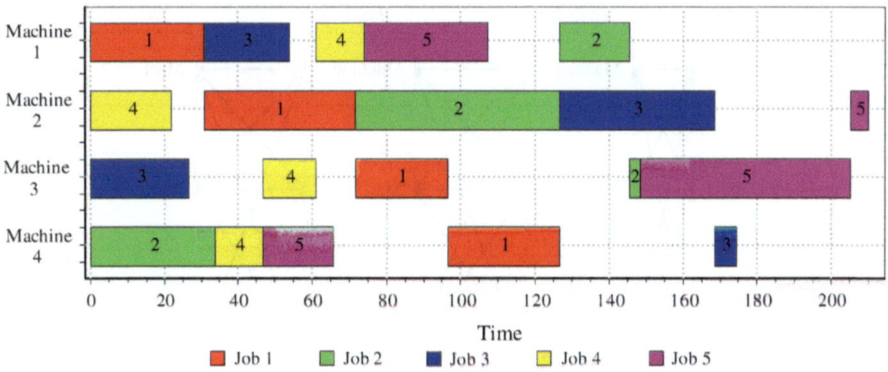

Fig. 3.11 Gantt chart with a possible sequence in a job shop with the processing times of Table 3.3 and the processing routes of Table 3.4

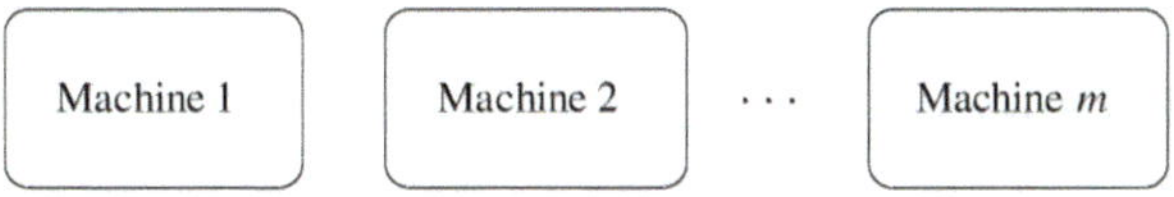

Fig. 3.12 Flow layout schematics of an open shop

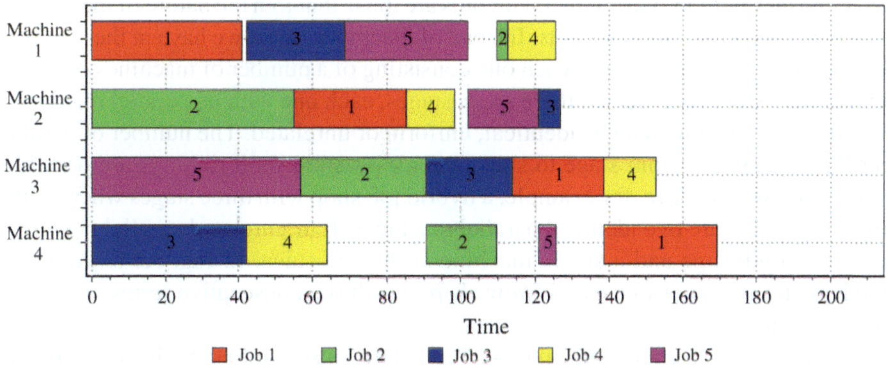

Fig. 3.13 Gantt chart with a possible sequence for an open shop with the processing times of Table 3.3

machines cannot process more than one job at the same time. One possible sequence, following the processing time data of Table 3.3 is given in Fig. 3.13.

As can be seen, the open shop is very flexible. The schedule has been 'squeezed' and there are less idle times in the machines. This is because of the total routing flexibility.

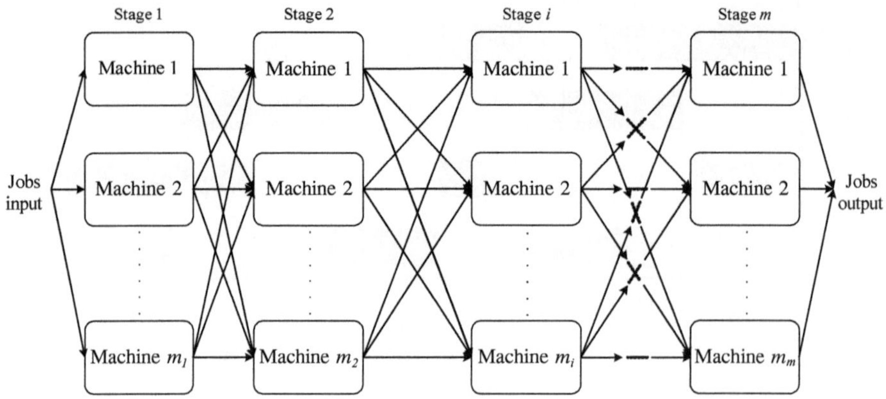

Fig. 3.14 Flow layout schematics of a hybrid flow shop

3.2.3.6 Hybrid Layouts

Normally, at production plants, important machinery is replicated. This was already commented in the parallel machine layout. However, in parallel machines, there is only one task per job. More general layouts are those that join the nature of the parallel machines and shop environments. Instead of machines, what we have in these hybrid layouts are production stages, each one consisting of a number of machines disposed in parallel. More specifically, there are m stages, each one with m_i parallel machines, where these machines can be identical, uniform or unrelated. The number of parallel machines can vary from stage to stage and all possible shop routings are possible. Therefore, we can have, for example, a hybrid job shop with three stages where at the first stage there are two identical parallel machines, four unrelated parallel machines at the second stage and a single machine at the last stage. In Fig. 3.14, we depict the layout schematics of a *hybrid flow shop*, which is a consecutive series of parallel machine layouts.

Hybrid layouts add the machine assignment dimension into the scheduling model as for each job and stage, an assignment decision to machines should be made. Let us give an example. We have a hybrid flow shop problem with three stages and 3, 4 and 3 machines respectively. There are five jobs to be processed. The processing times are given in Table 3.5.

As can be seen, the parallel machines at each stage are unrelated, as it does not take the same processing time to process job one on machines one, two or three of the first stage, for example. Now we need job assignments to machines at each stage and also the sequences for the jobs assigned to each machine at each stage. These are given in Table 3.6. The resulting Gantt chart can be seen in Fig. 3.15.

Several interesting conclusions can be drawn from this example. One may think that, since the parallel machines at each stage are unrelated, good schedules will consist on assigning each job to the fastest machine at that stage. However, this

Table 3.5 Processing times (p_{ij}) of the five jobs, three stages and 10 machines in a hybrid flow shop

Stage (i)	Machine (l)	Jobs (j)				
		1	2	3	4	5
1	1	27	45	56	40	30
	2	80	85	25	12	75
	3	44	17	39	29	25
2	4	25	80	12	16	25
	5	39	45	97	24	88
	6	44	71	28	25	96
	7	38	26	10	27	44
3	8	93	30	67	29	10
	9	45	63	66	88	89
	10	20	80	35	15	25

Table 3.6 Job sequence for each machine in the five jobs, three stages and 10 machines hybrid flow shop scheduling example

Stage (i)	Machine (l)	Job Sequence	
1	1	1	
	2	4	3
	3	2	5
2	4	1	5
	5	4	
	6		
	7	2	3
3	8	2	5
	9	1	
	10	4	3

greedy approach will not, in general, result in the best solution. Consider that the objective is to finish the set of jobs as soon as possible. The fastest machine for job 1 at the third stage is machine 10 as $p_{10,1} = 20$. However, job 1 has been assigned to machine 9, which has more than double the processing time. This is because jobs 3 and 4 have been already assigned to machine 10 (the fastest machine for them) and machine 10 is already overloaded. Therefore, if we want to finish job 1 as soon as possible, a slower machine has to be employed at the third stage. A second conclusion is also that machine 6 in the second stage is not used for this schedule. This is because the second stage has more machines than are actually needed.

3.2.3.7 Other Layouts

The previous layouts are not, by a long shot, the only possible ones. They should be seen as a broad classification inside which, with some degree of abstraction, some

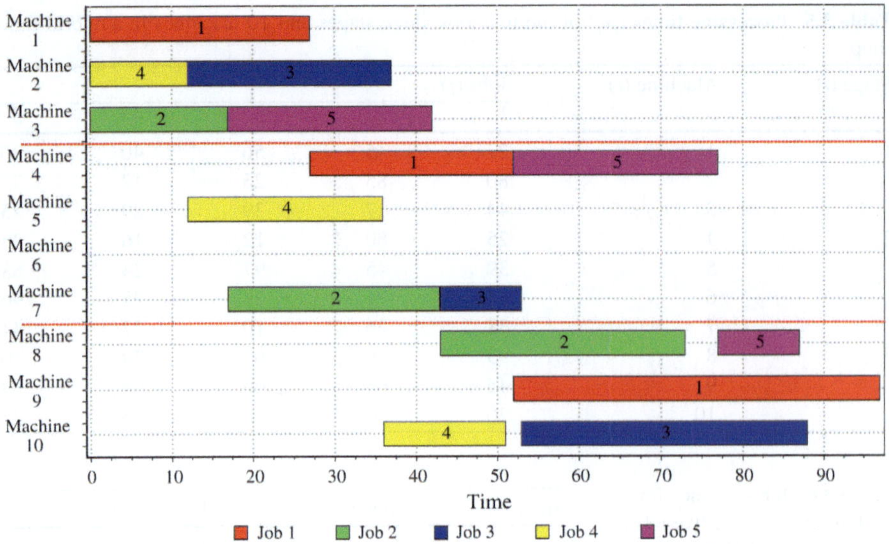

Fig. 3.15 Gantt chart with the a possible sequence for a hybrid flow shop problem with the processing times of Table 3.5 and the job sequences of Table 3.6

production environments could fit. Some other prototypical environments can be briefly summarised as follows:

- Manufacturing cells. They are a mixture between the high flexibility of the job shop and the mass production oriented low cost flow shop. Often referred to as cellular manufacturing, results in sets of machines grouped into cells. Each cell produces one type of product or family of products. Most of the time, machines at each cell are numerically controlled and are highly automated and it is not unusual to have robotic arms or automatic handling systems to move products from machine to machine inside the cell. Properly implemented cells have shown to be more flexible and responsive than more traditional mass production lines, like flow shops. Some basic references about cellular manufacturing are Hyer and Wemmerlöv (2002), Irani (1999).
- Assembly shops. These layouts model the bill of materials structure of several products whose intermediate products have to be manufactured first. Therefore, tasks of some jobs have to be finished in order to a subsequence assembly task to begin, which uses the previous tasks as raw materials. We will later see that these assembly operations can be modelled by precedence relationships.
- Production lines. A production line is very well understood when picturing a car manufacturing line. Production lines are often called assembly lines. Here, the products move in a linear way along conveyor belts or along any other possible means of transportation. Operations are performed over the products as they move as there is no buffer in between operations. Often, the places at the production line where operations are performed are referred to as stations. In general, an

assembly line is a medium- to high-volume, low product variety version of the assembly shop. An assembly line need not to be linear. It can have several feeder lines at certain points. Assembly lines are used, e.g. for manufacturing aircrafts, automobiles, appliances and farm equipment (see Morton and Pentico 1993).
- Batch shop. Batch processing exists when a specific number of jobs can be produced simultaneously on one machine at the same time. The completion time of a batch will be determined by the longest processing time of a job in its batch. A batch is characterised by homogenous properties (of its sub-jobs) and is processed as a whole through the shop, sometimes with additional no wait or other constraints. Batch shops occur e.g. in refinery, chemical, semiconductor or other process industries. Batch operations can be found at ovens where several jobs are collected to guarantee a high level of utilisation of the oven.
- Flexible manufacturing systems. Under this category, we find a very large number of possible layouts. We have included them here as in all other mentioned layouts, machines are more or less fixed and no flexibility is allowed in the short or medium term. Flexible manufacturing systems or FMS in short, are engineered with flexibility in mind so that new product types, or rapid changes in specifications of the products can be performed easily. FMS systems are also characterized by routing flexibility or by allowing different ways of manufacturing for the same products. FMS systems are commonly employed in the production of small lots of highly personalized products.

3.2.4 Additional Processing Constraints

Layouts are subject to countless situations, constraints, characteristics and events of all types. In the following, we provide just a short list of the number of processing constraints that real problems might have:

- Jobs might not be independent among each other as there might be a Bill of Materials (BOM) structure as commented in the assembly shops.
- Machines are actually never continuously available. Aside from economic/demand reasons, programmed maintenance activities, random breakdowns or other failures might prevent machines to operate at any time.
- In many situations, the processing of jobs can be interrupted and resumed at a later time. Pre-emption is allowed in most production settings.
- Setup times occur frequently in practice as well as many other operations on machines that are not just processing times of jobs on machines. Mounting or unmounting jigs or fixtures, cleanings, start-up or stop-down delays, etc.
- Buffers and storage areas for holding in-process semi-finished jobs are not infinite. For bulky and/or large products, storage areas might be even more limiting than machine availabilities.
- Jobs might not be allowed to wait indefinitely in between machines. Cooling down of products that have to be treated while hot (like steel rods) or products that might

dry up or evaporate are clear examples. In some extremes, products might not be allowed to wait at all in between stages. Frozen food is a good example of such no waiting allowed scenarios.

- Similarly, a minimum waiting time in between operations might be necessary since products might be too hot to handle after a furnace or kiln operation, to name an example.
- Assuming that all data is known in advance and that this data is deterministic is likely to be unacceptable. In controlled environments and tested manufacturing systems, the time needed for a given operation of a product might be known and pretty much fixed. However, the nature of some production processes is stochastic and processing times (as well as any other data) might be hard to estimate or might come in the form of a statistic distribution with known parameters.
- Production floors are highly dynamic. This means that jobs arrive over time, release dates change, due dates might be re-arranged with clients, weights or importances might vary. Basically, every piece of information, even if originally known in advance, is subject to change.
- In-process inventory, or semi-finished products do not instantly move from one machine to the other. Transportation times have to be taken into account. Sometimes, transportation is carried out by simple conveyor belts, other times, complex robotic or automated guided vehicles are used to move the products inside production plants. As a matter of fact, transporting the products can be seen as a routing problem in complex plant layouts and large factories.
- Machines are operated by personnel. This has huge implications at so many levels. From work shifts or timetables, to skill levels, learning effects and others. Machines might operate at different speeds depending on the number of persons attending to them. Similarly, more than one machine might be supervised by a single person depending on the shop layout.
- Fabrication is not perfect. Frequently, reworkings, fixings and reintroduction of jobs in the lines occur.
- In today's globalised economy, factories are no longer isolated elements. Distributed manufacturing and complex supply chains bring implications that reach into the scheduling decisions at manufacturing plants. The number of possible situations in such cases is very large.
- Jobs might have utterly complex processing routes, even alternative processing routes that might even change the bill of materials in dependence of the chosen manufacturing route.
- Processing times, even if known, can be worked on. Assigning more specialised personnel or additional tooling or any other resource could shorten, for an additional cost, the processing times.

As a matter of fact, the previous list is just a glimpse of situations that may arise in manufacturing scheduling. These will be treated in larger detail in Chap. 4.

3.2.5 Scheduling Criteria

As already discussed in Sect. 3.2.1, another element required to define a scheduling model is to establish the criterion (or criteria) employed in such decision problem. Usual criteria are based on some broad categories that can be summarised as follows:

- Utilisation-based criteria. These objectives are concerned with making the most out of the available productive resources: maximising machine utilisation, minimising idle times, reducing changeover or setup times, maximising the number of completed jobs per unit of time, etc.
- Customer's satisfaction. These include performance measures related with the fulfilment of the due dates, or the number of jobs that are completed before their due dates d_j or the equivalent of minimising the number of late jobs. Some other related criteria are those in which it is desired to minimise the number of 'late units' or the amount of time jobs are completed beyond their due dates.
- Storage oriented. When storage is a serious concern, minimising the work-in-progress, minimising the queue lengths in between machines or other similar procedures is beneficial. In order not to store finished products for too long, it might be interesting not to complete jobs way before their corresponding due dates.
- Cost oriented. Large batches are usually more economical due to economies of scale. Many other policies can be enforced and decisions can be made in order to reduce production costs. Usually, cost objectives are related with the previous items, specially with throughput and storage.
- Just in time, lean manufacturing and others. Just-in-time production requires jobs to be completed exactly at their due dates, not before and not after. In lean manufacturing, special emphasis is put in reducing waste and non-value adding operations throughout production. Other criteria could come in the form of maximising the number of accepted jobs for production subject that those jobs can be completed before their due dates (i.e. deadlines), etc.

Most of the scheduling literature is concerned about a single criterion or objective, and most of the time is of the minimisation type. As it happens with the processing constraints, scheduling criteria are as varied and numerous as possible real-life manufacturing scheduling scenarios one might encounter. The previous itemized list is just a brief attempt at structuring the various measures. As stated, these performance measures, along as many others, will be discussed in more detail in Chap. 5. Later, the consideration of more than one criteria under multiobjective approaches will be introduced in Chap. 10.

3.3 Classification of Scheduling Models

As already noted in several parts in this book, there is a sheer variety of production layouts, processing constraints and criteria in scheduling models. The traditional way to handle the diversity of scheduling models has been to classify them using taxonomies and notations that allow establishing the similarities and differences among the different models. As we will discuss in Chap. 7, this classification is extremely useful to build solution procedures for scheduling models based on these of similar models. In the following sections, we present the two most popular classification schemes.

3.3.1 Early Classification Schemes

Taxonomies or classifications for scheduling models started to appear in the 1960s with the book of Conway et al. (1967). These first taxonomies try to classify production scheduling models according to their main characteristics already mentioned, namely the floor layout, the processing constraints and the performance measure studied. Conway et al. (1967) proposed a quadruplet $A/B/C/D$ where each capital letter has a more or less precise meaning:

- A indicates the number of jobs that have to be scheduled in a given problem. In other words, A indicates the number of jobs n.
- B describes m the number of machines in the shop floor.
- C describes the processing layout in a simplistic way:
  - If $C = 1$ the layout is formed by a single machine. In some other related works, the field A is omitted in the case of one single machine.
  - If $C = P$ then we have a parallel machines layout.
  - If $C = F$ the problem is a flow shop.
  - $C = J$ indicates a job shop.
  - Finally, if $C = O$ an open shop is dealt with.
- D serves for indicating the type of optimisation criterion for the scheduling model.

For example, a problem denoted as $10/1/1/D$ or simply $10/1/D$ represents a single machine problem where 10 jobs have to be scheduled under the unspecified criterion D (performance measures will be precisely detailed in Chap. 5). Another example could be $n/m/F/D$ which represents the typical flow shop problem with a general number of jobs n and m machines disposed in series and an unspecified criterion D.

It is easy to see that the previous classification is sorely incomplete. For example, there are no distinctions between the three types of parallel machines. Also, processing constraints are neglected as the notation does not allow for their inclusion. Hybrid layouts and/or more complex layouts do not have a corresponding letter or expression way. For all these reasons, in the early 1970s, other much more comprehensive

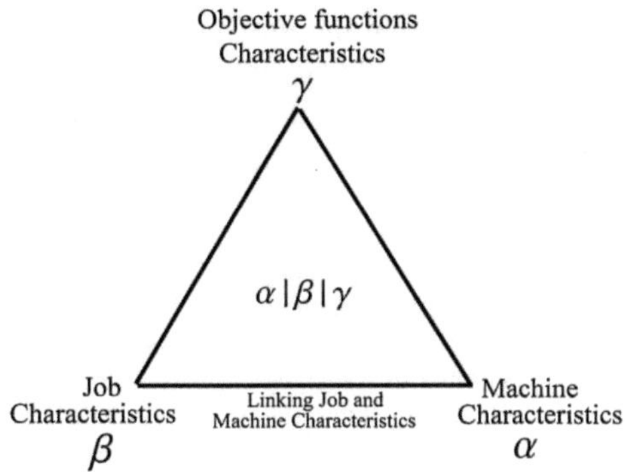

Fig. 3.16 $\alpha|\beta|\gamma$-classification of scheduling models

taxonomies or scheduling problem classification schemes were proposed. These are discussed in next section.

3.3.2 Current Classification Schemes

From the shortcomings of the previous classification of Conway et al. (1967), other authors proposed extensions, like RinnooyKan (1976) which proposed a notation based on the triplet $\alpha/\beta/\gamma$. This scheme was further extended and refined in the works of Graham et al (1979), Błazewicz et al. (1983), Lawler et al. (1993).

In the following, we highlight the most important aspects of this notation, which is pictured in Fig. 3.16.

- The first field in the triplet α defines the processing layouts. It is further split into two sub-fields $\alpha_1\alpha_2$ so that:

$\alpha_1 = \varnothing$: single machine layout, sometimes also denoted as $\alpha_1 = 1$.
$= P$: identical parallel machines.
$= Q$: uniform parallel machines.
$= R$: unrelated parallel machines.
$= F$: flow shop.
$= J$: job shop.
$= O$: open shop.
$\alpha_2 = 1, 2 \ldots, m$: fixed number of machines in the layout.
$= \varnothing$: the number of machines is not fixed to a precise number (i.e. the problem studied is not limited to a case with a specific machine count). Often this is also referred to as simply m. Note that in single machine problems ($\alpha_1 = 1$), α_2 has no meaning.

Therefore, with the α field, the machine layout can be easily defined. For example, $\alpha = P3$ means the considered layout is composed by three identical parallel machines. $\alpha = Fm$ or $\alpha = F$ represents a flow shop layout with any number of machines.

Note again that the previous α field suffers from some shortcomings as there is not a predefined way of representing hybrid layouts. The review paper of Vignier et al. (1999) introduced a significant extension. The authors split the α field into four sub-fields $\alpha_1\alpha_2\alpha_3\alpha_4$ as follows:

$\alpha_1 = HF$: hybrid flow shop.
$= HJ$: hybrid job shop.
$= HO$: hybrid open shop.
$\alpha_2 = 1, 2 \ldots, m$: the number of stages. Again, it could be a fixed number or a general number of stages m.
$\alpha_3 = P$: identical parallel machines at stage i.
$= Q$: uniform parallel machines at stage i.
$= R$: unrelated parallel machines at stage i.
$\alpha_4 = 1, 2 \ldots, m_i$: number of parallel machines at stage i.

Note that the two last sub-fields α_3 and α_4 are repeated for every stage i specified in the sub-field α_2. For example, a hybrid layout with $\alpha = HJ3, 1, P2, R3$ denotes a hybrid job shop with three stages where the first stage has one single machine, the second stage has two identical parallel machines and the third stage has three unrelated parallel machines. Another example could be $\alpha = HFm, \left((Pm_i)_{i=1}^m\right)$ which stands for a general hybrid flow shop with m stages where each stage i contains m_i identical parallel machines.

- The second field in the triplet notation, β indicates the processing constraints. This field is split into as many sub-fields as processing constraints there might be in the studied problem, separated by commas. As we exposed in previous sections, processing constraints are varied and numerous and are object of later study in Chap. 4, where we will define many possible β fields.
- Lastly, the third field in the triplet notation, γ, gives information about the single objective considered in the studied problem. As with the processing constraints, we have devoted a specific chapter to scheduling objectives which is Chap. 5. Furthermore, in Chap. 10 we will significantly extend the γ field in order to allow the notation of multiobjective problems.

We have to bear in mind that even with the hybrid layout extensions given by Vignier et al. (1999), it is simply not possible to capture every possible production scheduling layout. Not only this is not possible, but also it would be unrealistic to expect so. Notation schemes have to be succinct enough so to facilitate a quick and unambiguous identification of a layout. Complex layouts are simply approximated to the closest possible and still easily representable layout. Additional characteristics of complex layouts are given in plain text afterwards. As a result, other researchers and practitioners might grasp the most important aspects of the represented layout as if there are complex job routes, stages and so on.

3.4 Production Planning, Lot Sizing and Lot Streaming

In Sect. 1.5 we discussed already that the concept of a job in scheduling manufacturing, albeit probably clear in the scientific literature, is far from being a precise concept in real manufacturing. As a matter of fact, probably, the first barrier when having a scheduling practitioner at a company communicating with a scheduling researcher is to agree on what a 'job' is. Given the extreme importance of identifying this unit of information in order to build scheduling models, we devote this section to discuss first how jobs (as they are interpreted in the scheduling models presented in this chapter) result as the output of a production planning process (as it is described in Sect. 2.4) and, secondly, some alternatives (lot sizing and lot streaming) to aggregate/disaggregate job entities.

Although Sect. 2.4 has already provided additional insight on the planning process, production management is a large and broad field and it is outside the scope of this book and therefore and we will simply stress that, at least with respect to the implications over manufacturing scheduling, production planning is the process of obtaining a list of products to be manufactured, along with their quantities.

Some traditional production planning techniques, and more particularly, Materials Requirements Planning (MRP) suffer from several shortcomings that have been long documented in the literature and identified in practice. Some of these problems are:

- Data integrity. Unless high integrity of the data is assured, the result is unsatisfactory. Bill of Materials, inventories, orders and all related data has to be carefully checked in order to avoid what has been sometimes called a GIGO (Garbage In, Garbage Out) situation.
- MRP systems usually assume constant lead times and are inflexible as regards as other data like fixed reorder quantities, points, safety stocks, etc.
- The major problem is that the planning carried out by MRP systems does not take production capacity into account. Results might be impossible to implement.
- MRP typically does not issue early warnings for stock outs, demand spikes, shortages, etc.

Furthermore, some authors have criticised the MRP approach for its limited structure that basically breaks down the production demands into product families and later into individual products without actually considering the interactions between them and the underlying model structure. This myopic decomposition leads to important productivity and flexibility losses. Although more advanced techniques such as Manufacturing Resources Planning (MRPII) or Capacity Requirements Planning (CRP) overcome some of these problems, alternatives involving the use of linear and integer programming to model and to solve complex production planning problems in which high quality solutions (usually optimal) are of interest. One of the basic models is the well known Single-Item, Single Level, Uncapacitated Lot-Sizing problem, denoted usually as *SLULSP*, which can be formulated as a simple MILP model considering holding, fixed and unit production costs for the product.[2]

Although simplistic, *SLULSP* model takes into account variable demands, variable holding, fixed and unitary production costs over the periods and therefore, even in its simplest form, surpasses the capacities of the outdated MRP methodology. Furthermore, it can be easily extended in several ways. Maximum storage capacities can be added. Similarly, maximum production capacities or maximum number of production runs over the periods are also easy to add. Other more complex production planning optimisation models include the multiple-item, single level, uncapacitated lot-sizing problem, or the more general multi-level variants.

From a manufacturing scheduling perspective, the output of such models provides the value of the variables indicating the quantity to be produced for each periods in the planning period for each type of product, thus identifying the 'jobs' (the amount of each type of product to be produced) to be scheduled by aggregating the items to be produced at the same time (lot sizing), together with their corresponding due dates (the times in which the amount has to be completed).

In the view discussed in the paragraph above, jobs would be actually made up of lots of several discrete and identical items, like for example lots of microprocessors in a semiconductor industry or lots of metal sheets in a metallurgy firm. Consequently, the whole lot will be considered a single unit of processing, as thus will have to be finished in the upstream machine before starting in the subsequent downstream machine. *Lot streaming* breaks this assumption by allowing the job to be broken down into 'sublots'. The sublots, once completed on the previous machine, move on to the next machine, effectively resulting in different sublots of the same product or job being simultaneously produced on different machines. Obviously, this 'overlap' may be beneficial in many ways. Lot streaming within the scheduling context is the short-term optimisation approach to the previous production planning models. Recall that a production planning period t is usually not shorter than a week and there is a lot of production time in a week available for scheduling. Lot streaming has many different variants and characteristics, like for example a limit on the number of sublots, equal or variable size sublots, continuous or discrete size sublots and many others. On the other hand, lot streaming may not be interesting due to economic/technical reasons if the subsequent process has an operating cost which is not depending on the lot size (here the classical example is that of a furnace). The effect of lot streaming in a flow shop layout, considering the previous example given in Fig. 3.6 is shown in Fig. 3.17.

As we can see in Fig. 3.17, machines are better utilised as when compared with Fig. 3.6. The effect is particularly interesting in job 4 which is in the first position of the sequence. Lot streaming allows for units of job 4 to be available as soon as after 25 units of time have elapsed. This is much shorter when compared to the first units that could be available at time 50 in the flow shop pictured in Fig. 3.6.

---

[2] Note that, in some references, processing costs are time-independent and consequently, are removed from the objective function since processing costs are decision-irrelevant.

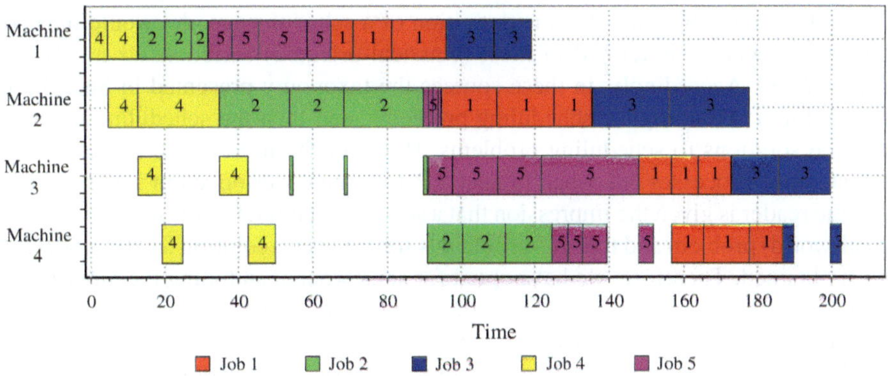

Fig. 3.17 Gantt chart for a streaming flow shop problem based in the example of Fig. 3.6

A final remark has to be done to discard the idea that, in principle, sublots could be simply modeled as different jobs. Although theoretically interesting, this approach quickly results in problems with a very large number of jobs (literally thousands) and it is therefore quite unpractical.

3.5 Conclusions and Further Readings

Manufacturing scheduling is a complex problem arising in many industries. As with any complex problem, a scientific approach is often needed in order to obtain satisfactory solutions. In this chapter, we have introduced the basic concepts of modelling as a necessity for scheduling problem solving, together with the main characteristics of models, types of models, common errors and shortcomings.

Next, some basic definitions for scheduling models have been provided, and special emphasis has been given to the processing layouts as defined in the rich scheduling literature. We have defined the layouts of single machine problems, parallel layouts, shop problems (flow, job and open shops) and the hybrid layouts often found in practice. A short introduction to other less frequent shop arrangements has also been given.

The main processing constraints and an introduction to scheduling criteria has been given in this chapter, although both will be significantly extended in the next two chapters, namely Chaps. 4 and 5. Of special interest are the links to real production scheduling problems not often highlighted elsewhere. Early as well as current scheduling model classification schemes, along with some extensions, have been introduced. Finally, we have devoted one section to discuss first how jobs (as they are interpreted in the manufacturing scheduling context) result as the output of a production planning process and, second, some alternatives (lot sizing and lot streaming) to aggregate/disaggregate job entities.

We have already noted that the definition of a scheduling model would more or less match what it is termed in most (classical-oriented) textbooks as a *scheduling problem*. Accordingly, in these sources the taxonomy presented in Sect. 3.3 is referred to a taxonomy for scheduling problems, and then the scheduling methods provide solutions to scheduling problems. However the need of the modeling and transference processes described in Sect. 3.2.1 is often overlooked in some sources, and the reader is given the impression that a scheduling model arises naturally from a (real-world) problem and that this problem is solved once a solution for the model has been obtained. Therefore, we have opted for somewhat abandoning the mainstream scheduling literature, but in our defense we can always state that our definition of scheduling model (and instance) is taken from a source as early as Coffman (1976).

Regarding further readings, the literature is full of comprehensive works on scientific modelling. Some examples are Gilbert (1991) or Mayer (1992) and also the classical book of Churchman (1984) or more modern volumes like DaCosta and French (2003). The formal definitions and notation given in Sects. 3.2 and 3.3 follow the mainstream scheduling literature, such as for example, Conway et al. (1967), Baker (1974), French (1982), Błazewicz et al. (2002), Brucker (2007) or more modern texts like Pinedo (2012), Baker and Trietsch (2009), Pinedo (2009). Production planning is a very rich field where lots of books and papers have been published. Some of these texts are Orlicky (1975), Plossl (1994), Wight (1995), Higgins et al. (1996), Toomey (1996) Artiba and Elmaghraby (1997), Drexl and Kimms (1998), Onwubolu (2002), Sheikh (2003), Voss and Woodruff (2003), Stadtler and Kilger (2005), Proud (2007). Specially interesting is the book of Pochet and Wolsey (2006), where the mathematical and optimisation aspects of production planning are studied in great detail. A good review paper for the Single-Item, Single Level, Uncapacitated Lot-Sizing problem is given by Brahimi et al. (2006). Regarding lot streaming, some basic references are Potts and VanWassenhove (1992), Trietsch and Baker (1993) or more recently, Chang and Chiu (2005). Additionally, in the case of the flow shop layout, there is a book published that solely deals with the lot streaming variant (Sarin and Jaiprakash 2007).

References

Artiba, A. and Elmaghraby, S. E., editors (1997). *The planning and scheduling of production systems: methodologies and applications*. Chapman & Hall, London.

Baker, K. R. (1974). *Introduction to Sequencing and Scheduling*. John Wiley & Sons, New York.

Baker, K. R. and Trietsch, D. (2009). *Principles of Sequencing and Scheduling*. Wiley, New York.

Błazewicz, J., Ecker, K. H., Pesch, E., Schmidt, G., and Węglarz, J. (2002). *Scheduling Computer and Manufacturing Processes*. Springer-Verlag, Berlin, second edition.

Błazewicz, J., Lenstra, J. K., and RinnooyKan, A. H. G. (1983). Scheduling Subject to Constraints: Classification and Complexity. *Discrete Applied Mathematics*, 5:11–24.

Brahimi, N., Dauzère-Pérès, S., Najid, N. M., and Nordli, A. (2006). Single item lot sizing problems. *European Journal of Operational Research*, 168(1):1–16.

Brucker, P. (2007). *Scheduling Algorithms*. Springer, New York, fifth edition.

Chang, J. H. and Chiu, H. N. (2005). Comprehensive review of lot streaming. *International Journal of Production Research,* 43(8):1515–1536.

Churchman, C. W. (1984). *The Systems Approach.* Dell Publishing Company, New York. Revised and upgraded edition from the 1968 original.

Coffman, E. G. (1976). *Computer & Job/shop Scheduling Theory.* John Wiley & Sons.

Conway, R. W., Maxwell, W. L., and Miller, L. W. (1967). *Theory of Scheduling.* Dover Publications, New York. Unabridged publication from the 1967 original edition published by Addison-Wesley.

DaCosta, N. and French, S. (2003). *Science and Partial Truth: A Unitary Approach to Models and Scientific Reasoning.* Oxford University Press, Oxford.

Drexl, A. and Kimms, A., editors (1998). *Beyond Manufacturing Resource Planning (MRP II): advanced models and methods for production planning.* Springer, New York.

Dudek, R. A., Panwalkar, S. S., and Smith, M. L. (1992). The lessons of flowshop scheduling research. *Operations Research,* 40(1):7–13.

French, S. (1982). *Sequencing and Scheduling: An Introduction to the Mathematics of the Job-Shop.* Ellis Horwood Limited, Chichester.

Gilbert, S. W. (1991). Model-building and a definition of science. *Journal of Research in Science Teaching,* 28(1):73–79.

Graham, R. L., Lawler, E. L., Lenstra, J. K., and Rinnooy Kan, A. H. G. (1979). Optimization and Approximation in Deterministic Sequencing and Scheduling: A Survey. *Annals of Discrete Mathematics,* 5:287–326.

Higgins, P., Le Roy, P., and Tierney, L., editors (1996). *Manufacturing planning and control: beyond MRP II.* Springer, New York.

Hyer, N. and Wemmerlöv, U. (2002). *Reorganizing the Factory: Competing Through Cellular Manufacturing.* Productivity Press, Portland.

Irani, S. A., editor (1999). *Handbook of Cellular Manufacturing Systems.* Manufacturing & Automation Engineering. John Wiley & Sons, New York.

Lawler, E. L., Lenstra, J. K., and RinnooyKan, A. H. G. (1993). Sequencing and Scheduling: Algorithms and Complexity. In Graves, S. C., RinnooyKan, A. H. G., and Zipkin, P. H., editors, *Logistics of Production and Inventory,* volume 4 of *Handbooks in Operations Research and Management Science,* Amsterdam. Elsevier Science Publishers, B. V.

MacCarthy, B. L. and Liu, J. (1993). Addressing the gap in scheduling research: A review of optimization and heuristic methods in production scheduling. *International Journal of Production Research,* 31(1):59–79.

Mayer, R. E. (1992). Knowledge and thought: Mental models that support scientific reasoning. In Duschl, R. A. and Hamilton, R. J., editors, *Philosophy of science, cognitive psychology, and educational theory and practice,* pages 226–243, Albany. New York State University.

Morton, T. E. and Pentico, D. W. (1993). *Heuristic Scheduling Sysmtems With Applications to Production Systems and Project Management.* Wiley Series in Engineering & Technology Management. John Wiley & Sons, Hoboken.

Onwubolu, G. C. (2002). *Emerging optimization techniques in production planning and control.* Imperial College Press, London.

Orlicky, J. (1975). *Material Requirements Planning.* McGraw-Hill, New York.

Pinedo, M. (2009). Planning and *Scheduling in Manufacturing and Services.* Springer, New York, second edition.

Pinedo, M. L. (2012). *Scheduling: Theory, Algorithms, and Systems.* Springer, New York, fourth edition.

Plossl, G. W. (1994). *Orlicky's Material Requirements Planning.* McGraw-Hill, New York, second edition.

Pochet, Y. and Wolsey, L. A. (2006). *Production planning by mixed integer programming.* Springer, New York.

Potts, C. N. and VanWassenhove, L. N. (1992). Integrating scheduling with batching and lot-sizing: A review of algorithms and complexity. *The Journal of the Operational Research Society,* 43(5):395–406.

Proud, J. F. (2007). Master Scheduling: *A Practical Guide to Competitive Manufacturing.* Wiley, New York, third edition.

RinnooyKan, A. H. G. (1976). *Machine Scheduling Problems: Classification, Complexity and Computations.* Martinus Nijhoff, The Hague.

Sarin, S. C. and Jaiprakash, P. (2007). *Flow Shop Lot Streaming.* Springer, New York.

Sheikh, K. (2003). *Manufacturing resource planning (MRP II): with introduction to ERP, SCM and CRM.* McGraw-Hill Professional, New York.

Stadtler, H. and Kilger, C., editors (2005). *Supply chain management and advanced planning: concepts, models, software and case studies.* Springer, New York, third edition.

Toomey, J. W. (1996). *MRP II: planning for manufacturing excellence.* Springer, New York.

Trietsch, D. and Baker, K. R. (1993). Basic techniques for lot streaming. *Operations Research,* 41(6):1065–1076.

Vignier, A., Billaut, J.-C., and Proust, C. (1999). Les problèmes d'ordonnancement de type flowshop hybride: État de l'art. *RAIRO Recherche opérationnelle,* 33(2):117–183. In French.

Voss, S. and Woodruff, D. L. (2003). *Introduction to computational optimization models for production planning in a supply chain.* Springer, New York.

Wight, O. W. (1995). *Manufacturing Resource Planning: MRP II: Unlocking America's Productivity Potential.* John Wiley & Sons, New York, second edition.

Chapter 4
Scheduling Constraints

4.1 Introduction

This chapter belongs to the part of the book devoted to scheduling models, one of the three elements (recall from Chap. 1 that the other two are methods and tools) that constitute an scheduling system. More specifically, this chapter discusses the constraints describing scheduling models. Already in Chap. 2 (Sect. 2.5), it was stated that manufacturing scheduling is subject to a large variety of constraints. These were refined in Sect. 3.2.4, where some constraints arising in real-life settings were introduced. This chapter gives further insight into this issue by bringing a characterisation and description of scheduling constraints. Since—as already mentioned in Sect. 3.3.2—the scheduling constraints are gathered in current scheduling classification schemes in the field β of the notation triplet $\alpha/\beta/\gamma$ the scheduling constraints, we will comprehensively work with the notation and with this field β in this chapter as well.

In order to provide a (hopefully) coherent view of scheduling constrains, we have necessarily simplified many of them, and broadly classified them into process, operations, transportation and storage constraints. Some other situations will be also addressed. Despite this classification, note that scheduling constraints are rarely independent from each other and that, even inside the same constraint, the treatment and the low-level details inevitably vary from one production floor to another. Last, it is to note that some constraints only make sense in some production layouts and/or under some specific criteria.

More specifically, in this chapter we

- describe constraints affecting the flow of operations in the shop floor (Sect. 4.2),
- introduce scheduling constraints that are more closely related to the operations or tasks of the jobs (Sect. 4.3),
- discuss restrictions attached to the movement of material in the shop floor (Sect. 4.4),

J. M. Framinan et al., *Manufacturing Scheduling Systems*,
DOI: 10.1007/978-1-4471-6272-8_4, © Springer-Verlag London 2014

- present constraints arising when considering that storage capacity is not unlimited (Sect. 4.5) and
- give some hints on further relevant constraints (Sect. 4.6).

4.2 Process Constraints

Under process constraints we list all situations that affect the otherwise 'standard' flow of operations in single, parallel or shop environments.

4.2.1 Job or Task Precedence Constraints

Precedence constraints started to be studied very early in the scheduling literature with some pioneering work in the early 1970s of Lawler (1973), Sidney (1975) and the already mentioned one of Lenstra and Rinnooy Kan (1978). Nowadays, it is still a very hot topic of research within the scientific community where literally hundreds of papers are being published with new results for many different production layouts. This type of scheduling constraint breaks the common assumption that all jobs to be scheduled and to be processed on machines are independent. It is very common for jobs to model actual products that have a Bill of Materials (BOM) structure. While all the items in the BOM might be just raw materials that are purchased from third companies, it is easy to find real cases where these products are intermediate items that also have to be processed.

Precedence constraints, in their simplest form, dictate that the first task or operation of a job cannot be started until the last job or operation of all of its predecessors are completed. The usual notation is the following: $\beta = prec$: Precedence relations exist among jobs. In the scientific literature, precedence constraints are divided into four different categories:

1. Chain precedence relations ($\beta = chains$), where each job might have, at most, one predecessor and at most one successor.
2. Intree precedences ($\beta = intree$), where each job might have many predecessors but, at most, one successor.
3. Outtree precedences ($\beta = outtree$), where each job might have, at most, one predecessor and many successors.
4. Tree precedences ($\beta = tree$), where there is no limit in the number of predecessors and successors.

These categories are depicted in Figs. 4.1 through 4.4.

Precedence constraints are not very intuitive. At a first glance, one could think that having them in a problem simplifies things. In some sense, this is true. For example, a five job, single machine problem will have $5! = 120$ possible different job processing sequences. This number drastically decreases as precedence constraints

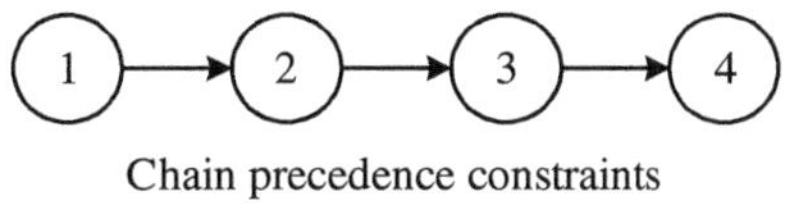

Fig. 4.1 Job precedence graph example. Chain precedence constraints

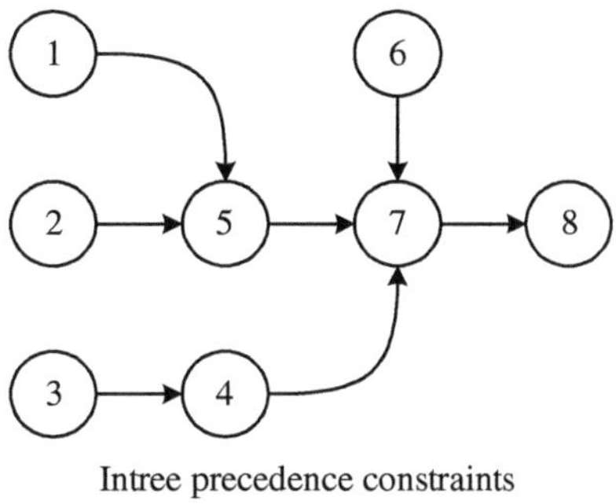

Fig. 4.2 Job precedence graph example. Intree precedence constraints

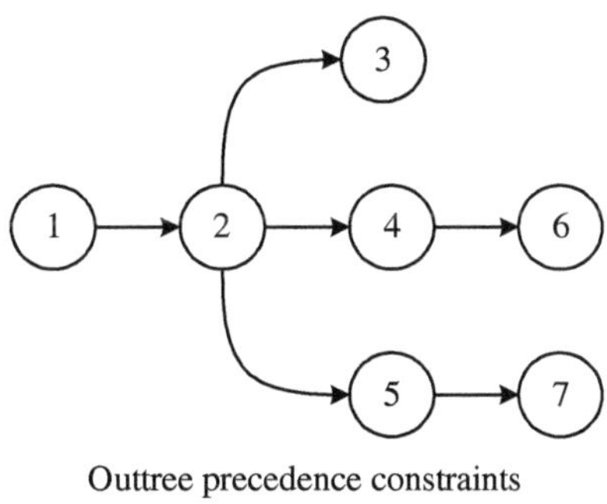

Fig. 4.3 Job precedence graph example. Outtree precedence constraints

are added. The most extreme case would be when every job has a predecessor except the first, for example, $1 \rightarrow 2 \rightarrow 3 \rightarrow 4 \rightarrow 5$. In this situation, only the sequence $\pi = (1, 2, 3, 4, 5)$ would be feasible. However, precedence constraints complicate matters entirely. Picture for example a parallel machines layout. This layout was presented in Sect. 3.2.3.2. There an example problem was introduced in Table 3.2 and pictured in the Gantt chart of Fig. 3.4. Now, let us assume that $7 \rightarrow 4$, $7 \rightarrow 3$ and $2 \rightarrow 1$. The resulting Gantt chart is shown in Fig. 4.5.

As we can see, idle times appear on machines. This might be seen as not a big deal, but in most parallel machine problems with no constraints, machines are not needed to lay idle, if we only consider semi-active schedules. As a matter of fact, the regular parallel machine problem under some performance criteria, is basically reduced to an assignment problem. Although more theoretically oriented, authors like Lenstra and Rinnooy Kan (1978) pointed out already long ago, that adding precedence constraints makes relatively simple problems much more difficult.

Precedence constraints also allow to capture more complex relationships. For example, the precedences might be among tasks of different jobs (as in a flow shop) and not only between the last task of a job and the first task of the succeeding job.

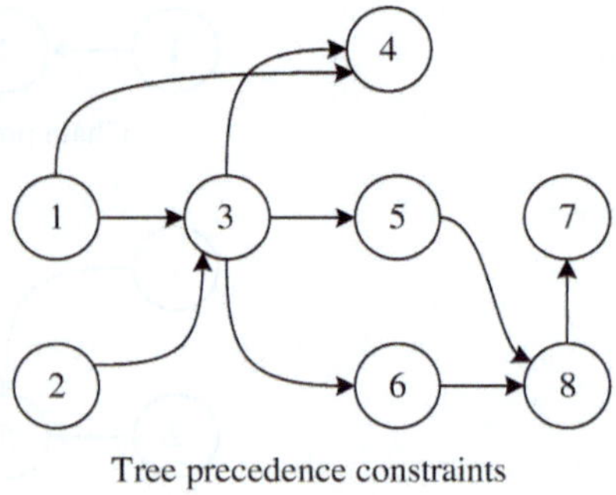

Fig. 4.4 Job precedence graph example. Tree precedence constraints

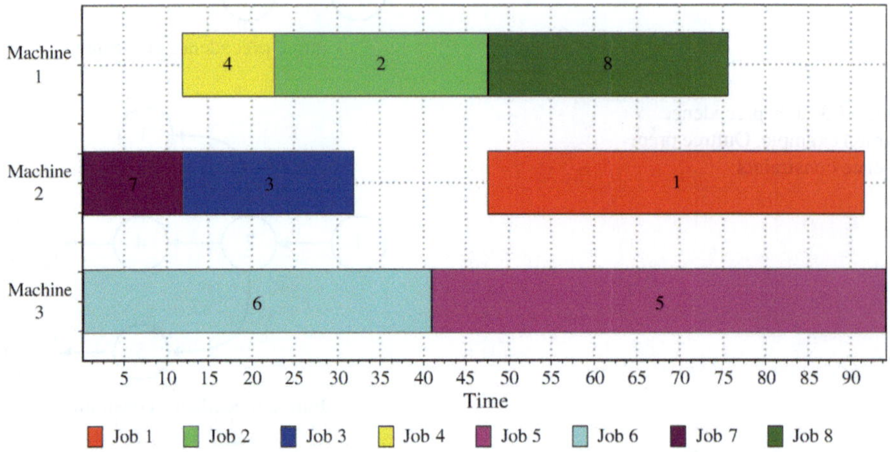

Fig. 4.5 Gantt chart with the result of the assignment and sequencing of eight jobs on three *parallel* machines with some precedence constraints

Precedences do not necessarily have to be what is called 'end-start' as some more complex situations could be tasks that have to start at the same time ('start-start') or finish at the same time ('finish-finish') and many others.

4.2.2 Changeovers or Setup Times

Generally speaking, operations carried out at machines that are not directly related with the processing of the jobs are commonly referred to as 'changeovers'. Changeovers include, but are not limited to, adjustments, changes, cleaning, testings, re-tooling, fixing or removing jobs, etc. Changeovers are also commonly known as setup times since they model the time that is needed to setup a machine prior and/or after processing a job.

Evidently, as changeovers model many different situations, a complete classification and taxonomy of the different setup times is clearly out of the scope of this book. However, and broadly speaking, there are some possible distinctions. First, setup times can be independent of the job sequence or dependent of the job sequence. Setup times are job sequence dependent if the amount of setup time depends both on the job that was just processed on the machine and on the next job to be processed in the sequence. After processing a job, the machine ends up in a 'configured state' suitable for the just finished job. The amount of needed reconfiguration or setup might be dependent of the type of job to be processed next. For example, if the next job to be processed is essentially similar in type to the previous one, only minor setups might be needed. However, if a completely different job is to be processed, large reconfigurations might be necessary.

Second, setup times might be anticipatory (also referred to as separable or detached setup times) or non-anticipatory (inseparable or attached). Anticipatory changeovers or setups are those that can be performed at the machine as soon as this machine finishes the previous job in the sequence. Note that due to the nature of the expected processing sequence, there might be an idle time in between processing of consecutive jobs. Therefore, an anticipatory setup can be carried out at any time during this idle period. A machine cleaning, for example, is a clear case of an anticipatory setup time. Conversely, non-anticipatory setup times require the next job in the sequence to be already present or to have arrived at the machine in order to carry out the setup. For example, imagine the process of adjusting an unpolished wood board in a polishing machine. This adjustment might require the wood board (job) to be already present in order to firmly fix it (setup) to the machine. If the wood board is not present, the setup cannot be carried out.

Let us ellaborate a simple example. Figure 4.6 shows a small two job, two machine flow shop problem where there is a single anticipatory setup between the jobs on the second machine. As can be seen, the setup time has no impact on the processing sequence as there is enough idle time at machine two between the two jobs to do the setup, which is depicted in solid black. Notice also that the setup time has been 'right justified' to the start of the second job but it could have been also left justified to be performed right after job one is finished or it could have been performed at any time during the idleness period of machine two.

A significantly different scenario arises in the case of a non-anticipatory setup, shown in Fig. 4.7. The setup can only be started as soon as job two is completed on the first machine and once it arrives to the second machine.

From the two previous classifications we have the following four combinations:

- Non-anticipatory and sequence independent setup times.
- Non-anticipatory and sequence dependent setup times.
- Anticipatory and sequence independent setup times.
- Anticipatory and sequence dependent setup times.

Notice that the first case, when the setup cannot be detached and when it is independent from the sequence, can be modeled by simply adding its duration to the

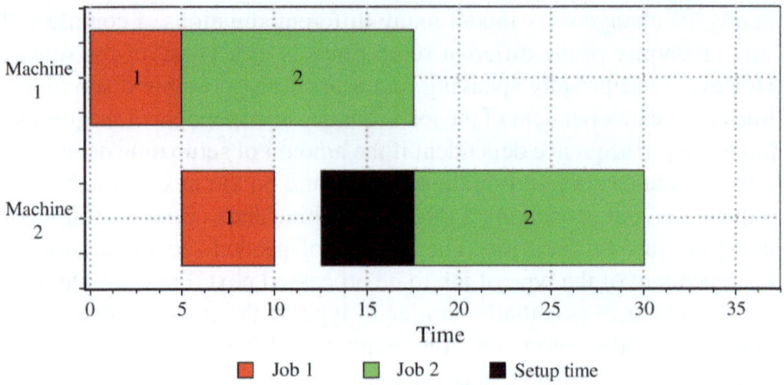

Fig. 4.6 Two job, two machine flow shop with an anticipatory setup on the second machine

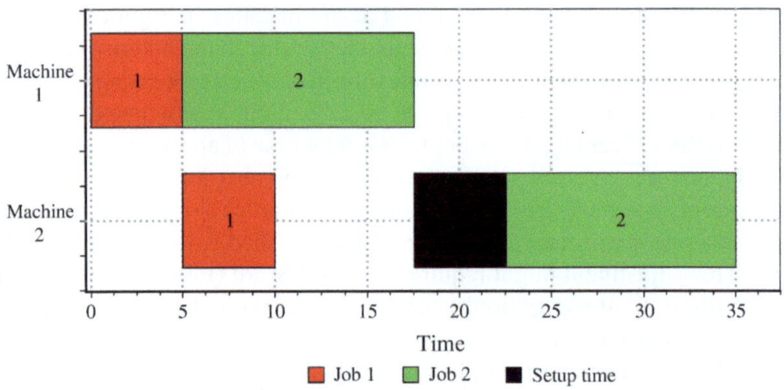

Fig. 4.7 Two job, two machine flow shop with a non-anticipatory setup on the second machine

processing time of the job in the machine. After all, the machine is going to be busy anyway. All other cases require special attention.

It is important to remark that in flexible manufacturing systems, as well as in many production environments, setup times can amount to a significant portion of the productive time if they are not handled with care. The time needed to clean, to name an example, a chemical reactor might be superior to the processing time needed to produce a batch of a given product.

The previous classifications are not the only possible ones. Other classifications further break down setups into what is referred to as family, class or batch setup times. In these cases, jobs are grouped into families so that setup times occur only when finishing production of one family and before starting production of the next one. Sometimes, there are minor setups in between jobs of the same family and major setups between families. Some other classifications include also removal times, which are seemingly identical to setup times, but that are performed after jobs are completed and not prior to the starting of jobs.

It is safe to say that most production systems include, in one way or another, setup times. In some cases, setup times might be sufficiently small to be considered negligible, but in many other cases, setups have to be considered explicitly. Real cases of setup times abound. A clear example comes from the paper cutting industry where the paper cutting machines (guillotines) need to be adjusted when changing from one cutting batch to another. Another example can be obtained from the ceramic tile manufacturing sector. Ceramic tiles are produced in identical processing lines composed of several processes; moulding press, dryer, glazing line, kiln, quality control and finally, packing and delivery. If a production line is set up for a given product format and colour, for example a 33×33 cm. black glazed ceramic tile, changing to a 45×45 cm. white glazed ceramic tile will need significant setup times to change the moulds in the moulding press and to clean and prepare the glazing line for the white glaze. However, changing to a 33×33 cm. yellow glazed tile will only require cleaning in the glazing line, as the moulding press is already set up. A more detailed case study taken from the ceramic tile sector is given in Chap. 15.

As far as the field β in the notation triplet $\alpha/\beta/\gamma$ is concerned, we denote setup and removal times as follows:

$\beta = S_{nsd}$: There exists sequence independent setup times. S_{ij} denotes the known and deterministic amount of setup time to be performed on machine i before processing job j.

$= S_{sd}$: There exists sequence dependent setup times. In this case, for each machine, we have the so-called 'setup time matrix', where S_{ijk} denotes the setup time needed on machine i for processing job k after having processed job j.

$= R_{nsd}$: Sequence independent removal times.

$= R_{sd}$: Sequence dependent removal times.

For families or batches, one can use, for example, SB_{nsd}, SB_{sd}, RB_{nsd} and RB_{sd}.

No specific notation is given to differentiate the anticipatory from the non-anticipatory setups cases. However, anticipatory setups are largely assumed in the scientific literature.

4.2.3 Machine Eligibility

In parallel machine problems, or in hybrid environments where stages have parallel machines, it is normal not to expect that all machines are able to process all jobs. In Chap. 3, at Sect. 3.2.3.2, it was already mentioned that in real life, the unrelated parallel machines model, where each machine is able to process each job at a specific speed, is the most probable layout. Factories evolve over time. Old machinery is replaced by new equipment and newer machines do not have the same speed and/or characteristics as old ones. However, usually constraints go further than just varying processing times. In some settings, some machines might be able to process just a subset of jobs and/or some jobs might be assigned to just a subset of machines. This is called machine eligibility.

When machine eligibility is present in a parallel machine layout, each job $j \in N$ has an associated set M_j so that $M_j \subseteq M$ which indicates the subset of machines where job j can be processed. For multistage hybrid problems, this set can be further indexed by the stage number, i.e. M_{ij}. The notation is as follows:

$$\beta = M_j : \text{Machine eligibility constraints.}$$

Once again, restricting eligible machines seems to simplify problems. However, this is not often the case (see, e.g. Urlings et al. 2010).

4.2.4 Permutation Sequences

In the flow shop layout, as commented in Sect. 3.2.3.3 is usually simplified to consider only permutation sequences. This is notated as follows:

$$\beta = prmu : \text{Permutation flow shop problem.}$$

Basically, most of the scheduling research is carried out in the permutation version of the flow shop. Also, in some hybrid environments, the term 'permutation schedules' indicates that there is a single permutation of jobs as a simplified solution for the problems. At each stage, jobs are launched to machines in the permutation order and assignment to machines at each stage is solved separately.

4.2.5 Machine Availability and Breakdowns

Almost throughout the whole book, we have always assumed that machines are continuously available throughout the time horizon. While this simplifying assumption is very convenient, it is really a long shot away from reality. Machines are subject to preventive maintenance, corrective maintenance (breakdowns) or just timetabling or working shift constraints, among other things. The result is the same: machines are not continuously available.

The scientific literature about machine availability and breakdowns is abundant. However, two main distinctions can be stated. The first is when the intervals of machine availability (or unavailability) are known in advance. In this case, for every machine $i \in M$ and for a given production scheduling horizon T, there is a set of h machine availability intervals $AV_i = \{\{BAV_1^i, EAV_1^i\}, \{BAV_2^i, EAV_2^i\}, \ldots, \{BAV_h^i, EAV_h^i\}\}$ where BAV_l^i and EAV_l^i are the starting and ending time of the machine availability period, respectively, and $BAV_l^i < EAV_l^i$ for $1 \leq l \leq h$ and for $i \in M$. A possible notation for this is as follows:

β = *brkdwn*: Machines are subject to breakdowns (general notation).
= NC_{zz}, NC_{inc}, NC_{dec}, NC_{inczz}, NC_{deczz}, NC_{sc}, NC_{win} model different patterns of of machine availabilities. These are zig-zag, increasing, decreasing, increasing in zig-zag, decreasing in zig-zag, staircase and arbitrary, respectively. These patterns are more or less explanatory and can be well understood in parallel machine settings. In a zig-zag pattern, machines with high availability are alternated with machines with low availability and increasing patterns denote settings in which each subsequent machine has less availability.

The second distinction in the machine availability literature is when some information of the machine availability intervals is not known a priori or when the machines simply breakdown with some probabilities or following some probability distributions.

4.2.6 Re-circulation, Re-processing and Skipping

Some important process constraints or special situations include the cases where jobs might visit one specific machine or stage more than once. A similar situation is when some jobs skip one or more machines or stages in the processing sequence. Some of the processing layouts commented in Chap. 3, like the flow shop or the job shop already consider what is called re-circulation. Re-circulation models, for example, double firing processes in ceramic tile manufacturing, the repeated polishing operations in furniture manufacturing and any other environment where a job has to visit a given machine or stage more than once in the processing sequence. Some examples of studies in the literature where re-circulation is allowed include, but are not limited, in the very least, to, Bertel and Billaut (2004) or Choi et al. (2005).

As far as re-circulation goes, the notation is as follows:

β = *recrc* : Re-circulation exists as a process constraint. Indicates that at least one job visits at least one machine more than once.

Some related scenarios arise when the outcome of the processing of a job is not perfect and there is a possibility of needing a re-processing. Note that this is in several ways different to re-circulation. First, there is no certainty in that a re-processing will be needed and therefore, re-processing is stochastic. Second, usually a re-process directly involves the same job leaving and entering again the same machine, whereas in re-circulation several machines could be visited in between two visits to the same machine. There is no accepted notation for re-processing but $\beta = reproc$ looks like a valid proposal.

Last, stage skipping could be a factor to consider in some problems. In some scenarios and layouts, like for example flow shops or hybrid flow shops, some jobs might not need processing in some machines or might not need to visit some stages. This is referred to as stage skipping and can be denoted by $\beta = skip$. Note that skipping (or synonymously passing of stations or missing operations) is often (e.g. in flowshops) modelled/interpreted by adding respective operations of jobs with processing time 0. However, this approach does not allow jobs simply to skip a station. Instead, the respective jobs have to queue up in front of the machine. And when they are to be processed, they immediately pass the machine because of their processing time 0. While this makes the consideration of the respective scheduling problem much simpler from an optimization method point of view, in fact this approach ignores the real-world potential for optimization where the respective situation is interpreted by 'real' skipping of a machine.

4.2.7 No-Idle Machines

An interesting situation arises when no-idle time is allowed at machines. This constraint models an important practical situation that arises when expensive machinery is employed. Idling on expensive equipment is often not desired or even not possible due to technological constraints. Some examples are the steppers used in the production of integrated circuits by means of photolithography. Other examples come from sectors where less expensive machinery is used but where machines cannot be easily stopped and restarted or is completely uneconomical to do so. Ceramic roller kilns, for example, consume a large quantity of natural gas when in operation. Idling is not an option because it takes several days to stop and to restart the kiln due to a very large thermal inertia. In all such cases, idling must be avoided. Notation goes as follows:

β = *no-idle*: No machine can stay idle after starting the process of operations.

Generally speaking, the processing sequence of jobs where a no-idle processing constraint is present in machines must ensure that once a machine starts processing, it is not stopped until the last job in the sequence is finished. This is graphically depicted in Fig. 4.8 for a five job, four machine flow shop example.

As one can see, the start of each job is delayed so to end up with complete blocks of consecutive jobs and no-idle times on machines.

The no-idle processing constraint has been seldom studied in the literature. Additional work thus remain to be done in this setting as much more complex processing layouts might benefit from the consideration of such processing constraints.

4.2.8 Batching Machines

High throughput machines usually accept more than one job at the same time. Actually, arriving jobs might be grouped into batches and then the machine simultaneously completes all the batch at the same time. This is very common in the semiconductor industry (Lee et al. 1992) or in tyre manufacturing (Bellanger and Oulamara 2009).

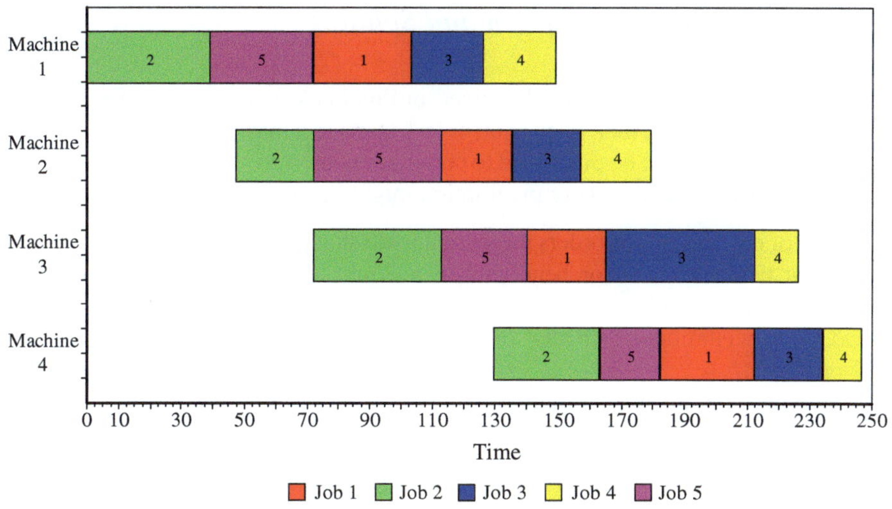

Fig. 4.8 Gantt chart example of a no-idle flow shop

Batching might be economical if the arriving jobs require the same setup and can be processed in batches. Usually, two types of batches are considered. $p - batch$ (parallel batching) is when the processing time of the whole batch in the machine is considered to be equal to that of the largest processing time of any job in the batch. $s - batch$ (serial batching) occurs when the processing time of the batch is equal to the sum of the processing times of all the jobs in the batch. Additionally, b denotes the *batch size*, or the maximum number of jobs that can enter a batch. If this number changes from machine to machine, then it is used b_i instead. As a result of all of the above, the notation is therefore:

$\beta = p - batch(b)$ (parallel batching) : All machines consider batches of maximum size b (or b_i) where the processing time of the batch is equal to that of the largest processing time among the jobs in the batch.

$= s - batch(b)$ (serial batching) : All machines consider batches of maximum size b (or b_i) where the processing time of the batch is equal to that of the sum of the processing times of the jobs in the batch.

The literature on batching scheduling is extremely large, as there are many possible sub-cases, generalisations and variations in the batching decisions.

4.3 Operations Constraints

This section deals with all the scheduling constraints that are more closely related to the operations or tasks of the jobs. Note that the distinction with the previous process constraints is, in some cases, not entirely clear cut and in many cases debatable. However, in some way or another, these constraints can be linked to jobs more than to machines.

4.3.1 Interruption, Preemption and Splitting

The concept of preemption was discussed in this book as early as in Sect. 1.5. In contrast, if it is assumed that, once a job or task starts in a machine, it cannot be interrupted and the job has to be processed to completion, then preemption is not allowed. There are many scenarios where preemption might be needed or even interesting:

- Arrival of new urgent job orders might require to stop jobs already being processed.
- It might be economical or beneficial to preempt a job and/or to continue it in another machine and/or at a later time.
- Sudden cancellations of clients' orders might require to stop processing a batch of products.
- Breakdowns on machines or any other unexpected event might result in preemption.

There are many possible types of preemptions and varied classifications. If, once interrupted, the preempted operation is lost and when resumed, processing has to restart from the beginning, the preemption is denoted as non-resumable. This is the case for example in a pre-heating phase in metallurgy processes. If one stops the pre-heating operation, it will be required to start pre-heating from the beginning again at a later time since the processed product might have cooled off by then. Note that setup time operations can also be preempted and most of the time, if a setup operation is aborted, a different setup is started (the next product in the sequence is changed) and therefore, the initial setup will have to be restarted from the beginning. Setup times are then denoted as non-resumable. Conversely, a preemption can be resumable if the interrupted job is just continued from the point where it was interrupted to completion. Sometimes, an intermediate situation exists where the operation can be resumed, but some penalty (time, costs, etc.) exists. This case is referred as semi-resumable.

Jobs might be preempted just once, a limited number of times or an unlimited number of times. Jobs that break down into operations, as in the shop layouts, might complicate things, since preempting a non-resumable operation might result in the whole job being repeated again (all operations). A preempted operation, for example in a parallel machine scenario, might have to return to the same machine or might be able to return to any other machine. As one can see, the possibilities are endless. However, not all scenarios are so complex. Sometimes, preemptions are as simple as stopping a machine because the working shift is over and the processing of the same job, on the same machine, continues the next working shift or after the weekend with no introduction of a different job in between.

As far as the notation goes, the basic preemption is referred to as follows:

$\beta = pmtn$	: Preemption allowed (general). In other works it is also referred to as *prmp*.
$= pmtn - non - resumable$	: Non-resumable preemption.
$= pmtn - semi - resumable$	: Semi-resumable preemption.
$= pmtn - resumable$	: Resumable preemption.

Another aspect that might affect the different tasks of jobs is splitting. In specific manufacturing environments, jobs/operations might be split into several sub-jobs/sub-operations. This problem is partially related to preemption of jobs/operations. However, the latter is often induced by processing requirements of other (more urgent) operations while job splitting is usually caused by lot sizing or related considerations.

4.3.2 Release Dates, Due Dates, Deadlines, Processing Windows

Release dates, as well as due dates and deadlines where already defined in Chap. 3 as input data for a scheduling problem. However, they also indicate constraints that affect operations. As commented, the first task of a job cannot start before the release date and the last task of a job should be ideally completed before the due date and forcibly completed before the deadline.

When such data exist in a production setting, it is denoted as follows:

$\beta = r_j$: There are release dates for the jobs. Omitting this field is the same as assuming that all release dates are zero, i.e. $r_j = 0, \forall j \in N$.

$\beta = d_j$: There are distinct due dates for each job.

$= d_j = d$: There is a single identical due date for all the jobs. This is referred to as the common due date case.

$= \overline{d}_j$: The due dates are compulsory and are referred to as deadlines. Some early research work denotes deadlines as $\triangle_j$.

Note that many times the due dates d_j are not explicitly stated in the β field of the notation since, as we will see in Chap. 5, many optimisation objectives implicitly consider the due dates.

If one finds in the same production setting release dates r_j and deadlines $\overline{d}_j$ the result is the so-called 'processing window' as the entire processing of a job must be completed within the $\overline{d}_j - r_j$ time interval. This is often referred to as 'time window'. Deadlines in general and processing windows in particular raise the issue of schedule feasibility, i.e. not all schedules or possible orderings of the jobs and tasks on the machines might be feasible. It might even be impossible to complete all jobs within their respective processing windows.

4.3.3 No Wait, Minimum and Maximum Time Lags and Overlapping

Normally, in shop or hybrid shop layouts, the next task of a given job can start as soon as the previous task is fully finished. However, if the next machine or stage is busy, then the task, in the form of semi-finished product or in-process inventory,

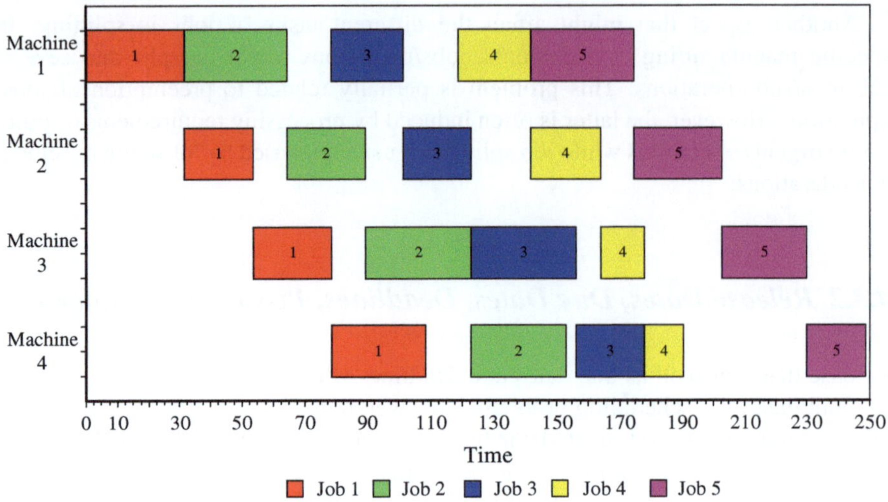

Fig. 4.9 Gantt chart example of a no-wait flow shop

waits at some place until the machine is free. In other words, tasks are allowed to wait indefinitely in between machines or stages.

There are many scenarios where waiting is not an option and the processing of some or all jobs needs to be carried out without interruptions between machines. This situation is commonly known as 'no-wait'. A typical example comes from steel production where the steel ingots are not allowed to wait between production stages or otherwise they would cool off. Most prepared food production lines also have, in some way or another, no-wait restrictions. A clear example is a frozen food line where no-wait must be enforced or otherwise the cool chain might break. Other examples can be found in chemical and pharmaceutical industries, among many others.

Constructing a no-wait schedule is not trivial. If the next machine needed by a job is busy, it is required to go back to the previous tasks of the same job and to delay their start, so that tasks never wait. Figure 4.9 shows a five job, four machine no-wait flow shop example. From the figure, we see that, for job four, the three initial tasks have to be delayed, even including an idle time on the first machine, to make sure that the job four does not find the last machine busy.

Note that in the flow shop layout, the no-wait constraint necessarily implies a permutation sequence. It is of paramount importance to consider that much of the constraints usually interact and a seemingly easy way to deal with a constraint, could end up being hard to solve when in conjunction with another special situation. As regards notation, the no-wait situation is easily described:

$\beta = nwt$: Jobs cannot wait in between stages. Mathematically, this implies that $SO_{ij} = EO_{i-1,j}, \forall j \in N, i \in N \backslash 1$. Note that this is sometimes referred to as *no-wait*.

In some works, like in the one of Hall and Sriskandarajah (1996), the authors extend this notation to clearly specify between which machines or stages waiting is not allowed. For example $no - wait(3, 4)$ indicates that no waiting is allowed for any job between stages or machines three and four.

The no-wait setting is in obvious contrast with the regular setting in which indefinitely waiting could be possible. It should be pointed out that the Gantt chart profiles of every single no-wait job is fixed, independent of the sequence of jobs. This information can be used when developing optimization approaches to no-wait problems.

Intermediate scenarios between assuming indefinite waiting times and the no-wait case are possible, as there are situations where jobs or tasks might wait some time, but not too much. For example, in a frozen food cold chain, 10 min waiting might be acceptable knowing that the temperature will not raise beyond a microbial contamination threshold in those 10 min. Therefore, a maximum waiting time or 'maximum time lag' is allowed. Jobs or tasks can wait, for no more than the maximum time lag. In general, we refer to lag^+_{ij} to the maximum time lag for job $j \in N$ after being processed on machine $i \in M$ and before proceeding to machine $i + 1$. In order to satisfy this maximum time lag, the following inequality must hold for all tasks: $EO_{ij} \leq SO_{i+1,j} \leq EO_{ij} + lag^+_{ij}$.

Similarly, there might be 'minimum time lags'. This implies that subsequent tasks of a job cannot start until a given minimum time has elapsed since the last operation. There are many situations where minimum time lags could appear. Picture for example painting operations that require a minimum drying time before subsequent operations can be performed. Products that require some type of furnacing or kiln firing might need some cooling off time before being handled. We denote by lag^-_{ij} to the minimum time lag for job $j \in N$ after being processed on machine $i \in M$ and before proceeding to machine $i+1$. Minimum time lags effective need jobs to satisfy the following: $SO_{i+1,j} \geq EO_{ij} + lag^-_{ij}$.

A straightforward extension is when both maximum and minimum time lags exist. This imposes a 'time window' interval inside which the next task of any job should be started. Some other authors, even consider negative time lags. Negative time lags allow for a subsequent task to start actually before the preceding task is finished. This can be used to model, for example, large batches of products that can effectively overlap in between stages, much like the lot streaming and lot sizing discussion of Sect. 3.4 of Chap. 3.

4.3.4 Special Processing Times

Nothing has been mentioned about the nature of the processing times so far apart from mentioning in Chap. 3 that processing times are supposed to be known and fixed in advance. However, there are many situations that might affect this otherwise strong assumption.

First of all, there is a sizeable part of the theoretical research in scheduling that deals with very specific processing times. These are referred to as the unit processing times and identical processing times cases, denoted as follows:

$\beta = p_{ij} = 1$: All processing times are equal to 1. This is referred to as unit processing times.
$= p_{ij} = \{0, 1\}$: Zero or unit processing times.
$= p_{ij} = p$: All processing times are equal to a value p.

Note that these special processing times cases might be erroneously classified as too theoretical. While it is true that the bulk of research in these cases is theoretical, there are some situations in practice that benefit from these special cases. For instance, in Lee et al. (1992), batching machines for the burn-in operations in semiconductor manufacturing are studied. In these machines, all the jobs that simultaneously enter the machine are modelled as having the processing time equal to the largest job that enters the batching machine. As we can see, such studies are highly relevant in practice.

Some other specific situations regarding processing times are also often studied in the literature. For instance, *proportional processing times* model the situation in which the processing times of the tasks of a given job are proportional to each other. For example, the first machine might be faster than the second machine, the third faster than the fourth machine and so on. These problems are usually noted as $\beta = p_{ij} = p_j/s_i$, where s_i is a speed factor of the machine i. There is a body of the literature about the proportionate flow shop, with some key works like the ones of Ow (1985) and Hou and Hoogeveen (2003).

So far, it has also been assumed that processing times, once determined, do not depend on anything else. In reality, processing times can be acted upon. Picture a painting operation of a large wood or metal piece. Painting might be carried out by personnel operating spray paint guns. The more personnel assigned to that painting operation, the shorter the processing time and the more expensive the operation will be. In other words, sometimes, processing times depend on other factors and can be somehow controlled. This situation is referred to in the scientific literature as controllable processing times and there is also a large body of research. The variety and diversity of controllable processing times scheduling is overwhelming as there are literally hundreds of possible variants. Note that controllable processing times are often also referred to as resource dependent processing times or even compressible processing times. The interested reader is advised to read through the previously cited reviews for more information.

No matter how the processing times are estimated, they are no more than that: an estimate. In some production systems, production or processing times cannot just be estimated by any amount. In these cases, if a fixed and deterministic processing time is not acceptable, some statistical distribution for each processing time has to be assumed. Under this situation, we enter the stochastic processing times case. The body of research under stochastic processing times is huge. Note that when one refers to stochastic scheduling, most of the time the meaning is stochastic processing times.

However, it has to be noted that analytical results are only possible for simplified production settings. For complex production scenarios, simulation is believed to be the only viable choice.

The above special cases for processing times are not the only possible ones. Time-dependent processing times occur when the actual processing time depends on the instant in which operations are started. A special case of the previous one is the so-called deteriorating processing times where the processing time of a given task increases as time passes by.

Last, it has been demonstrated that processing times decrease over time in repetitive production lines. This is the so-called learning effect. While it is debatable that processing times with learning considerations should be addressed in short-term scheduling, this has not deterred authors from proposing several models in which the processing times are subject to the most varied learning effects.

4.4 Transportation Constraints

Another situation that is very common in practice is that a task of a job cannot start just as soon as the previous task of the same job is finished. Doing so implies that the job, whatever product it models, arrives instantaneously from the output of the previous machine to the input of the next machine. Normally, the job will need to be carried over to the next machine. There are numerous ways of transporting products from machine to machine: conveyor belts, robotic cells, robotic arms, wagons, product boxes, automated guided vehicles, overhead cranes, as well as many others. As a matter of fact, the variety in product handling inside a factory is probably as high as the variety in manufacturing systems themselves. Alongside with this variety, there are many different approaches to transportation constraints in the scientific literature.

Basically, two types of transportation constraints can be considered. First, there is the transportation time, so that jobs do not instantaneously move from machine to machine. Should there be no more constraints, transportation times could be modeled simply as a minimum time lag with the time it takes to transport the job from one machine to the next. However, sometimes lot streaming (see Sect. 3.4) is employed so jobs are able to be splitted for transport from one machine to the next which might allow overlapping processing of jobs (sub-jobs or sub-lots of a job) on consecutive machines.

There is a second group of transportation constraints: the number of transporters is limited. This second situation results in a hard problem as the handling of products must be solved together with the scheduling. Obviously, if all transporters are busy transporting products, then the job has to wait somewhere until one is free. The immediate result of this is that the transportation time is not a fixed and constant term, but a time that depends on how the scheduling is being done and on the traffic in the transportation.

In the most complex scenario, the scheduling problem becomes a combined scheduling and routing problem in which countless routes for different transporters

Fig. 4.10 An automated guided vehicle (AGV) transporting a box full of ceramic tiles

have to be derived to transport all the goods inside the factory. Lately, some special emphasis is being observed on the routing of automated guided vehicles (AGVs) like those shown in Fig. 4.10.

The previous aspects of transportation constraints are not the only ones. Sometimes, transportation constraints go as far as the final delivery of products to clients. This is referred to as *product delivery coordination* and has been widely studied. Other scenarios include the necessary coordination of scheduling and transportation in complex supply chains.

4.5 Storage Constraints

In all previous sections, we have shown some alternatives for the many assumptions of the theoretical scheduling problems. However, there is still an important one remaining: storage. In-process inventory has to be stored inside the plant and obviously, space is not unlimited. When the storage areas are full, upstream machines have to stop while downstream machines work over the in-process inventory to free up some space in the storage areas.

Buffer capacity requirements occur per job between two consecutive operations of this job, i.e. after the previous operation has been finished and before the next operation starts. Often, unlimited buffers are supposed in manufacturing scheduling. However, sometimes physically buffer limitations are present explicitly or the goal of minimising working capital implicitly demands for limited in-process inventory. Buffer capacity requirement can be simply measured by 1 as buffer unit per job, i.e. by the number of jobs in a buffer (e.g. if jobs are buffered in standardised containers) or by some other measures (size, temperature category, etc.). This requirement or demand for buffer capacity has to be balanced with the available buffer capacity which, (a) might be limited by an upper bound with respect to the indicators determining jobs'/operations' buffer capacity type, and (b) can be provided centrally for all respective buffer requirements of the system, decentrally for a pair of consecutive or nearby located machines/stages or something in between, e.g. a subset of machines/stages. Additionally, e.g. in job shop environments, one might distinguish between buffers which are located physically and/or logically in front of a machine and those behind a machine.

Blocking is a special situation that results from the limited storage. It mainly affects the flow shop layout. If there is a limited storage capacity in between machines, it might happen that when there is no more capacity, the previous machine cannot finish and release a job since there is no place where to put it. As a result, the machine is 'blocked' with the job. As regards the notation, buffers and blocking are as follows:

$\beta = \text{buffer} = b$	: There are $m-1$ buffers of capacity b in between the m machines.
$= \text{buffer} = b_i$	: There are $m-1$ buffers of different capacities b_i which depend on the machines.
$= block$	: There is blocking. Implies a limited storage or buffer in between machines.

4.6 Other Constraints

At the beginning of the present chapter it was stated that the reality of the production systems is extremely complex and that there are countless possible constraints. In previous sections we have just scratched the surface of this issue. Now we proceed with a brief enumeration of many other possible constraints not mentioned before.

It has been mentioned that in supply chains, transportation operations have to be considered. However, there is one basic assumption that has been considered up to this point. There is one single production floor or factory, or, if there is more than one, scheduling is carried out separately at each one. Large companies with many factories have additional decisions on where to produce the products that form the production plan, i.e. there is a previous decision of deciding in which factory each product should be manufactured. Then, on each factory, a scheduling problem has to be solved. These situations have been recently referred to as 'distributed scheduling problems' and are recently being studied. One example is the study of the distributed permutation flow shop layout by Naderi and Ruiz (2010). Figure 4.11 clearly shows one full flow shop problem with 10 jobs and four machines that has to be distributed between two factories, each one also with four machines. Obviously, the job factory assignment decisions and the scheduling problem at each factory cannot be separated if a good quality result is desired.

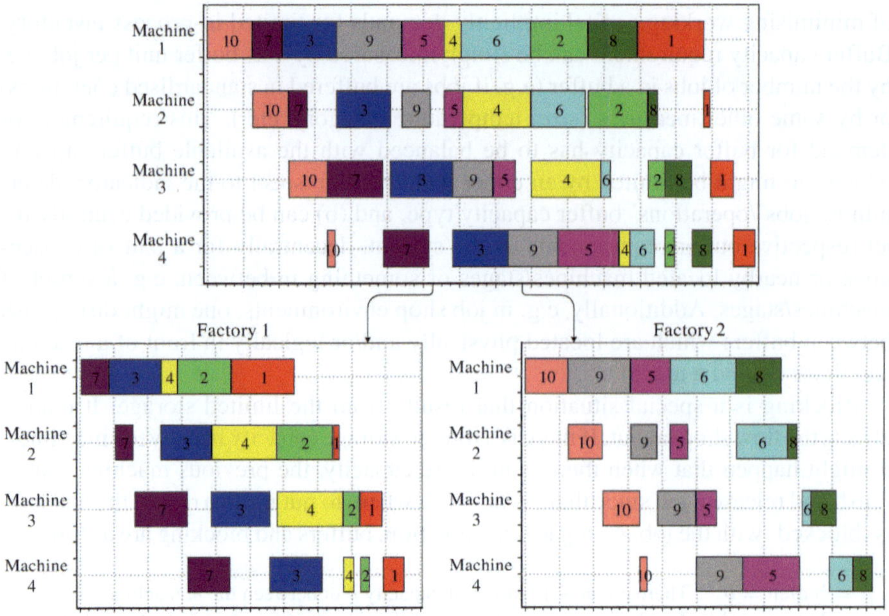

Fig. 4.11 An example of a distributed permutation flow shop problem

There are many other constraints. Setup times are a never-ending source of special situations. Some machines, for example, cannot be setup unless an specific technician is working and therefore, setups during the weekend, for example, have to be avoided. Furthermore, setups are often cleaning and adjustments of machines that are carried out by plant personnel. In this situation, setup times are also controllable.

4.7 Conclusions and Further Readings

During this chapter we have introduced some of the situations that can be found in production settings, grouped by process constraints, operations constraints, transportation and storage constraints. Note that this is just one possible grouping as the sheer variety of real situations easily overcomes any exhaustive classification attempt. This variety of scheduling settings and constraints results in a vast literature where most constraints are still studied in an isolated way most of the time. For example, the literature on no-wait scheduling is plenty, as it is the literature on setup times. However, in the intersection one finds very few research papers, i.e. no-wait scheduling with setup times. Moreover, it is safe to say that the intersection among three or more sets of constraints for any shop layout would return an empty set of references most of the time. Furthermore, the more constraints considered, the less complex the studied scheduling layout usually is. This is, there are some studies with many constraints considered simultaneously for the single machine problem, but just a handful of papers have been published where several constraints are simultaneously considered for complex hybrid job shop problems, to name an example.

The previous issue is actually a symptom that shows that more work needs to be done in the area of scheduling in order to cope with more realistic problems where several simultaneous constraints are considered. Any production scheduler working at a company has surely identified several of the previously commented constraints as important in his/her production setting. In the next chapters we will make an attempt at providing some general techniques and algorithms, as well as a case study, to help in bridging this gap between academic scheduling and real scheduling at production shops.

Further readings on the topic discussed in this chapter stem from the already commented main textbooks in scheduling. When describing the notation, we have sought to balance the simplicity against the thoroughness, also taking into account its acceptance. This is not possible in many cases: For instance, the notation on machine availability is not widely accepted (see as, for example, Lee et al. 1997 for a different notation).

Regarding specific works, apart from the references that have been cited in-line when describing specific sets of constraints in the corresponding sections, state-of-the-art and comprehensive reviews of the scheduling literature and setup times are available from Yang and Liao (1999), Allahverdi et al. (1999), Cheng et al. (2000) and Allahverdi et al. (2008). A good state-of-the-art review of scheduling problems with machine availabilities is available from Schmidt (2000). Reviews on flow shop are Park et al. (1984), Turner and Booth (1987), Framinan et al. (2004), Ruiz and Maroto (2005), Hejazi and Saghafian (2005) or Gupta and Stafford (2006), while for hybrid flow shops it is worth citing Linn and Zhang (1999), Vignier et al. (1999), Wang (2005), Ruiz and Maroto (2006), Quadt and Kuhn (2007), Ribas et al. (2010) and Ruiz and Vázquez-Rodríguez (2010). Stage skipping has been addressed by a number of authors, see, e.g. Leisten and Kolbe (1998) and Urlings et al. (2010) for the flow shop/hybrid flow shop settings, respectively. No-idle processing constraints have been seldom studied in the literature. To the best of our knowledge, the first papers are those of Adiri and Pohoryles (1982) and Vachajitpan (1982), and recent references on the topic are Pan and Wang (2008) and Ruiz et al. (2009). A comprehensive review of scheduling with batching decisions is due to Potts and Kovalyov (2000), although it only covers research that is already more than a decade old. In the last decade, more than double that number of papers have appeared for batching scheduling. Some recent examples of work in preemption are Agnetis et al. (2009) and Hendel et al. (2009).

Due dates are intensively studied in the research literature. Some examples are Sen and Gupta (1984), Baker and Scudder (1990) and more recently, Vallada et al. (2008). Some examples of processing windows research appear in Maffioli and Sciomachen (1997) or Yeung et al. (2009). Minimum, maximum and both types of time lags have been dealt with, among many others, by Riezebos and Gaalman (1998), Brucker et al. (1999) and Fondrevelle et al. (2006). Negative time lags and overlaps are discussed in network planning textbooks, being a recent reference Urlings et al. (2010).

Unit processing times are studied in Liua et al. (1999) and Averbakh et al. (2005) just to name a few. Zero and unit processing times are examined in Lushchakova and Kravchenko (1998), among others. Results for identical processing times have been published in Crama and Spieksma (1996) and Lee et al. (1992). A recent survey about scheduling with controllable processing times is that of Shabtay and Steiner (2007).

In the book of Pinedo (2012) there is a whole part devoted to stochastic scheduling. Regarding time depending processing times, a main source is Gawiejnowicz (2008). An interesting study for the time-dependent processing times single machine layout is given by Chen (1996), and a example of deteriorating processing times where the processing time of a given task increases as time passes by is Kubiak and van de Velde (1998). Some examples of processing times with learning effect are the papers of Lee et al. (2004) and Eren and Guner (2009) just to name a few.

Some recent references on lot streaming are Defersha and Chen (2012) and Feldmann and Biskup (2008), being Sarin and Jaiprakash (2007) a book on the topic. A good review of scheduling with transportation times is due to Lee and Chen (2001). Product delivery coordination can be found in Chang and Lee (2004). Excellent review papers on the coordination of scheduling and transportation are available from Potts and Hall (2003), Li et al. (2005) and Chen and Vairaktarakis (2005).

The main results of scheduling with blocking are due to Hall and Sriskandarajah (1996). Other relevant papers are those of Papadimitriou and Kanellakis (1980), Leisten (1990), Brucker et al. (2003) and Mascis and Pacciarelli (2005). Finally, an interesting application where setup times are considered controllable is Ruiz and Andres-Romano (2011).

References

Adiri, I. and Pohoryles, D. (1982). Flowshop no-idle or no-wait scheduling to minimize the sum of completion times. *Naval Research Logistics*, 29(3):495–504.

Agnetis, A., Alfieri, A., and Nicosia, G. (2009). Single-machine scheduling problems with generalized preemption. *INFORMS Journal on Computing*, 21(1):1–12.

Allahverdi, A., Gupta, J. N. D., and Aldowaisan, T. (1999). A review of scheduling research involving setup considerations. *OMEGA, The International Journal of Management Science*, 27(2):219–239.

Allahverdi, A., Ng, C. T., Cheng, T. C. E., and Kovalyov, M. Y. (2008). A survey of scheduling problems with setup times or costs. *European Journal of Operational Research*, 187(3):985–1032.

Averbakh, I., Berman, O., and Chernykh, I. (2005). The m-machine flowshop problem with unit-time operations and intree precedence constraints. *Operations Research Letters*, 33(3):263–266.

Baker, K. R. and Scudder, G. D. (1990). Sequencing with earliness and tardiness penalties: a review. *Mathematics of Operations Research*, 15:483–495.

Bellanger, A. and Oulamara, A. (2009). Scheduling hybrid flowshop with parallel batching machines and compatibilities. *Computers & Operations Research*, 36(6):1982–1992.

Bertel, S. and Billaut, J.-C. (2004). A genetic algorithm for an industrial multiprocessor flow shop scheduling problem with recirculation. *European Journal of Operational Research*, 159(3):651–662.

Brucker, P., Heitmann, S., and Hurink, J. (2003). Flow-shop problems with intermediate buffers. *OR Spectrum*, 25(4):549–574.

Brucker, P., Hilbig, T., and Hurink, J. (1999). A branch and bound algorithm for a single-machine scheduling problem with positive and negative time-lags. *Discrete Applied Mathematics*, 94 (1–3):77–99.

Chang, Y. C. and Lee, C.-Y. (2004). Machine scheduling with job delivery coordination. *European Journal of Operational Research*, 158(2):470–487.

Chen, Z.-L. (1996). Parallel machine scheduling with time dependent processing times. *Discrete Applied Mathematics*, 70(1):81–93.

Chen, Z. L. and Vairaktarakis, G. L. (2005). Integrated scheduling of production and distribution operation. *Management Science*, 51(4):614–628.

Cheng, T. C. E., Gupta, J. N. D., and Wang, G. Q. (2000). A review of flowshopn scheduling research with setup times. *Production and Operations Management*, 9(3):262–282.

Choi, S. W., Kim, Y. D., and Lee, G. C. (2005). Minimizing total tardiness of orders with reentrant lots in a hybrid flowshop. *International Journal of Production Research*, 43(11):2149–2167.

Crama, Y. and Spieksma, F. C. R. (1996). Scheduling jobs of equal length: Complexity, facets and computational results. *Mathematical Programming*, 72(3):207–227.

Defersha, F. and Chen, M. (2012). Jobshop lot streaming with routing flexibility, sequence-dependent setups, machine release dates and lag time. *International Journal of Production Research*, 50(8):2331–2352.

Eren, T. and Guner, E. (2009). A bicriteria parallel machine scheduling with a learning effect. *International Journal of Advanced Manufacturing Technology*, 40(11–12):1202–1205.

Feldmann, M. and Biskup, D. (2008). Lot streaming in a multiple product permutation flow shop with intermingling. *International Journal of Production Research*, 46(1):197–216.

Fondrevelle, J., Oulamara, A., and Portmann, M.-C. (2006). Permutation flowshop scheduling problems with maximal and minimal time lags. *Computers & Operations Research*, 33(6):1540–1556.

Framinan, J. M., Gupta, J. N. D., and Leisten, R. (2004). A review and classification of heuristics for permutation flow-shop scheduling with makespan objective. *Journal of the Operational Research Society*, 55(12):1243–1255.

Gawiejnowicz, S. (2008). *Time-Dependent Scheduling*. Springer.

Gupta, J. N. D. and Stafford, Jr, E. F. (2006). Flowshop scheduling research after five decades. *European Journal of Operational Research*, 169(3):699–711.

Hall, N. G. and Sriskandarajah, C. (1996). A survey of machine scheduling problems with blocking and no-wait in process. *Operations Research*, 44(3):510–525.

Hejazi, S. R. and Saghafian, S. (2005). Flowshop-scheduling problems with makespan criterion: a review. *International Journal of Production Research*, 43(14):2895–2929.

Hendel, Y., Runge, N., and Sourd, F. (2009). The one-machine just-in-time scheduling problem with preemption. *Discrete Optimization*, 6(1):10–22.

Hou, S. and Hoogeveen, H. (2003). The three-machine proportionate flow shop problem with unequal machine speeds. *Operations Research Letters*, 31(3):225–231.

Kubiak, W. and van de Velde, S. (1998). Scheduling deteriorating jobs to minimize makespan. *Naval Research Logistics*, 45(5):511–523.

Lawler, E. L. (1973). Optimal Sequencing of a Single Machine Subject to Precedence Constraints. *Management Science*, 19(5):544–546.

Lee, C.-Y. and Chen, Z.-L. (2001). Machine scheduling with transportation considerations. *Journal of Scheduling*, 4(1):3–24.

Lee, C.-Y., Lei, L., and Pinedo, M. (1997). Current trends in deterministic scheduling. *Annals of Operations Research*, 70:1–41.

Lee, C.-Y., Uzsoy, R., and Martin-Vega, L. A. (1992). Efficient algorithms for scheduling semiconductor burn-in operations. *Operations Research*, 40(4):764–775.

Lee, W. C., Wu, C. C., and Sung, H. J. (2004). A bi-criterion single-machine scheduling problem with learning considerations. *Acta Informatica*, 40(4):303–315.

Leisten, R. (1990). Flowshop sequencing problems with limited buffer storage. *International Journal of Production Research*, 28(11):2085.

Leisten, R. and Kolbe, M. (1998). A note on scheduling jobs with missing operations in permutation flow shops. *International Journal of Production Research*, 36(9):2627–2630.

Lenstra, J. K. and Rinnooy Kan, A. H. G. (1978). Complexity of Scheduling under Precedence Constraints. *Operations Research*, 26(1):22–35.

Li, C. L., Vairaktarakis, G. L., and Lee, C.-Y. (2005). Machine scheduling with deliveries to multiple customer locations. *European Journal of Operational Research*, 164(1):39–51.

Linn, R. and Zhang, W. (1999). Hybrid flow shop scheduling: A survey. *Computers & Industrial Engineering*, 37(1–2):57–61.

Liua, Z., Yua, W., and Cheng, T. C. E. (1999). Scheduling groups of unit length jobs on two identical parallel machines. *Information Processing Letters*, 69(6):275–281.

Lushchakova, I. N. and Kravchenko, S. A. (1998). Two-machine shop scheduling with zero and unit processing times. *European Journal of Operational Research*, 107(2):378–388.

Maffioli, F. and Sciomachen, A. (1997). A mixed-integer model for solving ordering problems with side constraints. *Annals of Operations Research*, 69:277–297.

Mascis, A. and Pacciarelli, D. (2005). Job-shop scheduling with blocking and no-wait constraints. *European Journal of Operational Research*, 143(3):498–517.

Naderi, B. and Ruiz, R. (2010). The distributed permutation flowshop scheduling problem. *Computers & Operations Research*, 37(4):754–768.

Ow, P. S. (1985). Focused scheduling in proportionate flowshops. *Management Science*, 31(7):852–869.

Pan, Q.-K. and Wang, L. (2008). A novel differential evolution algorithm for no-idle permutation flow-shop scheduling problems. *European Journal of Industrial Engineering*, 2(3):279–297.

Papadimitriou, C. H. and Kanellakis, P. C. (1980). Flowshop scheduling with limited temporary-storage. *Journal of the ACM*, 27(3):533–549.

Park, Y. B., Pegden, C., and Enscore, E. (1984). A survey and evaluation of static flowshop scheduling heuristics. *International Journal of Production Research*, 22(1):127–141.

Pinedo, M. L. (2012). *Scheduling: Theory, Algorithms, and Systems*. Springer, New York, fourth edition.

Potts, C. N. and Hall, N. G. (2003). Supply chain scheduling: Batching and delivery. *Operations Research*, 51(4):566–584.

Potts, C. N. and Kovalyov, M. Y. (2000). Scheduling with batching: A review. *European Journal of Operational Research*, 120(2):222–249.

Quadt, D. and Kuhn, D. (2007). A taxonomy of flexible flow line scheduling procedures. *European Journal of Operational Research*, 178(3):686–698.

Ribas, I., Leisten, R., and Framinan, J. M. (2010). Review and classification of hybrid flow shop scheduling problems from a production system and a solutions procedure perspective. *Computers & Operations Research*, 37(8):1439–1454.

Riezebos, J. and Gaalman, G. J. C. (1998). Time lag size in multiple operations flow shop scheduling heuristics. *European Journal of Operational Research*, 105(1):72–90.

Ruiz, R. and Andres-Romano, C. (2011). Scheduling unrelated parallel machines with resource-assignable sequence-dependent setup, times. 57(5–8):777–794.

Ruiz, R. and Maroto, C. (2005). A comprehensive review and evaluation of permutation flowshop heuristics. *European Journal of Operational Research*, 165(2):479–494.

Ruiz, R. and Maroto, C. (2006). A genetic algorithm for hybrid flowshops with sequence dependent setup times and machine eligibility. *European Journal of Operational Research*, 169(3):781–800.

Ruiz, R., Vallada, E., and Fernández-Martínez, C. (2009). Scheduling in flowshopns with no-idle machines. In Chakraborty, U. K., editor, *Computational Intelligence in Flow Shop and Job Shop Scheduling, volume 230 of Studies in Computational Intelligence*, pages 21–51, Berlin. Springer-Verlag.

Ruiz, R. and Vázquez-Rodríguez, J. A. (2010). The hybrid flowshop scheduling problem. *European Journal of Operational Research*, 205(1):1–18.

Sarin, S. C. and Jaiprakash, P. (2007). *Flow Shop Lot Streaming*. Springer, New York.

Schmidt, G. (2000). Scheduling with limited machine availability. *European Journal of Operational Research*, 121(1):1–15.

Sen, T. and Gupta, S. K. (1984). A state-of-art survey of static scheduling research involving due dates. *OMEGA, The International Journal of Management Science*, 12(1):63–76.

Shabtay, D. and Steiner, G. (2007). A survey of scheduling with controllable processing times. *Discrete Applied Mathematics*, 155(13):1643–1666.

Sidney, J. B. (1975). Decomposition Algorithms for Single-Machine Sequencing with Precedence Relations and Deferral Costs. *Operations Research*, 23(2):283–298.

Turner, S. and Booth, D. (1987). Comparison of Heuristics for Flow Shop Sequencing. *OMEGA, The International Journal of Management Science*, 15(1):75–78.

Urlings, T., Ruiz, R., and Sivrikaya-Şerifoğlu, F. (2010). Genetic algorithms with different representation schemes for complex hybrid flexible flow line problems. *International Journal of Metaheuristics*, 1(1):30–54.

Vachajitpan, P. (1982). Job sequencing with continuous machine operation. *Computers & Industrial Engineering*, 6(3):255–259.

Vallada, E., Ruiz, R., and Minella, G. (2008). Minimising total tardiness in the m-machine flowshop problem: A review and evaluation of heuristics and metaheuristics. *Computers & Operations Research*, 35(4):1350–1373.

Vignier, A., Billaut, J.-C., and Proust, C. (1999). Les problèmes d'ordonnancement de type flowshop hybride: État de l'art. *RAIRO Recherche opérationnelle*, 33(2):117–183. In French.

Wang, H. (2005). Flexible flow shop scheduling: optimum, heuristics and artificial intelligence solutions. *Expert Systems*, 22(2):78–85.

Yang, W.-H. and Liao, C.-J. (1999). Survey of scheduling research involving setup times. *International Journal of Systems Science*, 30(2):143–155.

Yeung, W.-K., Oğuz, C., and Cheng, T.-C. E. (2009). Two-machine flow shop scheduling with common due window to minimize weighted number of early and tardy jobs. *Naval Research Logistics*, 56(7):593–599.

Chapter 5
Objectives

5.1 Introduction

The framework for manufacturing scheduling envisioned in Chap. 1 presented three main blocks—models, methods and tools—that, together with the humans involved in the process, compose a scheduling system. This part of the book deals with scheduling models, and in Chaps. 3 and 4 we have reviewed the main elements of these models concerning systems layouts and processing constraints, respectively. Now a final element of the scheduling models—the objectives sought in the decision problem—are presented in this chapter before the building of scheduling models is properly discussed in Chap. 6.

Interestingly, most scheduling literature naturally assumes the existence of some objectives *intrinsic* to scheduling. However, we have already discussed that manufacturing scheduling decisions are part of a broader set of decisions collectively known as production management (see Chap. 2), and as a consequence, scheduling objectives should be aligned/integrated with those regarding other decisions in production management. Therefore, before introducing the most employed scheduling objectives, we provide a discussion of their choice in terms of the (broader) objectives considered in production management. Next, we classify most of the objectives employed in the scheduling literature depending on the existence or not of due dates, and will study the most important in detail. As we will discuss, reality can hardly be modeled with a single objective or criterion, therefore we will comment on multi-criteria or multi-objective problems as a pointer to a Chap. 10 where these topics will be treated in much more detail.

More specifically, in this chapter, we

- present a rationale for scheduling objectives within the broader context of production management (Sect. 5.2),
- introduce a notation and classification for scheduling performance measures (Sect. 5.3),
- discuss the main scheduling objectives (Sect. 5.4),

J. M. Framinan et al., *Manufacturing Scheduling Systems*,
DOI: 10.1007/978-1-4471-6272-8_5, © Springer-Verlag London 2014

- introduce ways to modulate the aforementioned objectives by using weights and/or rankings (Sect. 5.5),
- present an illustrative example (Sect. 5.6),
- point out the main issues regarding the evaluation of more than one criteria (Sect. 5.7).

5.2 A Rationale for Scheduling Objectives

As is well-known from general decision theory, quantifiable objective functions intend that the decision-maker wishes to determine extreme solutions (maximum, minimum), or solutions which achieve fixed and pre-determined objective function values, or intervals of objective function values which might be bounded from both sides or only from one side. The latter give minimum or maximum levels to be reached. In production management (and consequently in manufacturing scheduling), often the objectives are classified into the categories of costs (sometimes profit), time, quality and flexibility.

Since scheduling decisions are located in the operational, very short-term horizon level of the multi-level manufacturing planning and control framework (see Sect. 2.4.1), cost components are often not considered as objective functions in scheduling models, at least not explicitly. Instead, it is either supposed that on the scheduling level as well as on the shop floor level costs can only be influenced via time objectives (and therefore time-related objective functions are explicitly considered) or costs are proportional to time parameters and consequently, weighted time parameters are considered as a representation of costs. These time parameters may refer to machine and/or resource-related parameters, or to job-related parameters. The latter may be further classified into intrasystem objectives, e.g. maximising the utilisation of resources, and extrasystem objectives, such as minimising the delays with respect to the committed due dates. The category 'quality' in models of manufacturing scheduling is also usually exclusively regarded within the time framework, i.e. by service level indicators, being the reason why this category is not discussed further here as well. Referring to flexibility, as a well-accepted category of manufacturing goals, it has not been explicitly considered in manufacturing scheduling models for many years. However, during the last years several models and approaches have been published which explicitly consider variances and/or variabilities (usually measured by the coefficient of variation), e.g. of completion times, as objective functions, though usually in a multi-criteria setting. These indicators are intended to yield stable and robust schedules and are obviously directly related to flexibility aspects. Although flexibility is usually not primarily considered in static-deterministic models as described here, recognition of these objective functions reflects the fact that the models discussed here usually represent simplifications of dynamic-stochastic real-world settings, which require consideration of flexibility aspects at least indirectly.

Figure 5.1 summarises the above basic discussion of types of objective functions in manufacturing scheduling and gives examples for the different categories.

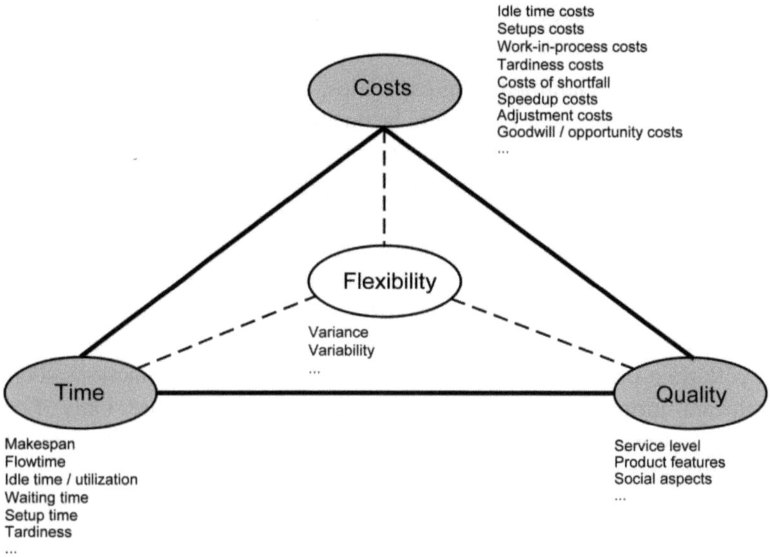

Fig. 5.1 (Extended) triangle of objective categories in manufacturing scheduling including flexibility

Note that the objective categories mentioned here are highly interdependent. Sometimes they act in parallel, such as in the case of idle time costs and utilisation. Unfortunately, most of them are often conflicting: The case of maximising machine utilisation is strongly against reducing work-in-progress or inventory levels. High machine utilisation is also conflicting with service level or with meeting due dates. If machine utilisation is low, there is a certain guarantee that machines will be free when needed for finishing a job that is about to be late. On the other hand, a very high inventory level practically guarantees that lead times will approach zero as all products will be readily available.

In addition, there is the issue of establishing a good level of performance in the shop floor for any of the previously cited criteria. A reference point, like for example, 'A cost threshold' or a 'customer's satisfaction level' has to be set. However, this is in fact a circular problem since usually measuring these criteria is, as mentioned, not easy. As a result, the performance of the schedules is often measured against past performance. Given that manufacturing systems are highly dynamic, comparing against past performance inevitably introduces a large degree of bias in the process. Another important issue is that, when measuring the performance of a schedule, the schedule has to be broken down into specific time horizons (e.g. weeks) and each time horizon is scheduled independently of others. The result might be that a very good optimisation for one period affects negatively the following periods. Other problems arise when breaking levels of management. For example, the schedule manager at a higher decision level and the plant manager at an operational level. While the scheduler might be interested in maximising customer's satisfaction, the plant manager will probably be centered around maximising machine utilisation or throughput.

In case of narrow manufacturing capacities, utilisation-oriented objective functions seem to be appropriate. This might be one reason why these objective functions became popular in the early days of manufacturing scheduling, i.e. in the mid of the twentieth century when after World War II every product which was produced could be sold and the only bottleneck in many industries was exclusively manufacturing capacity. However, looking at many industries during the last decades, surplus capacity was characteristic and the persistence of utilisation-oriented objective functions in many papers appears to be somewhat nostalgic. Instead, guaranteeing short and/or reliable delivery dates has become much more relevant for successful manufacturing. Thus, cycle-time-oriented and/or due-date oriented objective functions seem to be more relevant during the last years and in the foreseeable future. Only if capacity represents the true bottleneck for short-term business success, utilisation-oriented objective functions should be considered.

5.3 Performance Measures

Objectives employed in manufacturing scheduling models are related to the optimisation of some indicator of the performance of one schedule in the shop floor. We denote these indicators as *performance measures* and some of the most important will be defined next. Note that the notation corresponding to the data required below has been given in Sect. 3.2.2:

- C_j: Completion time of job j, i.e. time at which job j finishes its processing in the shop.
- F_j: Flowtime of job j. This models the time that the job stays in the shop while in production or waiting for processing. Clearly, $F_j = C_j - r_j$. Most importantly, in the lack of release dates, flowtime and completion time are equal, i.e. $F_j = C_j$.
- L_j: Lateness or slack of job j with respect to its due date d_j, or deadline $\overline{d}_j$. As such, it is calculated as $L_j = C_j - d_j$ or $L_j = C_j - \overline{d}_j$. The lateness function is shown in Fig. 5.2.
- T_j: Tardiness of job j. This measures only jobs that finish beyond their due dates and it is calculated as $T_j = \max\{L_j, 0\}$. Tardiness ignores negative lateness values, i.e. all jobs that are finished before the due date are not tardy and therefore not a reason to worry about. The tardiness function is shown in Fig. 5.3.
- E_j: Earliness of job j. Similarly to tardiness, it is only calculated for early jobs, i.e. $E_j = \max\{-L_j, 0\}$. Note that the earliness function is non-increasing with respect to C_j as shown in Fig. 5.4.
- U_j Tardy job or late job. This measure yields 1 if job j is late (i.e. $T_j > 0$ or equivalently $L_j > 0$, or equivalently $C_j > d_j$), and zero if not. Figure 5.5 depicts this function.

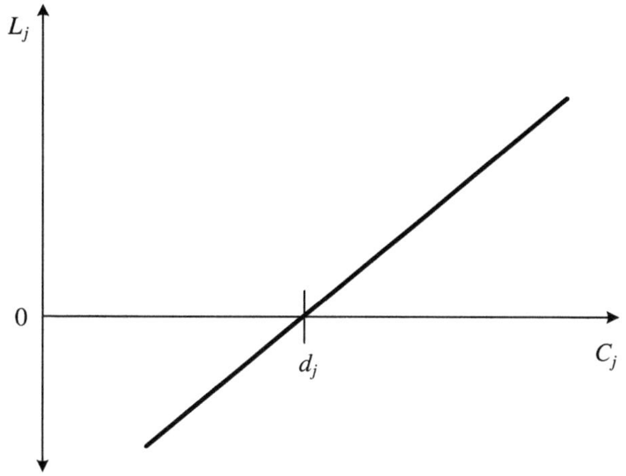

Fig. 5.2 Lateness function

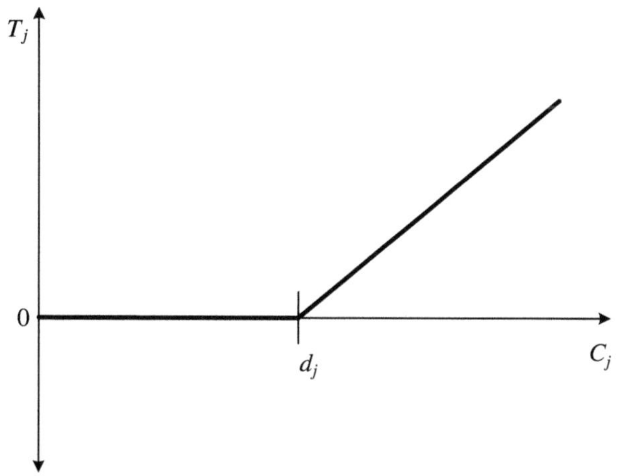

Fig. 5.3 Tardiness function

- V_j Early job. This measure yields 1 if job j is early (i.e. $L_j < 0$, or equivalently $C_j < d_j$), and zero if not. Figure 5.6 depicts this function.
- ET_j Earliness-Tardiness of job j. This measure is defined as the sum of the tardiness and the earliness of job j, i.e. $ET_j = E_j + T_j$. The function of ET_j is given in Fig. 5.7.
- JIT_j Just-In-Time job. This measure yields 1 if job j is neither early nor late (i.e. $L_j = 0$, or equivalently $C_j = d_j$), and zero if not. This is often referred to as 'just-in-time' or JIT. The function of JIT_j is given in Fig. 5.8.

Aside from the data in the model (most notably r_j and d_j), note that all above measures depend on the completion time of the job. The completion time of the job

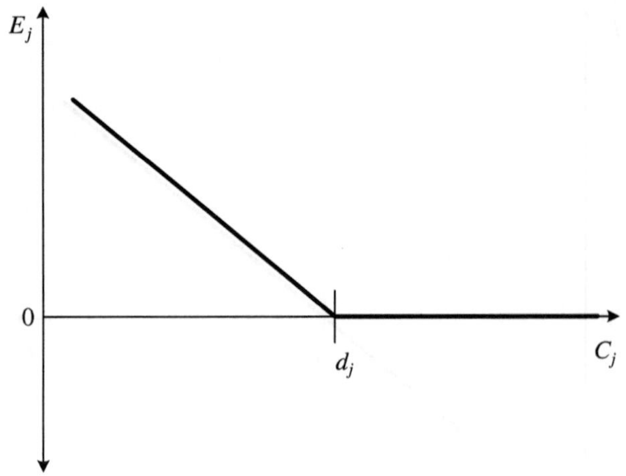

Fig. 5.4 Earliness function

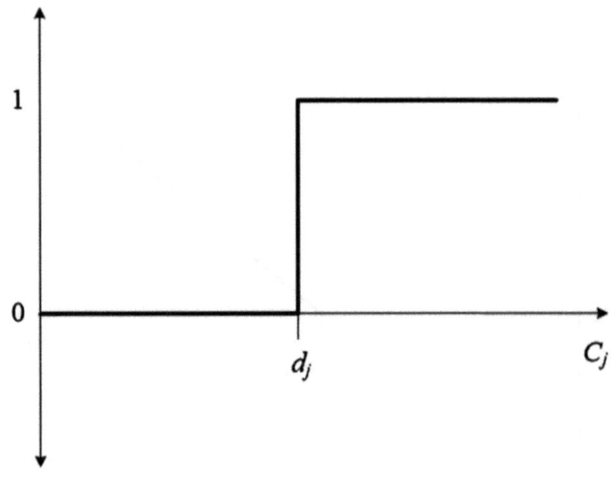

Fig. 5.5 Tardy jobs function

can be obtained for an particular schedule, i.e. the collection SO_{ij} of the starting times of operation for each job j on each machine i, according to:

$$C_j = \max_i \{SO_{ij} + p_{ij}\}.$$

When assuming some form of stochastic behaviour in the processing times, then the completion time of the job is a random variable that cannot be calculated according to a deterministic procedure. In these cases, statistical indicators of such random variable (such as its mean or variance) could be either obtained by a closed formula (unfortunately, only for specific cases), or at least estimated via samples.

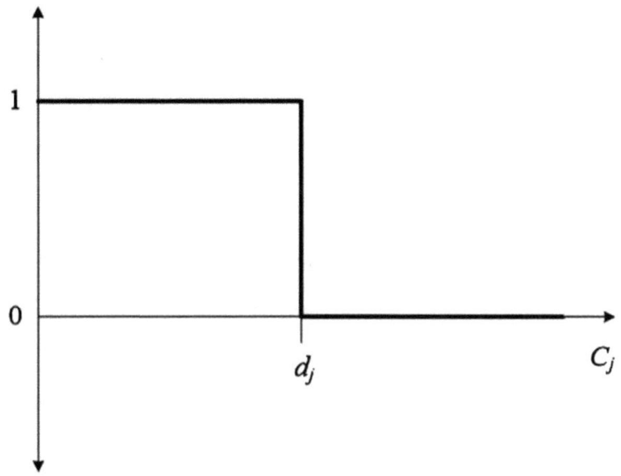

Fig. 5.6 Early job function

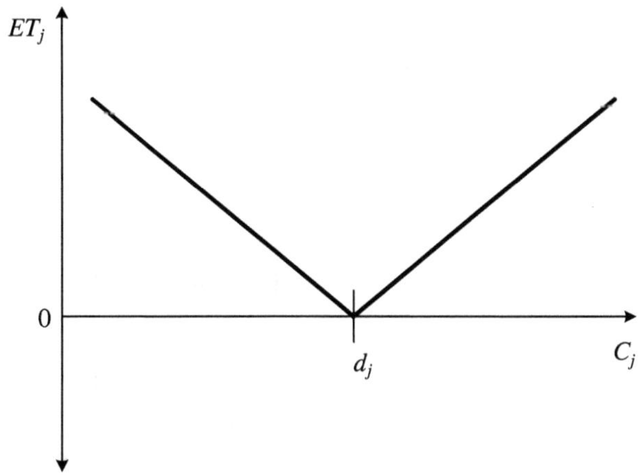

Fig. 5.7 Earliness-tardiness function

In addition, we will introduce two measures that are associated to rescheduling. Recall that, in rescheduling, the goal is to minimise disruptions from existing plans whenever new events occur. Therefore, we assume that there are two schedules Π_A and Π_B corresponding to points in time before and after the arrival of the new event, respectively. Two employed measures are:

- $D_j(\Pi_A, \Pi_B)$: Sequence disruption of job j. It indicates the difference (in absolute value) between the position of job j in schedule Π_A, and its position in schedule Π_B, i.e. $D_j(\Pi_A, \Pi_B) = |\Pi_A(j) - \Pi_B(j)|$.

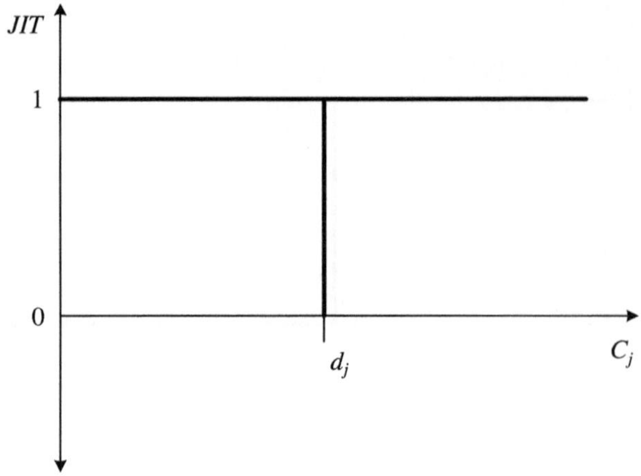

Fig. 5.8 JIT job

Table 5.1 Summary of performance measures

Measure	Notation	Formulae
Flowtime	F_j	$C_j - r_j$
Lateness	L_j	$C_j - d_j$
Tardiness	T_j	$\max\{0, C_j - d_j\} = \max\{0, L_j\}$
Earliness	E_j	$\max\{0, d_j - C_j\} = \max\{0, -L_j\}$
Earliness and tardiness	ET_j	$E_j + T_j = \|C_j - d_j\| = \|L_j\|$
Tardy job	U_j	$\begin{cases} 1 & C_j > d_j \\ 0 & \text{otherwise} \end{cases} = \begin{cases} 1 & L_j > 0 \\ 0 & \text{otherwise} \end{cases}$
Early job	V_j	$\begin{cases} 1 & C_j < d_j \\ 0 & \text{otherwise} \end{cases} = \begin{cases} 1 & L_j < 0 \\ 0 & \text{otherwise} \end{cases}$
Just-in-time job	JIT_j	$\begin{cases} 1 & C_j < d_j \text{or } C_j > d_j \\ 0 & \text{otherwise, i.e. if } C_j = d_j \end{cases} = \begin{cases} 1 & L_j \neq 0 \\ 0 & L_j = 0 \end{cases}$
Sequence disruption	$D_j(\Pi_A, \Pi_B)$	$\|\Pi_A(j) - \Pi_B(j)\|$
Time disruption	$\Delta_j(\Pi_A, \Pi_B)$	$\|C_j(\Pi_A) - C_j(\Pi_B)\|$

- $\Delta_j(\Pi_A, \Pi_B)$: Time disruption of job j. It indicates the difference (in absolute value) between the completion time of job j in schedule Π_A, and its completion time in schedule Π_B, i.e. $\Delta_j(\Pi_A, \Pi_B) = |C_j(\Pi_A) - C_j(\Pi_B)|$.

The above measures are summarised in Table 5.1.

5.4 Scheduling Objectives

From the previous notation, we are ready to define the most common objectives. A summary is provided in Fig. 5.2. According to the classification of scheduling models introduced in Sect. 3.3.2, the γ field of this notation will be used to represent the criteria under consideration. For practical purposes, we will classify them in five general groups:

- Feasibility. In this case, the objective is to find a feasible solution for the model.
- Non due-date related. There is no information (or it is not required) regarding due dates or deadlines.
- Due-date related. The objective involves the usage of information about due dates or deadlines.
- Rescheduling related. The objective involves information from two different schedules (usually associated to two different decision intervals).
- Additional objectives. Other objectives that can be considered.

Feasibility models indicate the absence of an objective (except that of finding a feasible schedule) and will not be discussed further. In such models, the γ field is indicated by a slash, i.e.:

$\gamma = -$: Feasibility problem.

Therefore, the model $1|prec, prmpt|-$ indicates the formal decision problem of finding (at least) one feasible schedule for the single machine layout with precedence constraints where pre-emption is allowed.

For the rest of the groups, a distinction can be done depending on the type of function g applied to the completion times, i.e. function f is of the max-form if $f = \max_{1 \leq j \leq n} g(C_j)$. In contrast, is of the sum-form if $f = \sum_{j=1}^{n} g(C_j)$. Furthermore, recall from Sect. 1.5 that g is said to be regular if it is a non-decreasing function of $(C_1, C_2, \ldots, C_n)$. Unless stated otherwise, the objectives presented in the following sections are regular.

A final remark can be done with respect to an stochastic settings, as there is an stochastic counterpart of all scheduling objectives considered here. In such cases, the otherwise deterministic objectives are now random variables and therefore the objective can be minimising some of their characteristics, most notably their mean or variance. These objectives can also be integrated in the model. For instance, the objective of minimising the expected value of objective O is notated as follows:

$\gamma = E[O]$: expected value of objective O.

5.4.1 Non Due Date Related Objectives

In this section, we discuss all objectives that do not consider due dates in their calculation. These are mainly concerned with the completion times and with the total time each job spends in the shop.

5.4.1.1 Max-Form Objectives

Makespan

The maximum completion time or makespan is defined as $C_{\max} = \max\{C_1, C_2, \ldots, C_n\}$. Regarding notation, makespan minimisation is denoted as:

$\gamma = C_{\max}$: Makespan minimisation.

Makespan can be seen as the time needed to finish the whole production plan since it measures from the time the first job starts processing (normally this is assumed to be zero, unless release times or other constraints exist), to the time the last job in the processing sequence is finished on the last machine it visits. It is thus connected with the idle time. However, no standard definition of idle time exists, as it depends if the 'heads' (i.e. the time that each machine is idle before starting processing its first job) and 'tails' (i.e. the time that each machine is idle after processing its last job) are both included, excluded, or just one of them is considered. For the definition of idle time including both 'heads' and 'tails', it can be proved that both objectives are equivalent and thus minimizing makespan and minimizing this definition of idle time are equivalent. Additionally, makespan is also closely connected to the production rate which can be calculated as $\mathrm{PR} = \frac{n}{C_{\max}}$. The utilisation of machine i is calculated as $\frac{\mathrm{TC}_i}{\mathrm{PR}}$ where TC_i is the total capacity of machine i.

Maximising machine utilisation is interesting as far as cost accounting is concerned. Capital equipment and expensive machinery have to be paid for. Therefore, using machines to produce products that increase the revenue of companies yield higher returns of the investments and, in the end, a higher profitability. On the contrary, idle machines are not generating revenue and are seen as a lost profit. However, it has to be noted that utterly increasing machine utilisation can result in undesired effects. For example, keeping machines busy churning out products that nobody demands will only reduce the level of service, increase the level of inventory and in the end, reduce benefits. In a make-to-stock environment, increasing machine utilisation makes all the sense. In addition, it should be mentioned that maximisation of utilization makes sense for bottleneck machines or bottleneck systems albeit high utilization might in turn result in larger waiting times as is well known from queuing theory.

Maximum flowtime

The maximum flowtime objective is defined as $\max_{1 \le j \le n} F_j$ and it is denoted as:

$\gamma = \max F_j$: maximum flowtime minimisation.

Maximum flowtime is a niche objective that has been seldom studied in the scientific literature. The reason is that with no release times, i.e. $r_j = 0, \forall j \in N$, we have that $C_{\max} = \max F_j$. Therefore, and specially for long production horizons, the effect of the release date r_j over the $F_{\max}$ decreases as $C_{\max}$ increases unless very large release dates are present. A weighted version of this objective can be also formulated (see Table 5.2).

5.4.1.2 Sum-Form Objectives

Total/average completion time or flowtime

Total flowtime and total completion time add up the flow time and completion times of all the jobs, respectively. Total flowtime and total completion times are seen as surrogates for cycle time, and it was already mentioned that WIP and cycle time (CT) are connected.

Minimisation of total/average flowtime and total completion times can be denoted as follows:

$\gamma = \sum C$: Total completion time, calculated as $\sum C = \sum_{j=1}^{n} C_j$.
$= \overline{C}$: Average completion time, calculated as $\overline{C} = \frac{1}{n} \cdot \sum_{j=1}^{n} C_j$.
$= \sum F$: Total flowtime, calculated as $\sum F = \sum_{j=1}^{n} F_j$.
$= \overline{F}$: Average flowtime time, calculated as $\overline{F} = \frac{1}{n} \cdot \sum_{j=1}^{n} F_j$.

Note that the minimisation of total flowtime/completion time is equivalent to the minimisation of the average flowtime/completion time.

5.4.2 Due Date Related Objectives

Meeting due dates is probably the highest regarded indicator of customer's satisfaction influenced by short-term operations management. As a result of this, many different objectives exist that in one way or another capture this importance. All these criteria are mainly based in the previous definitions of lateness (L_j), Tardiness (T_j) and Earliness (E_j).

At this stage, it is important to make some punctualisations. In a make-to-order production environment, every job usually corresponds to a client order (or to several orders from different clients that have been batched). Therefore, the due date set by or agreed with the client, is considered. Make-to-stock production systems are different as client orders are ideally served from the stock immediately as they are placed.

Table 5.2 Summary of objectives

Type	Objective	Notation	Formulae
Feasibility problem			
	Finding any feasible schedule	–	
Non due date related			
Max-form	Makespan	$\max C_j$ or $C_{\max}$	$\max_{1\leq j\leq n} C_j$
	Weighted makespan	$\max w_jC_j$	$\max_{1\leq j\leq n} w_jC_j$
	Maximum flowtime	$\max F_j$	$\max_{1\leq j\leq n} F_j$
	Maximum weighted flowtime	$\max w_jF_j$	$\max_{1\leq j\leq n} w_jF_j$
Sum-form	Total completion time	$\sum C_j$	$\sum_{j=1}^{n} C_j$
	Total weighted completion time	$\sum w_jC_j$	$\sum_{j=1}^{n} w_jC_j$
	Total flowtime	$\sum F_j$	$\sum_{j=1}^{n} F_j$
	Total weighted flowtime	$\sum w_jF_j$	$\sum_{j=1}^{n} w_jF_j$
Due date related			
Max-form	Maximum lateness	$\max L_j$	$\max_{1\leq j\leq n} L_j$
	Maximum weighted lateness	$\max w_jL_j$	$\max_{1\leq j\leq n} w_jL_j$
	Maximum tardiness	$\max T_j$	$\max_{1\leq j\leq n} T_j$
	Maximum weighted tardiness	$\max w_jT_j$	$\max_{1\leq j\leq n} w_jT_j$
	Maximum earliness	$\max E_j$	$\max_{1\leq j\leq n} E_j$
	Maximum weighted earliness	$\max w_jE_j$	$\max_{1\leq j\leq n} w_jE_j$
Sum-form	Total lateness	$\sum L_j$	$\sum_{j=1}^{n} L_j$
	Total weighted lateness	$\sum w_jL_j$	$\sum_{j=1}^{n} w_jL_j$
	Total tardiness	$\sum T_j$	$\sum_{j=1}^{n} T_j$
	Total weighted tardiness	$\sum w_jT_j$	$\sum_{j=1}^{n} w_jT_j$
	Number of tardy jobs	$\sum U_j$	$\sum_{j=1}^{n} U_j$
	Number of weighted tardy jobs	$\sum w_jU_j$	$\sum_{j=1}^{n} w_jU_j$
	Total earliness	$\sum E_j$	$\sum_{j=1}^{n} E_j$
	Total weighted earliness	$\sum w_jE_j$	$\sum_{j=1}^{n} w_jE_j$
	Number of early jobs	$\sum V_j$	$\sum_{j=1}^{n} V_j$
	Number of weighted early jobs	$\sum w_jV_j$	$\sum_{j=1}^{n} w_jV_j$
	Total earliness and tardiness	$\sum ET_j$	$\sum_{j=1}^{n} E_j + T_j$
	Total weighted earliness and tardiness	$\sum w_jET_j$	$\sum_{j=1}^{n} w_j(E_j + T_j)$
	Total weighted earliness and tardiness with different weights	$\sum w_jE_j + w'_jT_j$	$\sum_{j=1}^{n} w_jE_j + w'_jT_j$
	Number of just-in-time jobs	$\sum JIT_j$	$\sum_{j=1}^{n} JIT_j$
	Number of weighted just-in-time jobs	$\sum w_jJIT_j$	$\sum_{j=1}^{n} w_jJIT_j$
Rescheduling related			
Max-form	Maximum sequence disruption	$\max D(\Pi_A, \Pi_B)$	$\max_{1\leq j\leq n} \max D_j(\Pi_A, \Pi_B)$
	Maximum time disruption	$\max \Delta(\Pi_A, \Pi_B)$	$\max_{1\leq j\leq n} \max \Delta_j(\Pi_A, \Pi_B)$
Sum-form	Total sequence disruption	$\sum D(\Pi_A, \Pi_B)$	$\sum_{j=1}^{n} D_j(\Pi_A, \Pi_B)$
	Total time disruption	$\sum \Delta(\Pi_A, \Pi_B)$	$\sum_{j=1}^{n} \Delta_j(\Pi_A, \Pi_B)$

As a result, there are no due dates. However, stock levels decline as orders are served and reach a point where replenishment is needed. An 'internal' order is placed so to refill the stock with products. These internal orders are, as regards scheduling, identical to make-to-order client orders. The reason is that if the internal orders are produced late, client orders might go unserved due to stockouts.

In make-to-order systems, it is interesting to measure the service level which is just the fraction of client orders served by or before the due date. Alternatively, in make-to-stock production environments, the fill rate is the fraction of orders that are served from the inventory. Other measures include the amount of time in backorder or number of stockouts. All the following scheduling objectives are related to these real-life goals.

5.4.2.1 Max-Form Objectives

Maximum lateness

The objective of maximum lateness minimisation can be defined as follows:

$\gamma = L_{\max}$: Maximum lateness minimisation, calculated as $\max L = \max_{1 \le j \le n} L_j\}$.

This objective tries to minimise the maximum deviation from the due date or deadline in the schedule. The idea is to limit as much as possible the delay with respect to the committed date. Note that $\max L_j$ might be a negative number if all jobs in a given schedule are completed before their corresponding due dates. Indeed, if all jobs can be finished before their due dates (i.e. there are schedules for which $L_j < 0 \forall j$), then maximum lateness minimisation leads to finishing the jobs as early as possible).

Maximum tardiness

Maximum tardiness minimisation can be defined as follows:

$\gamma = T_{\max}$: Maximum tardiness minimisation, calculated as $\max T_j = \max_{1 \le j \le n} T_j$.

Minimising maximum tardiness has a direct meaning. For example, if $\max T_j =$ 15 then the most tardy job in the schedule is completed in 15 units of time (whatever these units might be) after its due date. However, $\sum T_j$ can quickly grow to large numbers if n is large in a given layout, which might be difficult to interpret.

Maximum earliness

In some scenarios, tardy jobs might be of no concern in comparison with having jobs finished before the due date. Finishing a product early has several added costs. First of all, the product has to be stocked if the due date is still far ahead in time, which results in inventory handling and costs, specially so if the product is cumbersome or voluminous. Furthermore, if the product is very expensive and/or if the manufacturing process is costly, finishing it early precludes the company to invoice the client, which can add a considerable financial burden. Some extreme cases could be perishable products, that have to be used shortly after production. Dairy foods or some chemical products are clear examples. In these cases, minimising earliness might be not just preferable, but mandatory.

Maximum earliness minimisation can be defined as follows:

$\gamma = \max E_j$: Maximum earliness minimisation, calculated as $\max E_j = \max_{1 \leq j \leq n} E_j$.

Earliness minimisation opens a very important aspect to consider when building schedules: idle time insertion (i.e. non semi-active schedules). It is usually desired to leave machines idle even when there are jobs to process. This is expected since doing otherwise, would result in jobs finishing before the due date. However, inserting idle time is not a straightforward issue at all since the amount of inserted time can be considered a continuous variable. It has to be decided before which tasks idle time is inserted. Furthermore, inserting idle time might in turn result in tardy jobs.

5.4.2.2 Sum-Form Objectives

Total/average lateness

Minimisation of total/average lateness can be denoted as follows:

$\gamma = \sum L_j$: Total lateness minimisation, calculated as $\sum L_j = \sum_{j=1}^{n} L_j$.
$= \overline{L}$: Average lateness minimisation, calculated as $\overline{L} = \frac{1}{n} \cdot \sum_{j=1}^{n} L_j$.

Total lateness and average lateness are not very appropriate objectives. The reason is that negative lateness values will compensate the positive ones and the final total or average lateness value might have no real meaning or interpretation. For example, a small average lateness $\overline{L}$ could be the result of all jobs meeting due dates, something that is desired, or it could be the outcome of a bad schedule where half of the jobs are very early (large negative lateness values) and the other half of the products are finished very late (large positive lateness values).

Total/average tardiness

As we have just discussed, total/average lateness minimisation does not seem to make sense in many settings. Instance, total/average tardiness minimisation can be used:

$= \sum T_j$: Total tardiness minimisation, calculated as $sumT_j = \sum_{j=1}^{n} T_j$.
$= \overline{T}$: Average tardiness minimisation, calculated as $\overline{T} = \frac{1}{n} \cdot \sum_{j=1}^{n} T_j$.

Note that as it was the case with the total and average completion time or flowtime, there is no real difference from the optimisation point of view in minimising total or average lateness or tardiness. As expected, the main drawback of the tardiness measure is that some jobs might be finished very early, which in many situations is not a desired result.

Total/average earliness

As with lateness minimisation, two variants are commonly studied:

$= \sum E_j$: Total earliness minimisation, calculated as $\sum E_j = \sum_{j=1}^{n} E_j$.
$= \overline{E}$: Average earliness minimisation, calculated as $\overline{E} = \frac{1}{n} \cdot \sum_{j=1}^{n} E_j$.

Number of tardy/early jobs

Sometimes, when a job is tardy or early, the damage is already done and it is not so important how much tardy or early the job is. In these cases, the interest lies in 'counting' how many jobs are early or late. Therefore, the following objectives can be defined:

$\gamma = \sum U_j$: Number of tardy jobs minimisation, calculated from : $\sum U_j = \sum_{j=1}^{n} U_j$.
$= \sum V_j$: Number of early jobs minimisation, calculated as $\sum V_j = \sum_{j=1}^{n} V_j$.

Note that $\sum U_j$ is a regular performance measure whereas $\sum V_j$ is non-regular.

An interesting feature of $\sum U_j$ and $\sum V_j$ is that they are very easy to interpret. As a matter of fact, they can be easily transformed to measures like the percentage of early or tardy jobs. From that perspective, it is very easy to assess the quality of a given schedule. Note however, that it might be impossible to finish all products before or after their due dates. As expected, the main drawback of these measures is that they neglect extremely early or tardy jobs. Normally, in real life, it is desired to have a small number of tardy (early) jobs and for the jobs that are tardy (early) a small tardiness (earliness) is desired.

Total earliness and tardiness

A straightforward extension of the tardiness and earliness minimisation is to add them both in a single expression. For all aforementioned reasons, it might be of interest to complete a job as close as possible to its due date, i.e. not early, not late. Earliness-tardiness minimisation is defined as follows:

$\gamma = \sum_{j=1}^{n} E_j + T_j$: Total Earliness and tardiness minimisation. Sometimes this is simply expressed as ET_j.

It is interesting to note that $ET_j = E_j + T_j = \max\{d_j - C_j, 0\} + \max\{C_j - d_j, 0\} = |C_j - d_j|$. This expression is equivalent to what is referred in some texts as the minimisation of the absolute deviation from the due date (Brucker 2007).

Number of JIT jobs

Similarly with $\sum U_j$ or $\sum V_j$, we can maximise the number of just-in-time jobs JIT as follows:

$\gamma = \sum JIT_j$: Maximisation of the number of just in time jobs, calculated as $\sum JIT_j = \sum_{j=1}^{n} JIT_j$.

Note that, in contrast to the rest of objectives, a maximisation criterion is the one that makes sense.

5.4.3 Rescheduling-Related Objectives

In Sect. 5.3, some rescheduling-related performance measures have been introduced, i.e. $D_j(\Pi_A, \Pi_B)$ the sequence disruption of job j, and $\Delta_j(\Pi_A, \Pi_B)$ the time disruption of job j. From these two performance measures, the following minimisation objectives can be formulated:

$\gamma = \max D_j(\Pi_A, \Pi_B)$: Minimisation of the maximum sequence disruption, calculated as $\max D_j(\Pi_A, \Pi_B) = \max 1 \leq j \leq n D_j(\Pi_A, \Pi_B)$.
$= \max \Delta_j(\Pi_A, \Pi_B)$: Minimisation of the maximum time disruption, calculated as $\max \Delta_j(\Pi_A, \Pi_B) = \max 1 \leq j \leq n \Delta_j(\Pi_A, \Pi_B)$.
$= \sum D_j(\Pi_A, \Pi_B)$: Minimisation of the total sequence disruption, calculated as $\sum D_j(\Pi_A, \Pi_B) = \sum_{j=1}^{n} D_j(\Pi_A, \Pi_B)$.
$= \sum \Delta_j(\Pi_A, \Pi_B)$: Minimisation of the total time disruption, calculated as $\sum \Delta_j(\Pi_A, \Pi_B) = \sum_{j=1}^{n} \Delta_j(\Pi_A, \Pi_B)$.

Note that these objectives correspond to performance measures related to the disruption induced by a new schedule. Nevertheless, more than one disruption can be considered, and additional objectives may arise.

5.4.4 Additional Objectives

In this section, we cover other potentially interesting albeit not as thoroughly studied objectives. One common characteristic of some previous objectives is that they are linear on C_j. Each unit of time that a job is completed after its due date, adds one unit to the tardiness value (or w_j units if we are talking about weighted tardiness). However, very large deviations from the due date are more important than small deviations. A simple way of penalising more large deviations from the due date is by simply squaring the deviations. A breed of new objectives and criteria arise:

$\gamma = \max D_j$: Maximum squared deviation from the due date. This is calculated as $\max\{(C_1 - d_1)^2, (C_2 - d_2)^2, \ldots, (C_n - d_n)^2\}$.
$= \sum D$: Total squared deviation from the due date. Calculated as $\sum_{j=1}^{n}(C_j - d_j)^2$.
$= \overline{D}$: Average squared deviation from the due date, calculated as $\overline{D} = \frac{1}{n} \cdot \sum_{j=1}^{n}(C_j - d_j)^2$.

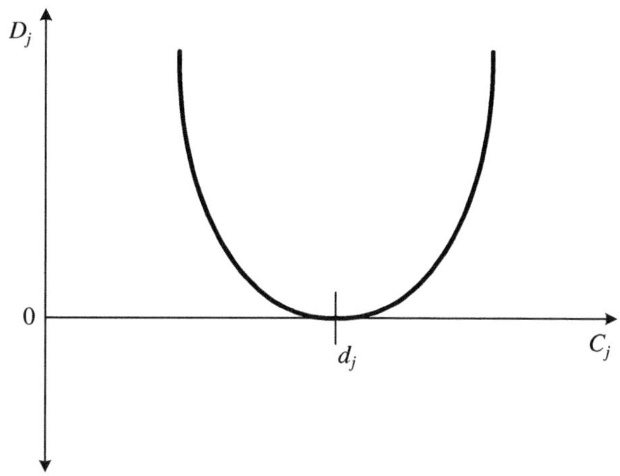

Fig. 5.9 Squared deviation from the due date, $(C_j - d_j)^2$

A typical D_j function is given in Fig. 5.9.

Note that the squared deviation is equal to the sum of the squared earliness and tardiness. However, if we are only interested in the earliness or in the tardiness all that we have to do is to add the square which results in objectives such as $\sum E^2$, $\sum T^2$ and all other alternatives. Also, it is possible to have weighted variants of these objectives, like $\sum w_j D_j^2$ or similar. The main drawback from the squared objectives is that they have little to no interpretation. It might be interesting to minimise the average squared earliness but the value of the objective itself will have no meaning. This is easily solved by considering other objectives, such as for example the percentage of early jobs, total earliness, etc.

In all tardiness measures, we have assumed that the due date d_j is a very specific point in time. Anything before or after this point in time is early or tardy, respectively. As expected, the defined objective of number of just-in-time jobs is a complicated one to maximise as it is hardly expected that all jobs finish exactly by their due date. As a matter of fact, for single machine or flow shop layouts, only a few jobs will exactly finish by their due dates, unless the due dates are very well spaced in time.

In practice, being a few hours early or late (even one or more days) is usually of no concern. As a result, it seems much more interesting to set the due date window or simply a due window, as seen in Sect. 3.2.2. Recall that d_j^- is the minimum due date, or 'earliest due date' before which the job will be considered early and d_j^+ is the maximum due date, or 'maximum due date' after which the job will be late. Under this scenario, the total earliness-tardiness is denoted and calculated as follows:

$\gamma = \mathrm{ET}_j^{ddw} = \sum_{j=1}^{n} E_j^{ddw} + T_j^{ddw}$: Total Earliness and tardiness minimisation, using a due date window, where $E_j^{ddw} = \max\{d_j^- - C_j, 0\}$ and $T_j^{ddw} = \max\{C_j - d_j^+, 0\}$.

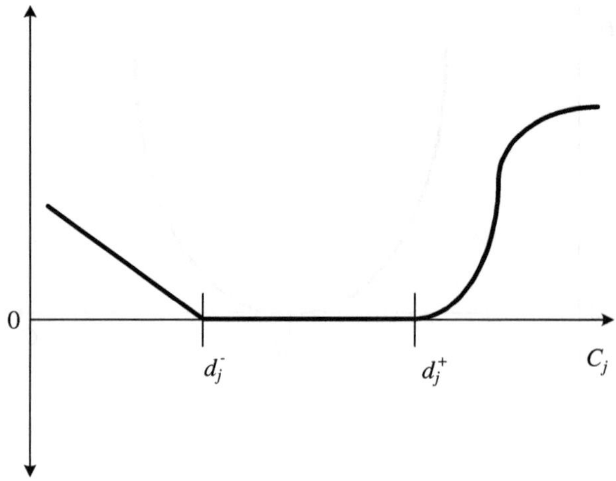

Fig. 5.10 A more realistic non-linear earliness and tardiness function with a due date window

Again, we could have all other possibilities as regards weights, different weights for earliness and tardiness or even some squared and weighted schemes. Picture for example the more realistic function of Fig. 5.10.

When some specific constraints are present in the system, lots of additional objectives can be derived. Setup times are a clear example. When setups require organisation or when they involve lots of time, money or both, it might be interesting to just minimise the number of setups, or to minimise the total setup time of the sequence. Instead of measuring time, one can measure costs. This is interesting when doing setups for a non-critical machine is cheap, for example, and carrying over setups for a critical or expensive machine might be more expensive. Furthermore, different setup operations might require more personnel or special tooling. Under these situations minimising the total setup cost is an interesting alternative.

Bringing up the cost minimisation objective raises a very complex and daunting issue. When doing scheduling, one is tempted to just state 'minimise the production costs'. However, this is easy to say, but very difficult to measure. While overall production costs are more or less accountable, specific production costs that actually might depend on how the products are scheduled is much more difficult. Furthermore, the costs of not meeting due dates are not easily measured. Some clients might be lost and measuring the lost revenue in these situations is all but easy. However, in some specific production companies, there might be an accepted and established production cost accounting. Under these settings, minimising production cost might be an interesting alternative to all other aforementioned performance measures.

5.5 Adding Weights, Priorities or Importance

We defined in Sect. 3.2.2 of Chap. 3 the weights of the jobs w_j. These weights are useful for expressing the importance, the priority, cost or whatever index of relative importance we might need. Given different weights for the jobs, we can modify all previous objectives to consider weights. Some examples are $\sum w_j C_j$ denoting $\sum_{j=1}^{n} w_j C_j$, or max $w_j T_j$ denoting $\max_{1 \leq j \leq n} w_j C_j$.

One big drawback of adding weights to the objectives is that, in some cases, the value of the objective cannot interpreted in an easy way. However, weighting allows for more richer and realistic objectives. Picture for example the earliness-tardiness minimisation. Not all jobs are equally important and it is not the same not being just in time for a big order of an important client that being just in time for a make-to-stock or stock-replenishment order. For this last case of earliness-tardiness minimisation, we have the following notation:

$\gamma = \sum w_j ET_j$: Weighted sum of earliness and tardiness minimisation, calculated as $\sum w_j ET_j = \sum_{j=1}^{n} w_j ET_j = \sum_{j=1}^{n} w_j E_j + w_j T_j$.

$= \sum w_j E_j + w_j' T_j$: Weighted sum of earliness and tardiness minimisation with different weights for earliness and tardiness, calculated as $\sum w_j E_j + w_j' T_j = \sum_{j=1}^{n} w_j E_j + w_j' T_j$

Notice how the last function is even more interesting. By giving different weights to each job and also different weights for earliness and tardiness, we can closely match up the importance of being early and being late. Normally, the consequences of being early are not as dire as those of being late. We could have a more realistic function given in Fig. 5.11.

5.6 An Illustrative Example

In this section, we introduce an example to illustrate some of the objectives discussed before. The example is given for a permutation flow shop with the five jobs and four machines with due dates, so we will compute the values for some objectives in the problem $F4|prmu|\gamma$ or $F4|prmu, d_j|\gamma$, depending on whether we refer to non-due date related or due-date related objectives. Table 5.3 provides the data of one instance, i.e.: p_{ij} the processing times of job j on machine i, weights (w_j), and due dates (d_j) for each job.

We will compute some objectives for a sequence, for example $(1, 2, 3, 4, 5)$. Figure 5.12 shows the Gantt chart for our example. Table 5.4 shows the completion times of each job j on each machine m_i for the given sequence $(1, 2, 3, 4, 5)$, the completion times C_j (note that they are the same than C_{4j} values), the differences between $C_j - d_j$, i.e the lateness of each job L_j, and the weighted lateness $w_j L_j = w_j(C_j - d_j)$.

In general, the following formulae describe a way to compute the completion time of a job j in the position [j] of a sequence π in a problem $Fm|prmu|\gamma$:

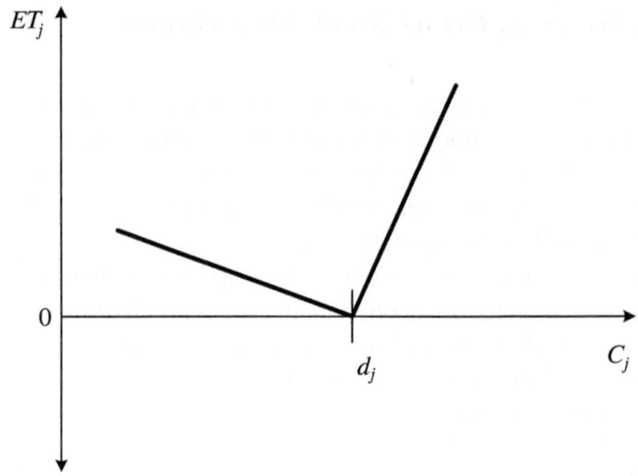

Fig. 5.11 Weighted earliness-tardiness function with different weights for earliness and tardiness

Table 5.3 Data for the illustrative example

Machine (m_i)	Job (j)				
	1	2	3	4	5
m_1	89	19	35	4	22
m_2	37	73	93	19	9
m_3	15	47	18	91	4
m_4	47	3	6	94	98
d_j	156	506	522	238	445
w_j	5	8	2	10	3

Table 5.4 Completion times of each job j on each machine m_i for the sequence (1, 2, 3, 4, 5)

c_{ij}	1	2	3	4	5
m_1	89	108	143	147	169
m_2	126	199	292	311	320
m_3	141	246	310	402	406
m_4	188	249	316	496	594

$$C_{i[j]} = \max\{C_{i-1[j]}, C_{i[j-1]} + p_{i[j]}\}$$

In Table 5.5, the first column includes all performance measures in Table 5.1. They are computed for each job. The last two columns show the maximum and the sum of the values, respectively, obtaining a great number of the objectives presented in the previous sections for the given sequence in this example. Note the completion times C_j for each job (note that they are the same than C_{4j} values of the previous table).

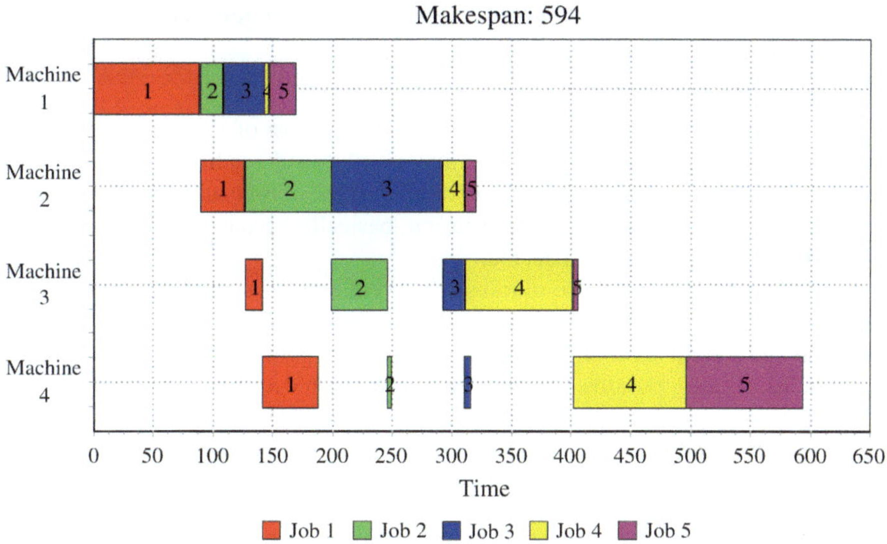

Fig. 5.12 Gantt chart for the permutation flow shop example with the sequence (1, 2, 3, 4, 5)

Table 5.5 Performance measures for each job j and objective values for the sequence (1, 2, 3, 4, 5)

	1	2	3	4	5	Max-form		Sum-form	
C_j	188	249	316	496	594	$C_{\max}$	594	$\sum C_j$	1843
w_jC_j	940	1992	632	4960	1782	$\max w_jC_j$	4960	$\sum w_jC_j$	10306
L_j	32	−257	−206	258	149	$\max L_j$	258	$\sum L_j$	−24
w_jL_j	160	−2056	−412	2580	447	$\max w_jL_j$	2580	$\sum w_jL_j$	719
T_j	32	0	0	258	149	$\max T_j$	258	$\sum T_j$	439
w_jT_j	160	0	0	2580	447	$\max w_jT_j$	2580	$\sum w_jT_j$	3187
E_j	0	257	206	0	0	$\max E_j$	257	$\sum E_j$	463
w_jE_j	0	2056	412	0	0	$\max w_jE_j$	2056	$\sum w_jE_j$	2468
U_j	1	0	0	1	1			$\sum U_j$	3
w_jU_j	5	0	0	10	3			$\sum w_jU_j$	18
V_j	0	1	1	0	0			$\sum V_j$	2
w_jV_j	0	8	2	0	0			$\sum w_jV_j$	10
ET_j	32	257	206	258	149	$\max ET_j$	258	$\sum ET_j$	902
w_jET_j	160	2056	412	2580	447	$\max w_jET_j$	2580	$\sum w_jET_j$	5655

Note that max-form objectives for U_j and V_j do not make sense, and they have not been included in the table.

5.7 Dealing with Conflicting Criteria: Multiobjective Scheduling

At the beginning of this chapter, we mentioned that most of the presented criteria may be conflicting. High machine utilisation usually results in high flowtime. Many other examples can be drawn. As a result of all this, the optimisation problem cannot be dealt with by optimising a single criterion, basically because by doing so, many other criteria will be severely affected.

The ideal situation would be to transfer everything into a 'profitability' measure and maximising this profitability would be the solution. The problem, as has been mentioned, is that this is an impossible task as it is not possible to calculate the costs and benefits of every possible schedule and production decision.

In any case, one has to consider the decision-making process as a more abstract entity. T'Kindt and Billaut (2006) state several items that make the decision-making process extremely complicated. Here we instantiate all these items for production scheduling:

- Usually it is not easy to define, without ambiguity, what is to be done or to be decided. In production scheduling problems, the very best one is to expect is an orientation or a vague indication of what is wanted.
- Put five top management individuals from the same company in a room and ask them about the production scheduling goal. Most probably, five different answers will be obtained and each one of them will be correct in its own right.
- Objectives, if rarely agreed upon, will evolve over time, i.e. they are not fixed.
- The best possible schedule can only be defined under some specific assumptions and normally for a simplified, approximated or theoretical scheduling problem. Therefore, it matters relatively not much if this solution is the best possible or just a very good one.
- There is a real possibility that a company, after using a scheduling system, changes objectives in view of 'what can be done', i.e. even if objectives are agreed upon, are measurable and there is no ambiguity, they might change.

Given all of the above, the only viable approach is to deal with several objectives *simultaneously*. By simultaneously we mean that several different performance measures have to be measured for every possible schedule and that a compromise decision has to be made. For example, let us assume that from five plan managers, three agree in that maximising machine utilisation is a key concern due to the very expensive equipment employed. The other two agree but insist that service level must not be sacrificed. One possible solution to this problem is to minimise the makespan $C_{\max}$ subject to a minimum service level, which could be defined as a maximum number of tardy jobs $\sum U_j$. Another possibility could be to minimise makespan and once the best makespan solution has been obtained, we could minimise the number of tardy jobs subject to no or moderate makespan deterioration. A third alternative would be to obtain several trade-off solutions between the best possible makespan value and the number of tardy jobs, and the lowest possible number of tardy jobs and the makespan value.

There are many different techniques aimed at multi-objective optimisation. Reality is indeed multi-objective, therefore it is needed to approach real problems from a multi-objective perspective. However, before delving into this, we need to first introduce single objective scheduling methods. Therefore, multi-objective scheduling will be later discussed in Chap. 10.

5.8 Conclusions and Further Readings

In the last two chapters, the different scheduling models were presented and many possible scheduling constraints were detailed. It was therefore expected that scheduling criteria were equally as rich and varied. We have made two main classifications of objectives: those based on completion times of the jobs and those based on the due dates. It is safe to say that a sizeable part of the scheduling literature is concerned with makespan minimisation. However, the practical relevance of this objective is debatable, specially with such a high prevalence.

In the long run, the best objective is the one that is trusted. This trust comes after the realisation of several schedules for which objectives where known and for which production and delivery went as expected. Advanced scheduling at production plants usually entail ad-hoc and evolved objectives that satisfy several criteria simultaneously.

Regarding further readings, a discussion about the relation between scheduling objectives and corporate objectives are given as back as in Gupta and Dudek (1971) or more recently, in more detail in Gary et al. (1995). This issue is also commented in the paper of Stoop and Wiers (1996). Some authors even apply quality control charts to measure the schedule performance (MacCarthy and Wilson 2001). The notation employed in Sect. 5.3 is rather standard and can be found e.g. in French (1982) or Pinedo (2012), among many others. Proofs that the maximisation of utilisation is equivalent to minimisation of makespan and equivalent to the minimisation of idletime, including heads and tails are commented in most textbooks, such as Conway et al. (1967), Baker (1974), French (1982), Błazewicz et al. (2002), Brucker (2007), Pinedo (2009, 2012), Baker and Trietsch (2009). The aforementioned textbooks usually commented and detailed the objectives presented here.

Regarding specific contributions, makespan minimisation is reviewed in many different papers, as the review papers are mostly linked with the processing layout or specific constraints. For the flow shop layout, some interesting reviews centered around makespan are those of Framinan et al. (2004), Ruiz and Maroto (2005), Hejazi and Saghafian (2005). Parallel machines have received attention in the reviews of Cheng and Sin (1990) and Mokotoff (2001) although these last review papers are not solely centered around makespan. The same is applicable to the hybrid shop layouts (mainly hybrid flow shops), which are reviewed in Linn and Zhang (1999), Vignier et al. (1999), Wang (2005), Ruiz and Maroto (2006), Quadt and Kuhn (2007),

Ribas et al. (2010), Ruiz and Vázquez-Rodríguez (2010). There is a whole body of scientific research dealing with inserted idle time in scheduling, being Kanet and Sridharan (2000) a good review on the topic. Good reviews of due-date related research are available from Sen and Gupta (1984), Baker and Scudder (1990) and more recently, Vallada et al. (2008), Jozefowska (2007) is a book solely devoted to Just-In-Time scheduling.

References

Baker, K. R. (1974). *Introduction to Sequencing and Scheduling*. John Wiley & Sons, New York.

Baker, K. R. and Scudder, G. D. (1990). Sequencing with earliness and tardiness penalties: a review. *Mathematics of Operations Research*, 15:483–495.

Baker, K. R. and Trietsch, D. (2009). *Principles of Sequencing and Scheduling*.Wiley, New York.

Błazewicz, J., Ecker, K. H., Pesch, E., Schmidt, G., and Węglarz, J. (2002). *Scheduling Computer and Manufacturing Processes*. Springer-Verlag, Berlin, second edition.

Brucker, P. (2007). *Scheduling Algorithms*. Springer, New York, fifth edition.

Cheng, T. C. E. and Sin, C. C. S. (1990). A state-of-the-art review of parallel-machine scheduling research. *European Journal of Operational Research*, 47(3):271–292.

Conway, R. W., Maxwell, W. L., and Miller, L. W. (1967). *Theory of Scheduling*. Dover Publications, New York. Unabridged publication from the 1967 original edition published by Addison-Wesley.

Framinan, J. M., Gupta, J. N. D., and Leisten, R. (2004). A review and classification of heuristics for permutation flow-shop scheduling with makespan objective. *Journal of the Operational Research Society*, 55(12):1243–1255.

French, S. (1982). *Sequencing and Scheduling: An Introduction to the Mathematics of the Job-Shop*. Ellis Horwood Limited, Chichester.

Gary, K., Uzsoy, R., Smith, S. P., and Kempf, K. (1995). Measuring the quality of manufacturing schedules. In Brown, D. E. and Scherer, W. T., editors, *Intelligent Scheduling Systems*, volume 4 of *Operations Research/Computer Science Interfaces*, pages 129–154, Dordrecht. Kluwer Academic Publishers.

Gupta, J. N. D. and Dudek, R. A. (1971). Optimality Criteria for Flowshop Schedules. *IIE Transactions*, 3(3):199–205.

Hejazi, S. R. and Saghafian, S. (2005). Flowshop-scheduling problems with makespan criterion: a review.*International Journal of Production Research*, 43(14):2895–2929.

Jozefowska, J. (2007). *Just-in-Time Scheduling*. Springer, Berlin.

Kanet, J. J. and Sridharan, V. (2000). Scheduling with inserted idle time: Problem taxonomy and literature review. *Operations Research*, 48(1):99–110.

Linn, R. and Zhang, W. (1999). Hybrid flow shop scheduling: A survey. *Computers & Industrial Engineering*, 37(1–2):57–61.

MacCarthy, B. L. and Wilson, J. R., editors (2001). *Human performance in Planning and Scheduling*. Taylor & Francis.

Mokotoff, E. (2001). Parallel machine scheduling problems: A survey. *Asia-Pacific Journal of Operational Research*, 18(2):193–242.

Pinedo, M. (2009). *Planning and Scheduling in Manufacturing and Services*. Springer, New York, second edition.

Pinedo, M. L. (2012). *Scheduling: Theory, Algorithms, and Systems*. Springer, New York, fourth edition.

Quadt, D. and Kuhn, D. (2007). A taxonomy of flexible flow line scheduling procedures. *European Journal of Operational Research*, 178(3):686–698.

Ribas, I., Leisten, R., and Framinan, J. M. (2010). Review and classification of hybrid flow shop scheduling problems from a production system and a solutions procedure perspective. *Computers & Operations Research*, 37(8):1439–1454.

Ruiz, R. and Maroto, C. (2005). A comprehensive review and evaluation of permutation flowshop heuristics. *European Journal of Operational Research*, 165(2):479–494.

Ruiz, R. and Maroto, C. (2006). A genetic algorithm for hybrid flowshops with sequence dependent setup times and machine eligibility. *European Journal of Operational Research*, 169(3):781–800.

Ruiz, R. and Vázquez-Rodríguez, J. A. (2010). The hybrid flowshop scheduling problem. *European Journal of Operational Research*, 205(1):1–18.

Sen, T. and Gupta, S. K. (1984). A state-of-art survey of static scheduling research involving due dates. *OMEGA, The International Journal of Management Science*, 12(1):63–76.

Stoop, P. and Wiers, V. (1996). The complexity of scheduling in practice. *International Journal of Operations and Production Management*, 16(10):37–53.

T'Kindt, V. and Billaut, J.-C. (2006). *Multicriteria Scheduling: Theory, Models and Algorithms*. Springer, New York, second edition.

Vallada, E., Ruiz, R., and Minella, G. (2008). Minimising total tardiness in the m-machine flowshop problem: A review and evaluation of heuristics and metaheuristics. *Computers & Operations Research*, 35(4):1350–1373.

Vignier, A., Billaut, J.-C., and Proust, C. (1999). Les problèmes d'ordonnancement de type flow-shop hybride: État de l'art. *RAIRO Recherche opérationnelle*, 33(2):117–183. In French.

Wang, H. (2005). Flexible flow shop scheduling: optimum, heuristics and artificial intelligence solutions. *Expert Systems*, 22(2):78–85.

Chapter 6
Construction of Scheduling Models

6.1 Introduction

After the definition of the main elements constituting scheduling models, this chapter is devoted to the process of actually building scheduling models. Manufacturing scheduling models are intended to simplify and to map the relevant real-world problem settings in a (usually formal) model which will be used to yield—via scheduling methods—a solution to this formal model. This solution would be then transferred to and implemented in the real-world setting.

Mathematical (i.e. formal) models include a combination of logical dependencies, mathematical relationships such as equations and inequalities, data structures and criteria. The quality of the answers to the real-world problem's questions which are produced by a model obviously depends on the accuracy of the structure and of the data of the model. Here, we describe and discuss the most prominent approaches to construct scheduling models.

More specifically, in this chapter we

- present basic approaches to construction of scheduling models (Sect. 6.2),
- describe decomposition and reintegration as a basic tool for handling model complexity (Sect. 6.3),
- address aggregation and disaggregation in manufacturing scheduling to reduce a model's complexity (Sect. 6.4),
- sketch out what validation and verification means in the context of (scheduling) models (Sect. 6.5).

6.2 Basic Approaches to Construction of Scheduling Models

Recall from Fig. 1.2 in Chap. 1 that we mapped the flow of a decision process where we started from a real-world problem for which a decision is required. Then a formal model is derived by means of simplification and formalisation. By some formal

J. M. Framinan et al., *Manufacturing Scheduling Systems*,
DOI: 10.1007/978-1-4471-6272-8_6,

procedure, this formal model is solved and this formal solution is transferred to a real-world solution which is implemented. In this section, we mainly address the 'upper right corner' of Fig. 1.2, i.e. we will focus on simplification and formalisation, especially the transfer of real-world relations into a formal, mathematical decision model and the techniques to set up a respective formal model.

Formulation of a decision model for manufacturing scheduling requires adequate real-world assumptions about constraints, objectives, tasks and logical dependencies and their transfer to the formal sphere. Both, decision variables and parameters/coefficients of the model have to be defined precisely and on an adequate level of detail (see also Chap. 2). 'Adequate' here means that both, the real-world aspects of the problem are sufficiently represented in the model *and* the model itself is expected to be tractable. Complexity aspects, therefore, are of eminent importance in model construction.

In Sect. 2.3.2.2 a classification of complexity has been given. According to this classification, note that model-oriented complexity in manufacturing scheduling might result both from mass aspects (multiplicity and variance) and chaos aspects (ambiguity and changeability) such as:

- number and heterogeneity of operations/jobs (including (non-) pre-emptability, etc.),
- number and heterogeneity of resources,
- number and heterogeneity of objectives,
- number and heterogeneity of possible decisions/decision variables (including, e.g. in flow shops the modelling decision whether only permutation flow shop solutions are considered or whether the permutation might vary from machine to machine),
- type, number and heterogeneity of interactions between operations, resources, objectives and/or decisions (including both, real-world and formal aspects such as (non-) linearity, etc.),
- ...

General approaches to manage these complexity drivers include abstraction (including deletion of respective aspects from the model) and coarsening (including decomposition, aggregation/disaggregation, e.g. with respect to time, operations, resources, objectives, ...).

While the above aspects of model building refer to both, the real-world and the formal sphere, simultaneously the formal type of model for the scheduling problem under consideration has to be determined. Several approaches based on different techniques are available. In Sect. 6.2.2 some of these approaches will be addressed, i.e. linear programming, mixed-integer programming, agent-based approaches, stochastic approaches and fuzzy logic and Petri nets.

Before going more into detail, it has to be mentioned that several standard modelling approaches are available for certain types of problems. These types of models are often somewhat like a reference point if a new (or modified) problem is addressed. Figure 6.1 shows some indications for the choice of the model type according to some classification criteria.

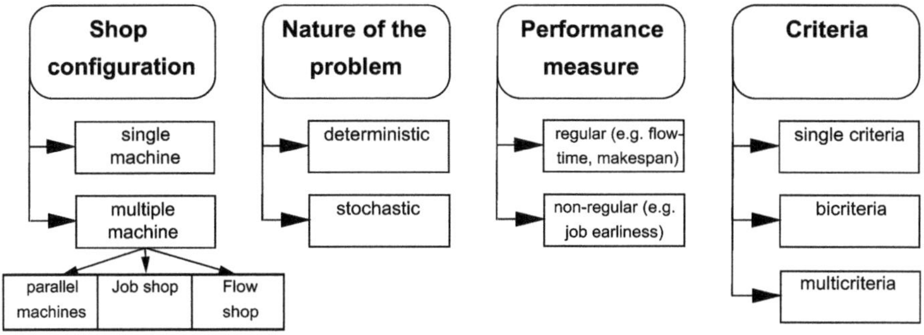

Fig. 6.1 Framework for selection of scheduling model types (Nagar et al. 1995)

6.2.1 Variables, Constraints and Objectives

The goal of a manufacturing scheduling model is to generate a valid schedule, i.e. a feasible and good, efficient or even optimal assignment of operations to the time scale. Schedules can be characterised via their activities/jobs/operations, machines/resources, hard and soft constraints, objectives and possibly other problem-specific aspects.

As already mentioned in Chap. 2 and later developed in Chap. 3, the basic element (atom) of every scheduling problem is the single operation. In its simplest version, this operation is assigned to a pair of exactly one job j and one machine i with a pre-specified and deterministic processing time. In consequence, the operation is fixed with respect to its machine index i, its job index j and its duration p_{ij}.

However, in more complex settings, an operation might also be split (or, on the contrary, be clustered) to batches of one or several jobs' overall processing requirements on the machine (i.e. more than one operation per job exists on one machine or several operations of several jobs on one machine are clustered and processed as one 'macro' operation) and/or to several machines on a certain manufacturing stage (e.g. in hybrid structures with multiple, possibly different machines on one stage of the manufacturing system). Or 'macro' operations, with respect to their processing time and/or resource requirements, might be fixed in advance or are variable themselves. Therefore, on one hand, depending on the specific setting, an operation might be broken down to several sub-operations on several machines. In turn, these sub-operations must add up to the overall operation under consideration. Every sub-operation must be dimensioned according to the resources/machine capacity it needs, i.e. according to its resource's/machine's production function. On the other hand, several operations might be processed together, in parallel (e.g. in an oven) or in sequence (because of their same setup status) and may or must be separated after this joint processing for the subsequent processing of the respective jobs. Formal approaches to settings like these are only partly standardised (batch processing, batch scheduling, simultaneous lot sizing and scheduling) and are often rather specific for the respective problem under consideration. This is the reason why we do not further specify these aspects here.

With respect to the availability of resources, every operation (or sub-operation) on one hand has to regard the availability of the machine(s) it is assigned to, i.e. constraints have to guarantee that only a limited number of operations (usually at maximum one operation) are executed simultaneously on the respective machine. On the other hand, limitations on possibly required additional resources (manpower, etc.) might have to be considered as well if this is relevant for the problem under consideration.

With respect to the availability of jobs (or operations), constraints relative to predecessors and successors of the operation under consideration have to be included. Usually, all predecessor operations have to be finished before the current operation can start. This is what is called end–start relation with minimum time distance (usually, in scheduling problems, this minimum time distance is 0) in network planning models. Supposing SO_{ij} being the starting time of the operation of job j on machine i and C_{ij} the respective finishing time, this constraint can be expressed as $SO_{ij} \geq C_{i'j'} + \text{MinTimeDist}\big((i, j), (i', j')\big)$, for all operations (i', j') being (immediate) predecessors of operation (i, j). Both, SO_{ij} and C_{ij}, are continuous-type variables of the scheduling model.

However, recall from Sect. 4.2.1 that other type of time distance, i.e. apart from the above described end–start relation, also can occur. Usually, not both, predecessor and successor relations have to be mapped in a model since they result from each other. (Jobs' or operations' release times and deadlines can be easily included in the SO_{ij}, C_{ij} notation.)

Basically, there are two different types of predecessors for every operation, i.e. fixed and changeable ones. Fixed predecessors result from physical constraints of operations or jobs, e.g. from pre-specified job/product routings. (MinTimeDist in the above formula might represent, for fixed predecessors, transportation times between two consecutive operations of one job.) Changeable predecessors result from the different decisions/schedules which represent the different solutions of the scheduling problem. A best choice of these changeable predecessors represent the optimal solution to the manufacturing scheduling problem under consideration. (MinTimeDist in the above formula might represent, for changeable predecessors, sequence-dependent setup times.)

Standard scheduling objective functions obviously can also easily be described by using the above SO_{ij}, C_{ij} formulas (see also Sect. 5.3). For a given solution (i.e. a complete assignment of operations to the time scale, ending times of jobs' last operations immediately define makespan, flowtime, utilisation, etc.).

Many other aspects might be included in a manufacturing scheduling model, such as:

- predictable or unpredictable non-availability periods of machines/resources,
- non-availability of materials, fixtures or tools (being also relevant with a short-term horizon, e.g. in just in time setting),

- pre-emption of jobs (with restarting of jobs/operations later on from scratch or from the status when pre-emption occurred or something in between),
- logical dependences of variables/jobs/... (e.g. joint production settings, cyclic production calendars, joint machine (non-) availability times),
- ...

All of these problem extensions require specific model expansions. The detailed description of these expansions is beyond the scope of this chapter.

The decision problem itself, however, is not covered completely by the above descriptions as is already indicated by the above discussion of fixed and changeable predecessors. In general, for all combinations of changeable predecessors (i_1, j_1) and (i_2, j_2), a decision has to be taken whether an operation (i_1, j_1) will precede (not necessarily immediately) an operation (i_2, j_2) or vice versa. In scheduling models, this is either included implicitly, e.g. if a specific permutation defines this precedence configuration in a permutation flow shop problem. Or the decision is explicitly designed, usually by defining binary variables, e.g. as

$$x_{(i_1,j_1),(i_2,j_2)} = \begin{cases} 1 & \text{if } (i_1, j_1) \text{ precedes } (i_2, j_2) \\ 0 & \text{if } (i_2, j_2) \text{ precedes } (i_1, j_1) \end{cases}$$

for all pairs of operations (i_1, j_1) and (i_2, j_2) for which both relative sequences are relevant, i.e. which might return an infeasibility if they are at least partly processed in parallel. (This is obviously the case if both operations take place on the same machine, which is opposed to be able to process no more than one operation at a time, but might happen sometimes also otherwise.)

Obviously, a necessary constraint for feasibility of a schedule then is

$$x_{(i_1,j_1),(i_2,j_2)} + x_{(i_2,j_2),(i_1,j_1)} = 1.$$

Additionally, a transitivity relation must hold, i.e.

$$x_{(i_1,j_1),(i_2,j_2)} = 1 \wedge x_{(i_2,j_2),(i_3,j_3)} = 1 \Rightarrow x_{(i_1,j_1),(i_3,j_3)} = 1$$

for all relevant triples, to avoid circles of precedence.

If such an explicit representation of precedence alternatives is used, the number of binary variables will usually be very large, and sophisticated approaches to reduce this number are very welcome as one approach to complexity reduction. Another way to reduce complexity, which might be applied in combination with the first one, is the design and the application of heuristic solution procedures which are addressed later in this book.

Time variables SO (and at least implicitly also C) and precedence variables x can be linked as follows. Suppose that only operations on a single machine i (maybe in a multi-machine setting) are considered. Then

$$M \cdot (1 - x_{(i,j_1),(i,j_2)}) + SO_{ij_2} \geq SO_{ij_1} + p_{ij_1}$$
$$M \cdot x_{(i,j_1),(i,j_2)} + SO_{ij_1} \geq SO_{ij_2} + p_{ij_2}$$

for all i, j_1 and j_2 with $j_1 \neq j_2$ and a large number M represents both alternatives of precedence for the operations (i, j_1) and (i, j_2), see Williams (2013). This representation of precedence alternatives on one machine requires only half of the binary variables as compared with the full table since (i, j_1) preceding (i, j_2) means automatically (i, j_2) not preceding (i, j_1) and vice versa.

Using (and possibly expanding or modifying) the approaches described in this section, generates a decision model for manufacturing scheduling problems, defining variables, constraints as well as objective function(s) in one model which is the starting point for the optimisation or search process to generate a good, efficient or even optimal solution.

So far, in this section we exclusively addressed the design of scheduling models with constraints that were strictly to be kept, i.e. hard constraints. However, in real-world settings sometimes constraints are not as strict, may it be because of the existence of processing alternatives not mapped in the model, for (time) buffers included only implicitly in the model or for other reasons. In such cases, a relaxation might be included in the respective constraints. E.g., in the above constraints, a time relaxation of size u (as a variable) might be included, e.g. as

$$M \cdot (1 - x_{(i,j_1),(i,j_2)}) + SO_{ij_2} \geq SO_{ij_1} + p_{ij_1} - u$$
$$M \cdot x_{(i,j_1),(i,j_2)} + SO_{ij_1} \geq SO_{ij_2} + p_{ij_2} - u$$

meaning here that the strict precedence requirement between the operations is replaced by some allowance of parallelity, e.g. in real-world terms by using a time buffer. However, to avoid unlimited violation of the precedence requirement, u has to be bounded either primarily by an additional constraint (addressing the size of u) or by dually penalising a positive value of u in the objective function. An alternative way of handling such soft constraints is by using fuzzy set approaches where the degree of membership corresponds to the amount by which the constraint is violated.

Another setting demanding for similar approaches are conflicting constraints in the original model, resulting in the non-existence of a feasible solution. Relaxing these constraints primarily or dually or using goal programming approaches might be adequate as well because of implicit buffers and/or technological or process flexibility not being included in the model.

Many other aspects of real-world scheduling settings can be mapped in formal scheduling models. Lead times, release or processing patterns, late modifications of jobs (with respect to time, size, design, ...), quality aspects and many other might increase the complexity of the real-world scheduling task under consideration. Also variability aspects, e.g. with respect to processing times might be a severe issue in many situations, may it be influenced by a human component, by planning deficiencies, by technological variability, etc. The enormous diversity of these aspects does not allow an comprehensive description of the respective modelling approaches which are by far not standardised.

Concluding this section, we point out:

1. In addition to the above-mentioned modelling aspects, the modelling process itself has to be executed regarding also the following. 1) The model must be (easily) understandable for everybody who is basically able to understand quantitative modelling approaches and gets into contact with the model. 2) The model must be easy/adequately to solve, to maintain and to update with respect to possible different future real-world settings (see also Williams 2013). 3) And the model—sometimes in contrast to solution efforts with respect to quantitative models—additional insight in the real-world setting gained from the (often iterative) modelling process itself is frequently even more valuable than solving the model to optimum.
2. The quality of the solution of a scheduling problem depends both on model formulation *and* on the solution method. Both these levels of dealing with manufacturing scheduling problems are highly interdependent and therefore have to be considered jointly, may it be simultaneously or in an iterative loop.

6.2.2 Some Basic Modelling Techniques Used in Manufacturing Scheduling

A real-world manufacturing scheduling problem can often be (quantitatively) modelled in more than one way. This results in different model and solution complexity as well as different solution quality. We therefore sketch out several typical modelling approaches in the sequel.

6.2.2.1 Binary/Integer Programming Modelling

As already indicated by the formulas in the previous section, mixed-binary formulations of scheduling problems are very popular. Because of the wide dissemination of both, mixed-integer model formulation as well as respective solution techniques, these approaches contribute a lot to understanding and solution of manufacturing scheduling problems. However, because of the enormous effort required to solve MIP models to optimality, only small problem instances might be solved accordingly.

Early basic mixed-binary formulations of scheduling problems are presented, among others by Bowman (1959), Wagner (1959), Dantzig (1960), Manne (1960), Greenberg (1968). Some of these models will be addressed in Chap. 8. Here, we will only present the model of Greenberg (1968) as a straightforward formulation of the standard n job, m machine job shop scheduling problem with makespan objective.

Table 6.1 (Modified) Greenberg model: notation

Data	
n	Number of jobs, $j = 1, \dots, n$
m	Number of machines, $i = 1, \dots, m$
R_j	Vector that specifies the machine sequence of the operations of job j (supposed to be a permutation of $1, \dots, m$ which means that every job visits every machine)
p_{ij}	Processing time of job's j operation on its i-th machine (i.e. on machine $R_j(i)$)
M	Large number
Variables	
C_{max}	Makespan of the schedule
SO_{ij}	Starting time of processing job j on machine i (continuous variable)
x_{ijk}	Binary variable, being 1 if, on machine i, job j starts before job k, 0 otherwise

(Modified) Greenberg model (Greenberg 1968)

Objective function:

$$\min \ C \tag{6.1}$$

Constraints:

$$SO_{R_j(i-1),j} + p_{i-1,j} \leq SO_{R_j(i),j} \qquad \forall\, j = 1, \dots, n,\ i = 2, \dots, m \tag{6.2}$$

$$SO_{ij} + x_{ijk} \cdot p_{ij} \leq SO_{ik} + x_{ikj} \cdot M \qquad \forall\, i, j, k \text{ with } j \neq k \tag{6.3}$$

$$x_{ijk} + x_{ikj} = 1 \qquad \forall\, i, j, k \text{ with } j \neq k \tag{6.4}$$

$$SO_{R_j(m),j} + p_{ij} \leq C \qquad \forall\, j \tag{6.5}$$

$$x_{ijk} \in \{0, 1\} \qquad \forall\, i, j, k \tag{6.6}$$

Combining constraints (6.5) and objective function (6.1), the minimum value for C is the minimum makespan since the left-hand side of (6.5) defines the finishing time of every job's last operation. Constraint set (6.2) guarantees the machine sequence per job, saying that job's $j\, R_j(i-1)$-th operation must be finished before its $R_j(i)$-th operation can start. Constraint set (6.3) guarantees that no two jobs are simultaneously processed on the same machine. Finally, constraint set (6.4) guarantees that, on machine i, either job j is processed before job k or vice versa.

Clearly, this model can be easily adjusted to some well-known other settings, such as, e.g. flowtime as objective function or including due dates (and related objective functions) or deadlines.

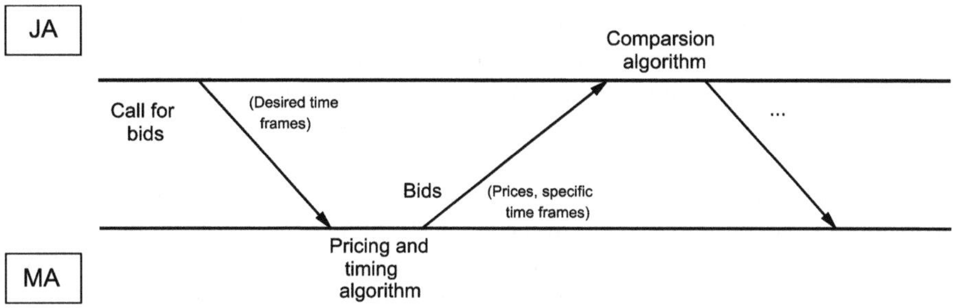

Fig. 6.2 Overview of bidding and pricing process (Pinedo 2012)

6.2.2.2 Agent-Based Approaches

In contrast to 'true' optimisation approaches to scheduling as used, e.g. in mixed-integer linear programming, Artificial Intelligence (AI) aims on satisfying aspiration levels, i.e. it is sufficient to generate solutions which are accepted by the decision-maker(s). (Therefore, 'true' optimisation approaches might be interpreted somewhat like an extreme case in the context of AI.)

Agent-based approaches (ABA) represent a special kind of AI methods. They first decompose the overall model into sub-models (agents) with specific structure, specific individual tasks and specific procedures for their 'local', agent-specific decision-making. And, second, these agents iteratively communicate with each other according to pre-specified procedures intending to give improved information to the counterpart for solving its local decision problem and, finally, to generate an accepted solution to the overall decision problem. The iterative process will be stopped when some pre-specified stopping criterion is fulfiled (e.g. with respect to solution quality or computation time).

Referring to manufacturing scheduling problems, in a most simple version of an ABA, agents might be defined as job agents (JA) or machine agents (MA). JAs internally assign a job's operations to the time scale based on the machine resources (time frames) received from the MAs, and compare the solution with former or alternative ones (comparison algorithm). On the level of communication, JAs demand machine capacity from the MAs intending to improve their current local performance. MAs locally, i.e. for the specific machine, bring jobs' operations into sequence and, in turn as feedback in communication, offer time windows for a job's operation on this machine to the JAs. Information exchange between the agents might be primal (e.g. explicit time information) and/or dual (i.e. (opportunity) cost and pricing information). This process is mapped in Fig. 6.2.

MAs will locally often use simple or more sophisticated dispatching rules or one-machine optimisation procedures, keeping in mind the requirements communicated to them by the JAs. Bottleneck machines will play a dominant role while non-bottleneck machines will often only be regarded as subordinated in the communication process. JAs will, e.g. locally determine job's earliest availability time for a certain operation on a specific machine (according to the machine sequence of operations of the specific job), and might determine a priority indicator of the specific job (e.g., according to the tightness of job's due date or the importance of the respective customer) and will give this information to the MA(s). Exclusive priority (rule) scheduling as a job-oriented and exclusive bottleneck scheduling as a resource-oriented approach represent the two extremes for rule-based ABA. Every plausible combination of these extreme cases might be considered of course as well (see Schmidt 1992, or Shen and Norrie 1998 for an overview). Obviously, ABA is rather flexible and can easily be applied to open, dynamic scheduling environments which makes them attractive for respective real-world settings while many (most) 'true' optimisation-oriented scheduling approaches focus on more or less 'closed' static settings. However, on one hand, the higher flexibility of ABA has to be traded-off with the lower solution quality (with respect to (sub-) optimality) as compared with optimisation-oriented approaches. On the other hand, it has to be regarded that an intrinsic problem of ABAs, not only in scheduling, is the myopic perspective of the agents. Generating an overall scheme for an ABA which, as far as possible includes the overall perspective of the problem under consideration by an adequate communication scheme is the challenge of modelling ABAs.

The basic principles of ABAs in manufacturing scheduling have been described above. More specific information can often only be given with respect to the specific problem setting under consideration. We will therefore not go into detail further here. However, in addition, we point out that the concept of ABAs to manufacturing scheduling presented so far supposes that the required information is

- available wherever it is required (or even all information is available everywhere in the system—which is not a true decentralised approach to the problem) and
- transferred truly between the agents, without any opportunistic behaviour.

We only mention that it might be necessary to check these requirements in real-world settings of manufacturing scheduling and, in consequence, adjust the communication procedures accordingly.

6.2.2.3 Modelling Approaches Including Uncertainty

For several reasons non-deterministic settings might occur in manufacturing scheduling problems. Processing times might be known only by their probability distribution and/or by some range, machine/resource availability might be stochastic with respect to time and/or size, availability (release dates) and priority of jobs might be non-deterministic, etc. Sometimes, not even probability data or ranges are known but only partly quantifiable relative information is given, such as high, medium and low with respect to, e.g. processing times.

There are several ways to deal with uncertainty in manufacturing scheduling problems. The simplest way is just to ignore uncertainty and to set up and solve the model with 'some kind of' expected values for the uncertain parameters. From a formal point of view, this might be an important drawback of respective approaches.

However, keeping in mind the often rather short-term horizon perspective of real-world scheduling problems, ignoring uncertainty might be not as critical from an application point of view. Nevertheless, the extent of uncertainty's influence is often not known in advance. Therefore, this approach seems to be appropriate in situations where uncertainty will have only little influence on the problem and its solution. Only slightly more elaborated are models which include buffers for uncertainty, e.g. by expanding processing times or by reducing capacities. The 'correct' determination of the size of these buffers is a most challenging task. Both approaches might run into feasibility and/or optimality problems. In addition to the application of deterministic approaches to non-deterministic problems, sensitivity analyses can be executed, may it be by using some kind of simulation or by applying explicit formal approaches (if available).

However, if uncertainty is expected to be an important issue with respect to the problem under consideration, is has to be considered explicitly—if possible. Two basic approaches are intensively discussed in the literature, i.e. stochastic scheduling approaches and fuzzy logic approaches.

Restricting to modelling issues in this chapter, *stochastic models* (in manufacturing scheduling) might be relatively easy formulated by just defining certain parameters of an originally deterministic model to be stochastic. Problems then result 'only' from model solution and not from model formulation. However, this interpretation is profoundly naive. Modelling and solution aspects usually influence each other and should be regarded jointly. Nevertheless, we will not deal with solution methods for stochastic scheduling problems here. Instead we point the reader to respective references, e.g. the book of Pinedo (2012), where stochastic scheduling problems and solution methods are discussed. Without going into detail and well-known for many stochastic problem settings also outside of manufacturing scheduling, stochastic aspects significantly increase the complexity of a model and the solution process. A treatable deterministic model will often become intractable if stochastic model components occur. Therefore, on one hand, valid results for stochastic scheduling problems will usually only be deducible for fairly simple problem settings, e.g. single-stage problems with one or more machines or simple flow shop or job shop problems. On the other hand, results and interpretations for stochastic scheduling problems will often differ from those for deterministic problems: (Stochastic) dominance results, deriving hints for scheduling policies or shop capacities (machines, buffers, ...) instead of detailed solutions might be the focal aspect instead of deriving single good or optimal solutions.

If uncertain problem data are available in some simple kind of distribution (e.g. triangular or trapezoidal) or only in some qualitative or linguistic categories (e.g. as low, medium or high, which is supposed to be representable in some trapezoidal distribution), fuzzy set or fuzzy logic approaches might be applied to manufacturing scheduling problems. For example, a due date for a job might be not as strict as it is often supposed in basic models. Traditionally, due dates are included in scheduling problems by respective objective functions such as lateness and tardiness. Missing the due date is then penalised. However, if due dates are not as strict, they might be represented by a trapezoidal distribution, representing the categories early, normal

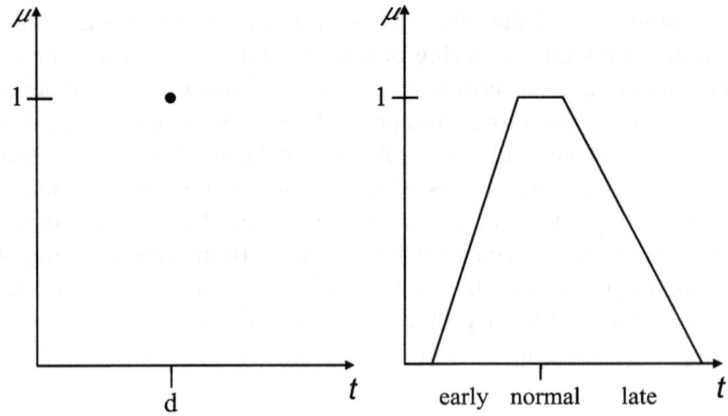

Fig. 6.3 Fixed due date and fuzzy due date

and late as is shown in Fig. 6.3, where μ indicates the so-called membership function with values between 0 and 1, and 1 indicating that the relationship under consideration is completely fulfilled while other values of μ can be interpreted accordingly.

Apart from single parameter data, also the objective function(s) and or the information on constraint satisfaction might be available only in linguistic terms. For example, makespan might be classified into short, normal and long. The same holds for the classification of trade-offs between multiple objective functions which, depending on the specific problem might be classified only linguistically. Flexible resources can be addressed accordingly (see, e.g. Slowinski and Hapke 2000).

As compared with stochastic programming approaches to manufacturing scheduling, fuzzy approaches on one hand are obviously somewhat simpler, also because of the rather simplified distributions supposed in fuzzy approaches. On the other hand, solving a fuzzy scheduling model usually also yields fuzzy result information, giving the decision-maker some impression of the expected effect of fuzziness, also with respect to the robustness of a solution subject to the fuzzy components of the model.

6.2.2.4 Other Modelling Approaches

Apart from those described above, there are several other approaches to model manufacturing scheduling problems, e.g. Petri nets and object-oriented approaches. We will just sketch out these approaches here.

From an application point of view, the main features of *Petri net-based approaches* for scheduling are twofold. First, they are easily able to handle multiple lots for complex relations that may exist among jobs, routes, machines and material handling devices, i.e. Petri nets provide an efficient method for representing concurrent activities, shared resources, precedence constraints and multiple lot sizes. Second, the generated schedule is event-driven (this facilitates real-time implementation), deadlock-free (since the Petri net model of the system can be a detailed representation of all the operations and resource-sharing cases, a generated schedule is the one from the system's initial condition to the finally desired one; it thus avoids any deadlock) and optimal or near-optimal with respect to makespan.

Within the Petri net approach, a bottom-up method is used to synthesise the Petri net model of a system for scheduling. First, a system is partitioned into sub-systems according to the job types, then sub-models are constructed for each sub-system, and a complete net model for the entire system is obtained by merging Petri nets of the sub-systems via the shared resources.

Although the underlying problems of scheduling tasks are highly complex a simplified model can sometimes be obtained using *object-oriented techniques*. As described by Błażewicz et al (2007), Schmidt (1996), object-oriented modelling attempts to overcome the disadvantage of modelling data, functions, and interactions between both, separately. The different phases of the modelling process in object-oriented approaches are analysis, design and programming. Analysis serves as the main representation formalism to characterise the requirements from the viewpoint of the application; design uses the results of analysis to obtain an implementation-oriented representation, and programming means translating this representation using some programming language into code. Comparing object-oriented modelling with traditional techniques, its advantages lie in data abstraction, reusability and extensibility of the models, better software maintenance and direct compatibility of the models of different phases of the software development process. A model built by object-oriented analysis consists of a set of objects communicating via messages which represent dynamic relations of pairs of them. The main static relations between pairs of objects are generalisation/specialisation and aggregation.

6.3 Complexity Reduction in Manufacturing Scheduling Using Decomposition and Re-integration

Complexity of manufacturing scheduling models has already shortly been addressed in Chap. 2. It can result from many different aspects and may be classified into real-world related aspects (including organisational ones) and formal aspects. Complexity reduction with respect to the solution procedure is discussed in Chap. 7. Here, we will focus on complexity reduction of the scheduling model itself.

There are several brute force approaches to reduce a scheduling model's complexity:

- The problem as a whole is simplified with respect to the number of possible solutions. For example, a flow shop problem is considered as a permutation flow shop problem.
- Only 'relevant' jobs and/or machines are considered, i.e. only a subset of all jobs and/or machines is included in the model. Important jobs/machines remain in the

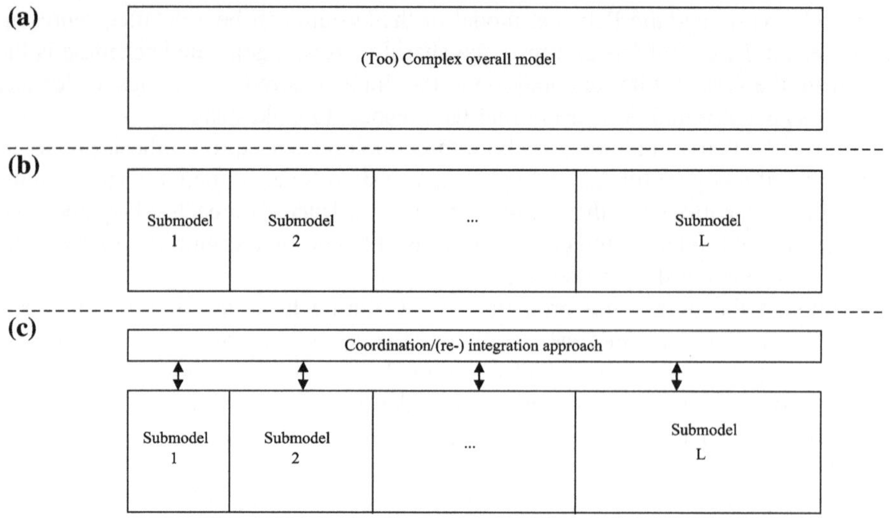

Fig. 6.4 Coordination approaches in complexity-reduced manufacturing scheduling problems: **a** Too complex overall model, **b** Model decomposition, **c** Decomposed model and coordination/integration

model, less important ones are added subsequently after a solution for the important items has been generated.
- Some constraints are relaxed or even ignored, e.g. preemption is excluded from consideration, sequence-dependent setup times are interpreted to be sequence-independent (and are therefore included in the processing times), transmission to a next machine is only possible for the whole job (although some sublot could also be transferred earlier to the next machine allowing overlapping processing, ...).
- The time horizon is cut off earlier which usually results in less operations/jobs to be considered.
- ...

All these and also more sophisticated approaches intend to reduce the number of scheduling objects (operations, jobs), resources (machines), variables and/or constraints to receive a model complexity which can be handled.

For a given manufacturing scheduling problem, suppose that we have either implicitly an impression of a too complex model or an explicit model which cannot be handled (solved) as is. To reduce this model's complexity, decomposition and subsequent coordination of the decomposed components of the model is required (see Fig. 6.4). The basic idea of decomposition is to separate the overall model into sub-models which are

1. tractable individually per sub-model (e.g. one machine problems or low number of jobs), and
2. connected with each other by advantageously few and clearly defined relations between the sub-models (to allow the subsequent integration of the partial solution to a solution of the overall problem).

(Decomposition might take place in several steps, on several levels and even iteratively. However, here we only describe aspects of a one-step decomposition approach.)

In most *resource-based* decomposition approaches, the overall scheduling problem is split into several sub-problems where these sub-problems contain each a subset of all machines. The subsets of machines are usually disjunct and in many cases consist only of a single machine or of the machines of a single production stage. Each sub-problem involves developing a schedule for the particular machine subset. By clustering resources/machines, from a formal point of view, resource-based decomposition approaches will primarily partition constraints referring to the respective resources/machines.

In *operation-based* decomposition approaches, an orthogonal view to the resource-oriented one is adopted. A complete shop schedule is interpreted as the integration of several job schedules. If n jobs are to be scheduled, e.g. (up to) n sub-problems are created where each involves the development of a schedule for a particular job (job group). Afterwards, these single job (job group) schedules are integrated to an overall schedule. By clustering operations in this type of decomposition approach, both constraints and variables are partitioned into the respective sub-problems.

In *event-based* decomposition approaches, each sub-problem involves making a single scheduling decision, e.g. which job/operation to schedule on a particular machine, or deciding to send a job to which machine. The schedule is developed, e.g. by an event-based simulation of the shop, as used in dispatching approaches to scheduling. In event-based decomposition, information is chronologically grouped. All information available at the time an event takes place is provided in a sub-problem.

These three basic decomposition strategies may be used isolated or in combination. The degree of sub-problem interdependencies also depends on the scheduling objective and the tightness of constraints in the specific problem instance. Therefore, bottlenecks should be anticipated before a decision for a specific decomposition approach is taken. If particular machines or jobs or time-periods/events are known to be crucial or expected to be crucial for the problem under consideration and its solution, then a focus should be laid on the respective aspects when decomposing the problem. From both, empirical evidence in more general cases as well as formal analysis for smaller problem sizes, it is well-known that the bottleneck type (machine, operations, events) indicates the direction of decomposition while in addition, first the bottlenecks themselves should not be clumped together with non-bottlenecks and second the respective other dimensions of decomposition/clustering approaches could be applied as well. For example, a problem with a clear bottleneck machine could separate this bottleneck machine from the other machines. The respective one machine problem for this bottleneck machine should be solved in detail while the other machines might be grouped together (and possibly aggregated) in the decomposition approach as well as all or some jobs on these non-bottleneck machines.

When constructing the sub-models, these models might include some anticipating information from the other sub-models as well—if this is possible, e.g. as used in the well-known shifting bottleneck procedure, machine by machine sub-models might include job availability data (i.e. machine-specific release dates) resulting from information provided by the other sub-models. Also the classical coordination mechanisms as (primal) budgeting (e.g. of time) and/or (dual) price mechanisms might be included in the coordination schemes. The whole tool set of centralised or decentralised coordination might be applied to manufacturing scheduling models in one way or another. It should also be mentioned that the structure of the sub-models might be induced by the organisational structure of the manufacturing setting under consideration.

After the overall model has been decomposed into sub-models according to the above description, every single sub-model has to be solved. We suppose that these sub-models are solvable in one way or another without any severe problems—that is the main reason why they have been built and how they should be constructed. Finally, the solutions of the sub-models have to be (re-) integrated using an integration or coordination approach. This step cannot be described comprehensively. Basically, the local information from the sub-models' solutions will usually consist of some (local) schedule. Therefore, the individual schedules have to be integrated on the coordination level according to the requirements of the overall (original) model. If the resulting 'total' solution does not meet the decision-maker's requirements, the process of decomposition, local solution and (re-) integration of local solutions might be repeated subject to modified information delivered from the coordination process to the sub-models. For example, this information could consist of modified release times of jobs, different assignment of jointly used resources or different prices for these resources, etc.

6.4 Aggregation and Disaggregation

Aggregation and disaggregation is a well-known tool for reducing complexity in planning approaches and planning models. It can also be used for manufacturing scheduling models according to the two-level approach presented in Sect. 6.2. Aggregation in manufacturing scheduling is intended to reduce the number of objects to be considered in a respective model.

As mentioned earlier, the basic object of consideration in scheduling models is the single operation of a job on a machine. Therefore, aggregation of operations is a nearby approach to reduce the number of objects to be scheduled. Similar operations (e.g. requiring the same setup status) can be fixed as consecutive operations and therefore be added up to one 'macro' operation. Similar machines (and the respective operations of jobs on these machines), may they be in parallel on one production stage or in sequence, can be interpreted as one machine. (This approach is usually only adequate for non-bottleneck machines.) Of course, as in every aggregation approach, the aggregates coarsen the model and induce some loss of information which has to be taken explicitly into account when the solutions for this complexity-reduced models are re-transferred to a solution of the original problem.

On one hand, the basic guideline for designing aggregation approaches is the more or less trivial declaration that no 'important' or 'relevant' information should be aggregated in a way that it will significantly influence the quality of the solution derived. However, on the other hand, there is (almost) no situation where there is no loss of information by using aggregation approaches. Therefore, a (negative) influence on the solution quality will always exist when using aggregation approaches. This influence may not be ignored but might be taken into account when designing an aggregation approach, e.g. increasing granularity of the aggregate model will usually increase the solution's quality but has to be traded-off with the increasing complexity of the resulting model.

Vancza et al (2004), although focusing on the integration of production planning on one hand and manufacturing scheduling on the other, give some more detailed hints which should be regarded when anticipating detailed scheduling models in some aggregate production planning counterpart which can also be adopted to complexity reduction in a pure scheduling context when using aggregation approaches:

1. Aggregate models must respect the main temporal constraints of jobs (e.g. due dates) as well as resource capacity constraints.
2. The aggregate model should also handle precedence relations that result from complex product structures (e.g. assemblies) and technological routings. However, resource assignment problems with finite capacities and precedence constraints are in general extremely hard to solve.
3. Albeit aggregation removes details—due to the differences between the planning and scheduling model—it may introduce new constraints that distort the original disaggregate problem. The effect of such constraints should be kept as small as possible.
4. Planning and scheduling models should be built by using common product and production data (e.g. bills of materials (BOMs), routings and resource calendars). Open orders should be addressed at both levels.

Concluding this paragraph, we list some standard subjects of aggregation and disaggregation in manufacturing scheduling:

1. An operation usually belongs to a job and a job might belong to a larger customer order or consists itself of several customer orders. Therefore, the aggregation of jobs to orders or vice versa and the respective influence on the respective operations' structure includes an aggregation/disaggregation approach. This aggregation/disaggregation is sometimes not seen as part of the scheduling model itself but is seen as external input to the scheduling model. However, clustering or decomposing jobs and/or orders accordingly might have a more or less severe influence on the quality of a schedule.

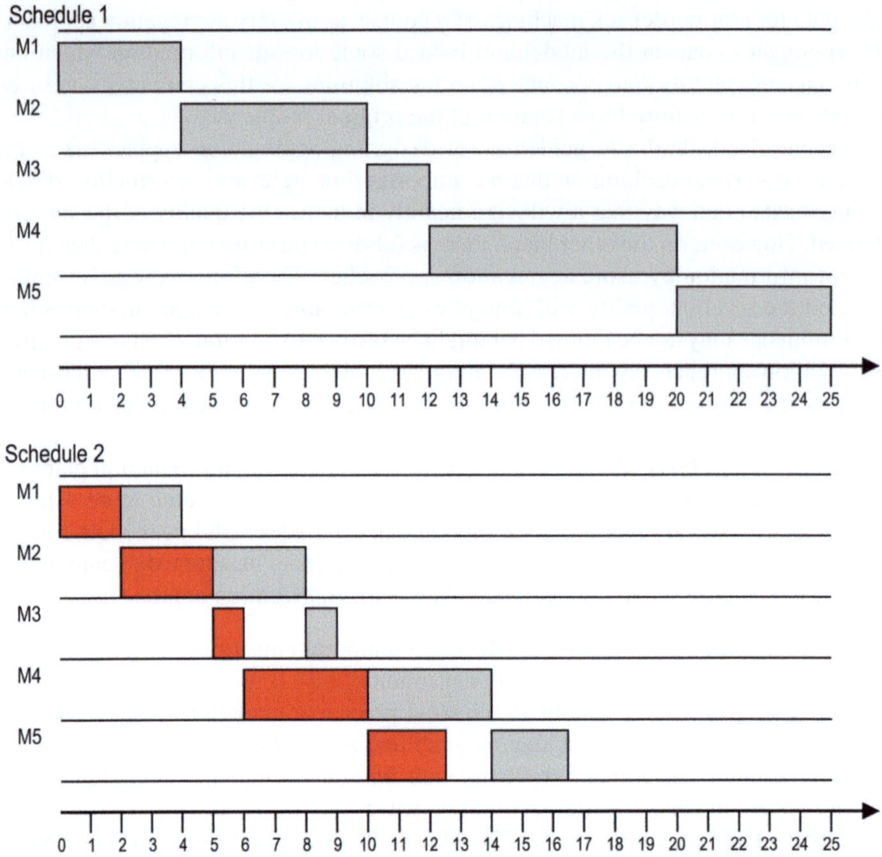

Fig. 6.5 Effect of lot streaming on makespan (cycle time)

2. Parallel and/or sequential batching (pooling) of operations (or whole jobs) on one machine might be decided outside of the scheduling problem (on an 'aggregate' decision level) or a subject of the scheduling model itself. Such questions occur, e.g. in process industries, if setups are relevant (with respect to time and/or costs, for product family scheduling problems), if ovens should not run below a certain utilisation rate (parallel batching), ...
3. Opposite to batching, disaggregation of operations (lot streaming or lot splitting) might become an issue, e.g. if due dates are tight and customers are satisfied with partial shipments or if a next operation of a job is intended to start before the previous operation is finished for the complete job. Lot streaming was introduced in Chap. 3, Sect. 3.2.2 and further detailed in Sect. 3.4. Figure 6.5 shows the effect of lot streaming for a small example.

4. Transportation times from one stage to another might be added to the processing times of the respective operations if a separate consideration of transport operations is not required.
5. Considering every single machine of the shop structure in detail might be rather complex on one hand. On the other, organisational issues might imply an aggregate view on machines in manufacturing scheduling, e.g. if group technologies are applied. There the (macro) scheduling problem consists of assigning (macro) operations/tasks to pre-determined groups of machines while the detailed scheduling task within each group is decentralised and executed by the group itself.
6. Scheduling settings with hybrid structures on specific stages might be simplified by aggregating the capacities of a hybrid stage into one macro-machine and by adjusting the operations' processing times accordingly. Of course, disaggregating a solution from the aggregate macro-model needs to regard the hybrid structure explicitly.
7. Additional resources (apart from the machines, e.g. flexible workers) which are limited and have to be assigned to different machines competing for these resources might be considered in a detailed or atomic manner on one hand. On the other, if these resources can be pooled and can be partitioned more flexible, the solution space will be enlarged and better schedules might be generated, e.g. by improvements of balancing of production rates, of idle time reduction etc. This type of resource aggregation also includes the determination of the number of resource groups and the assignment of resource groups to machine (groups) as well as an adequate expression for the aggregate capacity itself (Burdett and Kozan 2003; Le Pape 1994).
8. Aggregation of time periods, i.e. reducing the granularity of the time grid might reduce the complexity of a scheduling model. Although ceteris paribus the number of operations remains unchanged by time aggregation, dealing with weeks or days instead of hours and minutes usually reduces the differences between the objective function values for different schedules and probably increases the number of good schedules since the differences of the objective function are reduced by this aggregation process. However, this must be in line with the requirements of the real-world scheduling problem, e.g. a schedule with a time horizon of 1 month usually will not require an hour time grid while a day's schedule will.
9. Connected with 8, the length of the time horizon itself might be an issue for complexity reduction although it is not a 'true' aggregation approach.

The above-listed approaches may be used isolated or in combination. Anyway, in most models they have to be designed individually and, as mentioned before, because of their coarsening of information, they will generate some feasibility and/or optimality lag which has to be evaluated with respect to the detailed model or, better, with respect to the real-world problem under consideration.

6.5 Validation and Verification of Modelling Approaches/Models

Model verification and validation are essential components of every modelling process. They are crucial for the acceptance of a model. Roughly spoken and according to modelling approaches to manufacturing scheduling in this chapter, the difference between verification and validation can be described as follows: While model verification tries to proof whether a model formulation (and, in addition, its solution procedure) fulfils the requirements of the model within the model's sphere, model validation is intended to check the model's adequacy with respect to the underlying real-world problem.

Model verification is intended to confirm (as far as possible) *ex ante* that the model and jointly its solution procedure performs as required. This includes that the (formal) model is constructed and mapped correctly and the algorithms are implemented properly. However, we can almost never guarantee a 100 % error-free (formal) mapping and implementation of any scheduling model. Only statistical statements on model's reliability are possible. Consequently, a properly designed model testing approach is a necessity for assessing model adequacy.

Complementarily to model verification, model validation intends to ensure that the model meets its intended requirements in terms of the methods employed and the results obtained relative to the underlying real-world scheduling setting. The goal of model validation is to make the model useful in the sense that the model addresses the right problem and to provide accurate information about the (real-world) system being modelled. Model validation includes the following components: requirements validation, face validity, calibration, process validation, data validation, theory validation, veridical (= the degree to which an experience, perception or interpretation accurately presents reality) and validation of emergent structures and processes.

It is often too costly and time consuming to validate a model comprehensively over its complete domain of applicability. Instead, tests and evaluations are conducted until sufficient confidence is obtained that a model can be considered valid for its intended application. The trade-off between confidence requirements and validation costs has to be considered case specific.

With respect to the complexity reduction approaches described in earlier sections of this chapter, a more general perception of model validation should be shortly addressed: Model validation can be interpreted also as validating a model's adequacy with respect to another model. This other model must not necessarily represent the real-world setting. It might also be the basic (too) complex model. Then, the complexity-reduced formal model is validated relative to the underlying complex model (which is supposed to be a formal model too). That is, in terms of validation, complexity reduction apart from real-world adequacy validation includes a second validation aspect, namely between the two different formal models, the complex one and the one with reduced complexity.

More operational, model validation and model verification contain one or more of the following aspects:

- The restrictions (constraints) of the model must be feasible, redundancy-free and in line with the real-world problem.
- The objective function(s) must be in line with the objectives for the real-world problem.
- The decision space in the model must be in line with the real-world decisions.
- The formulated model must be error-free.
- The subjective influence of the decision-maker(s) must be minimised (according to a purely rational point of view) or in line with the real-world structural conditions, e.g. when giving weights to job, operations and/or objectives.

Summarising, model verification and model validation are two steps being applicable to all modelling approaches where a real-world problem is mapped into one or more formal models or model levels. With respect to manufacturing scheduling, both approaches have to be specified according to the specific setting under consideration.

6.6 Conclusions and Further Readings

In this chapter, we described and discussed several typical approaches to construct scheduling models. We also addressed complexity reduction techniques in manufacturing scheduling, also including model decomposition and, inseparably connected, re-integration of information from decomposed sub-models. Aggregation and disaggregation as a prominent method for complexity reduction was discussed as well. Finally, we sketched out basic aspects of model verification and validation.

It is beyond the scope of this chapter to address the aspects mentioned before and/or to formalise the approaches in more detail. The interested reader has been referred to more specialised references in each section. The main bottom lines of this chapter are:

- Manufacturing scheduling models can be formalised in different ways with different tools. Decision-makers have to choose the adequate approach(es) carefully and according to the real-world requirements.
- Complexity reduction is a main and permanent issue in modelling scheduling problems.
- There is a voluminous tool box to address complexity reduction issues, including decomposition and re-integration as well as aggregation and disaggregation.
- Issues of complexity reduction should undergo a valid analysis instead of being treated on a gut level.
- Specific validation and verification for manufacturing scheduling models are missing to a large extent. Nevertheless, both inner-model verification and real-world oriented validation should be addressed adequately and explicitly for every scheduling model.

Regarding further readings, a key text on mathematical modelling is Williams (2013). Basic information on modelling approaches including uncertainty can be found in Zadeh (1975, 1999), Dubois et al (1996, 2003), Dubois and Prade (1988), Fayad and Petrovic (2005), Kuroda and Wang (1996), Vlach (2000). The first five references are specifically devoted to fuzzy approaches. An overview of previous work on using Petri nets for scheduling problems is given by Zhou and Venkatesh (2000), while different methods for generating object-oriented models can be found in Schmidt (1996). Finally, for further information on model validation and verification, the interested reader is referred to Macal (2005), Gibson (2001), Sargent (2007).

References

Błażewicz, J., Ecker, K. H., Pesch, E., Schmidt, G., and Weglarz, J. (2007). *Handbook on scheduling: from theory to applications*. Springer, Berlin/Heidelberg/New York.

Bowman, E. H. (1959). The schedule-sequencing problem. *Operations Research*, 7(5):621–624.

Burdett, R. L. and Kozan, E. (2003). Resource aggregation issues and effects in mixed model assembly. In Kozan, E., Beard, R., and Chattopadhyay, G., editors, *Proceedings 5th Operations Research Conference of the Australian Society for Operations Research Queensland Branch on Operations research into the 21st century*, pages 35–53, Brisbane, Queensland - Australia. Queensland University of Technology.

Dantzig, G. B. (1960). A machine-job scheduling model. *Managemant Science*, 6(2):191–196.

Dubois, D., Fargier, H., and Fortemps, P. (2003). Fuzzy scheduling: Modelling flexible constraints vs. coping with incomplete knowledge. *European Journal of Operational Research*, 147(2):231–252.

Dubois, D., Fargier, H., and Prade, H. (1996). Possibility theory in constraint satisfaction problems: Handling priority, preference and uncertainty. *Applied Intelligence*, 6(4):287–310.

Dubois, D. and Prade, H. (1988). *Possibility theory*. Plenum Press, New York.

Fayad, C. and Petrovic, S. (2005). A fuzzy genetic algorithm for real-world job shop scheduling. *Innovations in Applied Artificial Intelligence: 18th Int. Conf. on Industrial and Engineering Application of Artificial Intelligence and Expert Systems, IEA/AIE 2005. Bari, Italy, 22–24 June 2005*, 3533:524–533.

Gibson, J. P. (2001). Formal requirements models: simulation, validation and verification. *Technical, Report NUIM-CS-2001-TR-02*.

Greenberg, H. H. (1968). A branch-bound solution to the general scheduling problem. *Operations Research*, 16(2):353–361.

Kuroda, M. and Wang, Z. (1996). Fuzzy job shop scheduling. *International Journal of Production Economics*, 44(1–2):45–51.

Le Pape, C. (1994). Implementation of resource constraints in ILOG schedule: A library for the development of constraint-based scheduling systems. *Intelligent Systems Engineering*, 3(2):55–66.

Macal, C. M. (2005). Model verification and validation. *Workshop on "Threat Anticipation: Social Science Methods and Models"*.

Manne, A. S. (1960). On the job-shop scheduling problem. *Operations Research*, 8(2):219–223.

Nagar, A., Heragu, S. S., and Haddock, J. (1995). A branch-and-bound approach for a two-machine flowshop scheduling problem. *Journal of the Operational Research Society*, 46(6):721–734.

Pinedo, M. L. (2012). *Scheduling: Theory, Algorithms, and Systems*. Springer, New York, fourth edition.

Sargent, R. G. (2007). Verification and validation of simulation models. In Henderson, S. G., Biller, B., Hsieh, M.-H., Shortle, J., Tew, J. D., and Barton, R. R., editors, *Proceedings of the 2007 Winter Simulation Conference*, pages 124–137, Piscataway, NJ. IEEE Operations Center.

Schmidt, G. (1992). A decision support system for production scheduling. *Revue des Systemes de decision*, 1(2–3):243–260.

Schmidt, G. (1996). Modelling production scheduling systems. *International Journal of Production Economics*, 46–47:109–118.

Shen, W. and Norrie, D. (1998). An agent-based approach for dynamic manufacturing scheduling. *Working Notes of the Agent-Based Manufacturing Workshop, Minneapolis, MN.*

Slowinski, R. and Hapke, M. (2000). Foreword. In Slowinski, R. and Hapke, M., editors, *Scheduling under Fuzziness*, Heidelberg, New York. Physica-Verlag.

Vancza, J., Kis, T., and Kovacs, A. (2004). Aggregation: the key to integrating production planning and scheduling. *CIRP Annals of Manufacturing Technology*, 3(1):377–380.

Vlach, M. (2000). Single machine scheduling under fuzziness. In Slowinski, R. and Hapke, M., editors, *Scheduling under Fuzziness*, pages 223–245, Heidelberg, New York. Physica-Verlag.

Wagner, H. M. (1959). An integer linear-programming model for machine scheduling. *Naval Research Logistics Quarterly*, 6(2):131–140.

Williams, H. P. (2013). *Model building in mathematical programming*. Wiley, Chichester, 5th ed.

Zadeh, L. A. (1975). Calculus of fuzzy restrictions. In Zadeh, L. A., Fu, K. S., Tanaka, K., and Shimura, M., editors, *Fuzzy sets and their applications cognitive and decision processes*, pages 1–39, New York. Academic Press.

Zadeh, L. A. (1999). Fuzzy sets as a basis for a theory of possibility. *Fuzzy sets and systems*, 100:9–34.

Zhou, M. C. and Venkatesh, K. (2000). *Modelling, simulation and control of flexible manufacturing systems: A petri net approach.* World Scientific, Singapore a.o.

Part III
Scheduling Methods

The Part III of the book consists of Chaps. 7–10 and is devoted to scheduling methods. Here we will review the main approaches related to solution procedures of scheduling models. We emphasize general approaches rather than specific solutions, and provide a general scheme to build and validate new procedures for scheduling problems. We pay attention to the integration of heuristic and exact approaches and devote a chapter to multiobjective procedures.

Chapter 7
Overview of Scheduling Methods

7.1 Introduction

In the previous part of the book, we have presented the concept of a scheduling model as a way to formalise the decision-making scheduling problem. This part of the book is devoted to present the methods to provide (good or even optimal) solutions for these scheduling models. In this chapter, we give an overview of scheduling methods, leaving for the next chapters the detailed discussion of specialised methods.

After a few basic definitions related to scheduling methods (most notably, the concept of algorithm), we discuss, in a rather intuitive way, why scheduling problems are tremendously hard to solve to optimality, thus giving a rationale for the traditional use of scheduling policies, which try to ease these difficulties at the expense of an unimpressive performance, at least as compared with 'true' scheduling algorithms. We also give an overview on the main types of scheduling algorithms, and discuss how the adequacy of a scheduling method for solving a manufacturing scheduling decision problem can be assessed.

More specifically in this chapter, we

- provide the main definitions regarding scheduling methods, together with a dicussion on the main assumptions behind these (Sect. 7.2),
- discuss the concept of computational complexity and its consequences when designing manufacturing scheduling methods (Sect. 7.3),
- review the main scheduling policies along with their rationale (Sect. 7.4),
- present an overview of the main types of scheduling algorithms (Sect. 7.5) and
- introduce the principal issues related to the (formal and real-world) assessment of manufacturing scheduling methods (Sect. 7.6).

J. M. Framinan et al., *Manufacturing Scheduling Systems*,
DOI: 10.1007/978-1-4471-6272-8_7,

7.2 Basic Concepts

In this section, we provide the standard definitions of scheduling methods and related terms. It is important to note that these terms refer exclusively to formal/computational aspects for solving mathematical models, and cannot be, in general, translated into the real-world decision problem in a straightforward manner. Therefore, along with these definitions, we provide a list of the main implicit and explicit assumptions behind, and leave for a later section (Sect. 7.6) the discussion on assessing the adequacy of a scheduling method for solving a real-world scheduling decision problem.

7.2.1 Formal Definitions

In the context of this book, a scheduling method is a formal procedure that can be applied to any instance of a scheduling model in order to obtain a feasible schedule that (presumably) obtains good scores with respect to a sought objective. Note that, according to Chap. 3, a scheduling model is assumed to be a representation of a well-defined manufacturing scheduling decision problem. Therefore, since the procedure can be applied to any instance of a scheduling model, it can be said (rather informally, as it has been noted in several places throughout the book) that a scheduling method *solves* a scheduling problem.

A scheduling method can thus be viewed as a procedure taking an instance of a scheduling model as an input in order to produce (at least) one schedule as an output. Since the procedure has to be formal, it can be described using a finite number of steps, and it is thus amenable to be coded and eventually executed by a computer. The technical name for such finite procedure is *algorithm*.

It is customary to classify algorithms into exact and approximate. Exact algorithms guarantee that no other schedule performs better than the one obtained with respect to the objective sought. The so-obtained solution is named *optimum* or *optimal solution*. In contrast, approximate algorithms do not guarantee that the solution is optimal although, in some cases, we will see that it is possible to estimate the maximum deviation from the optimum.

7.2.2 Assumptions

As mentioned above, there are several explicit and implicit assumptions in the previous definition of scheduling methods that are worth to be discussed:

1. Well-defined scheduling model. As already mentioned in Chap. 2, sometimes real-world manufacturing scheduling problems lack of precise description. Machine availability considerations, rework and many other aspects are excluded from the

model formulation. So it might be that some of the constraints of the problem are not included into the definition of the model or are even not known to the decision maker although they might be relevant for the real-world problem's solution. Relative to this exclusion/ignorance, the terminus 'solution' may become inadequate.

2. Formal model. It is supposed that the problem is given as a mathematically strictly formulated formal model. However, as already mentioned in Chap. 2 as well, these formal models always represent a more or less significant simplification of the real-world problem. Therefore, deriving a solution for this formal model, even if it is an optimal one, means nothing more than deriving an approximate solution for the respective real-world problem. Hence, with respect to the real-world problem, there will be neither an exact problem formulation nor an exact solution procedure or an exact solution in the strict sense. For the moment, we will confine our perspective to the formal sphere of the problem under consideration and later, in Sect. 7.6 we will address the assessment of the solutions provided by the formal model with respect to the real-world problem.
3. Objectives. Usually, the formal model intends to derive the best feasible solution of a manufacturing scheduling problem. However, there are many other plausible types of objective functions, which could represent the goals of the formal problem, e.g.

    - Generation of a feasible solution. This is especially relevant if the set of constraints does not obviously imply feasible solutions.
    - Generation of a good solution. That is, e.g., a solution being at most a certain percentage away from the optimal objective function value or being not worse than some predefined value, etc.
    - 'Dualising' constraints and/or objective functions. Constraints of an 'original' formal model might be dualised to the objective function and vice versa. With respect to the 'exactness' of this dualisation, it is usually supposed that these dualisation steps have been performed before exactness is an issue and are input into the formulation of the formal optimisation model, or these dualisations are part of the (maybe exact) solution procedure itself.
    - Number of solutions. In manufacturing scheduling models, often more than one optimal solution occurs, i.e. the same best (or satisfactory) objective function value is achieved by more than one solution. Most (also exact) solution procedures obtain only one of these solutions. However sometimes, it is required to determine more than one or even all best solutions.

    Note that defining the objectives in the previous manner would mean that an exact procedure would just fulfil these types of requirements or regards the additional model modifications as input, respectively, sometimes even not necessarily looking for an extreme (maximum or minimum) solution.
4. Single objective function. As will be discussed in Chap. 10, and is well known from multi-objective optimisation in general, the 'optimality' in the multi-objective context will often be expressed by some efficiency calculus, usually

by the concept of Pareto set which requires, for an 'exact' procedure, to determine the whole efficient boundary of solutions.
5. (Deterministic) exact procedure. Finally, to be called 'exact', a method for fulfilling the requirements of the formal optimisation problem (i.e. here usually to derive the optimal solution of every possible problem instance for a formal manufacturing scheduling problem) must deliver the solution within a finite (but possibly large) number of computation steps and in a completely deterministic manner. Completely deterministic here means that the result should be identically replicable in another run of the procedure for the same problem instance.
6. Stochastics setting. It should be mentioned that exact procedures might also be defined for stochastic decision problems, including stochastic manufacturing scheduling models. The exact nature of the algorithm is not in conflict with possible stochastic components of the model formulation as long as the criterion of the model is not a random variable. For instance, one may speak of an exact procedure for the $\alpha|\beta|C_{max}$ model as well as for the $\alpha|\beta|E[C_{max}]$ model where processing times are random variables.

So, in the sequel of this chapter, we will suppose that we have a precisely formulated (i.e. also completely quantified) manufacturing scheduling problem (e.g. a flow shop problem) with a unique objective function (e.g. makespan) which is intended to be minimised, i.e. one (not all) optimal solution is to be determined by a well-defined deterministic procedure, i.e. an exact procedure. These assumptions will allow us to go forth until Chap. 10, where several objectives would be discussed, and Chap. 14, in which the relaxation of these hypotheses will be discussed in the context of a manufacturing scheduling system.

Once these assumptions have been established, a natural question is why employing approximate algorithms (without performance guarantee) if there are exact methods at hand. The answer lies in the computational complexity of most scheduling models, which will be discussed in the next section, so we will see that approximate algorithms are rather often the only suitable alternative given the long times that some exact procedures take to complete. Indeed, the complexity of most real-world scheduling models is so high that in practice, a particular case of scheduling methods are the so-called scheduling policies (which will be studied in Sect. 7.4) which are widely employed.

7.3 Computational Complexity of Scheduling

Computational complexity theory is a field inside the theory of computation that focuses on classifying *computational problems* according to their underlying structure and difficulty. In our context, the computational problem is the scheduling model at hand, for which we intend to build formal procedures (i.e. algorithms) to find the best solution.

Rather informally, we can define two types of computational complexity:

- Algorithm complexity. Algorithm complexity informs us on the performance of a specific algorithm for solving a model by assigning to it an estimation of its cost (in time). This estimation is known as the computational or time complexity (also referred to as running time). Intuitively, the actual time to run the algorithm will depend, among other factors, on the specific instance of the model for which the algorithm is applied.
- Model complexity. Model complexity informs us about the complexity of the 'best algorithm' able to find the optimal solution this model by classifying the models into *complexity classes*.

It is interesting to note that the complexity of an algorithm can be calculated for any algorithm, either exact or approximate. However, model complexity refers to the complexity of the best exact algorithms to solve the model. In the next section, we will discuss these rather abstract concepts first by introducing an example (Sect. 7.3.1) and then by deepening into the two types of computational complexity (Sects. 7.3.2 and 7.3.3). Note that the treatment we are going to give to all these disciplines is going to be necessarily succinct, as a precise discussion is completely out of the intended scope of this book.

7.3.1 An Example

In this section, we illustrate the computational complexity of scheduling with a very simple model, i.e. a single machine, for which we try to find out a schedule that minimises the total tardiness of the jobs. This model is denoted as $1||\sum T_j$. For simplicity, we will focus on finding the optimal semi-active schedule; therefore, each possible semi-active schedule can be represented by specifying the sequence of the jobs for each machine, i.e. the order in which jobs are to be processed on the machines. Then, in our setting, $n!$ different sequences exist, which represent job permutations. Let us define by Π the set of those $n!$ sequences and by π any of those permutations.

Note that $1||\sum T_j$ is not trivial. Depending on the order in which the jobs are processed, some of them will finish before their due dates and some others will finish late, adding to the total tardiness value. Unless there is enough time to finish all jobs before the earliest due date, i.e. $d_j \geq \sum_{j=1}^{n} p_j$, there might be a possibility that some jobs will finish late.

Now, let us assume for the moment that the only method we have at our disposal is the complete enumeration of all the possible $n!$ job sequences in order to determine the minimum total tardiness. Let us also suppose that we want to solve different instances of the model with $1, 2, \ldots, 30$ jobs each one. Notice that 30 jobs is a small problem size for most real-life settings. Therefore, there would be $1!, 2!, \ldots, 30!$ possible solutions as a function of the number of jobs. We are going to assume that we have an 'old' computer capable of doing one billion (in short scale, i.e. 1,000,000,000) basic operations per second. This would be the equivalent of a CPU running at 1.0 GHz. For simplicity, we are also assuming that one full schedule can be calculated in just one basic operation. Normally, for calculating total tardiness, we would need at least n basic operations as the individual tardiness of all jobs has to be added. We can then calculate the time it would take for this 'old' computer to calculate all possible sequences. In order to put these CPU times into perspective, we compare these CPU times against those of a theoretical computer that is 5 million times faster, i.e. this computer would be running at 5 PHz or 5,000,000 GHz. A computer running at such incredible speeds is way beyond current computers that, at the time of the writing of this book, rarely operate at frequencies higher than 6 GHz. All this data is given in Table 7.1, where AU stands for 'Age of the Universe' which has been estimated to be 14 billion years.

As we can see, for trivial problems with about 12 jobs, both computers would give the optimal solution in less than half a second in the worst case. However, increasing just 3 jobs to 15 jobs (a 25 % increase in the size of the problem) results in a CPU time of 21.79 min for the slow computer (2,730 times slower). From 15 jobs onwards, the CPU times of the slow computer quickly turn unfeasible, as anything longer than 12 h or so is unpractical for daily use. It can be observed that applying these complete (or further elaborated implicite) enumeration approaches to combinatorial optimization problems (such as scheduling problems) often results in quickly finding the optimal solution but in long term proving that this solution is indeed an optimal one.

Conversely, a computer no less than 5 million times faster would be expected to solve problems of much larger size than the slow computer. However, if $n = 17$ is already unpractical for the slow computer, for the fast computer we would be able to solve problems of up to $n = 21$. In other words, a 5 million times faster computer allows us to solve problems that are less than 24 % larger.

This single machine scheduling problem is an example of a problem with an exponential number of solutions. Needless to say, realistically sized problems of $n = 150$ are completely out of the question. Saying that enumerating all possible sequences of a single machine problem with 150 jobs is impossible is actually an underestimation. Consider that 150! results in $5.71 \cdot 10^{262}$ sequences. This number is extremely huge. To have an idea, the number of atoms in the observable universe is estimated to be between 10^{78} and 10^{80}. Therefore, a simplistic problem with one machine and some 150 jobs is impossible to enumerate. The permutation flow shop problem also has a solution space of $n!$ However, other more complex problems, like the regular flow shop or job shop have an even much larger number of solutions with $(n!)^m$ or the parallel machine layout with m^n. A job shop with, let us say, 50 jobs and 10 machines, will have a solution space of $(50!)^{10} = 6.77 \cdot 10^{644}$ sequences.

The fact that a problem has an extraordinarily huge number of possible sequences does not necessary imply that no efficient methods exist for obtaining a good solution or even an optimal solution. For example, picture again the $1||C_{max}$ problem which also has $n!$ possible sequences but any of them will return the minimum makespan. The example given in Table 7.1 has an obvious caveat: evaluating all possible sequences in the search for the optimum is actually the worst possible method.

Table 7.1 Total CPU times for two types of computers for evaluating all solutions in single machine problems as a function of the number of jobs n

n	$n!$	Computer time (slow)		Computer time (fast)	
1	1	1	ns	0.20	fs
2	2	2	ns	0.40	fs
3	6	6	ns	1.20	fs
4	24	24	ns	4.80	fs
5	120	120	ns	24	fs
6	720	720	ns	144	fs
7	5,040	5.04	μs	1.01	ps
8	40,320	40.32	μs	8.06	ps
9	362,880	0.36	ms	72.58	ps
10	3,628,800	3.63	ms	726	ps
11	39,916,800	39.92	ms	7.98	ns
12	479,001,600	479	ms	95.80	ns
13	6,227,020,800	6.23	s	1.25	μs
14	87,178,291,200	87.18	s	17.44	μs
15	1,307,674,368,000	21.79	min	262	μs
16	20,922,789,888,000	349	min	4.18	ms
17	355,687,428,096,000	98.80	h	71.14	ms
18	6,402,373,705,728,000	74.10	days	1.28	s
19	121,645,100,408,832,000	3.85	years	24.33	s
20	2,432,902,008,176,640,000	77.09	years	8.11	min
21	51,090,942,171,709,400,000	16.19	centuries	170.30	min
22	1,124,000,727,777,610,000,000	356	centuries	62.44	h
23	25,852,016,738,885,000,000,000	819	millennia	59.84	days
24	620,448,401,733,239,000,000,000	19.66	million years	3.93	years
25	15,511,210,043,331,000,000,000,000	492	million years	98.30	years
26	403,291,461,126,606,000,000,000,000	12.78	billion years	25.56	centuries
27	10,888,869,450,418,400,000,000,000,000	345	billion years	690.09	centuries
28	304,888,344,611,714,000,000,000,000,000	690	AU	1.93	million years
29	8,841,761,993,739,700,000,000,000,000,000	20,013	AU	56.04	million years
30	265,252,859,812,191,000,000,000,000,000,000	600,383	AU	1.68	billion years

Slow computer running at 1 GHz. Fast computer running at 5 PHz

The previous example serves to illustrate two aspects earlier mentioned in this section: on the one hand, not all algorithms are equally efficient with respect to solving a scheduling model, so it is of interest analysing its performance. On the other hand, giving these incredible computer times, it seems that there are models for which it is difficult to obtain algorithms with a reasonable (not to mention good) performance.

7.3.2 Algorithm Complexity

Let us start with the running time of an algorithm. The running time is an estimation of the time needed by an algorithm (usually coded into a computer) to complete as a function of the input size. Measuring crude time is a tricky business since each computer architecture runs at different speeds, has different frequencies and many other specificities that make the comparison of running times of the same algorithm in two different architectures a daunting task. To overcome this problem, the running time is not actually given in time units but as a number of elementary operations performed by the algorithm instead. These elementary operations or 'steps' are an independent measure since each elementary operation takes a fixed amount of time to complete. As a result, the amount of time needed and the number of elementary operations are different by at most a constant factor.

For example, if we have an unsorted list of n elements, what is the time needed to locate a single item in the list (assuming that all items are different and that they are unsorted)? A basic search algorithm will require, in the worst case, to analyse each element and would require n steps to complete. These n steps are n steps with independence of the computing platform used, but obviously, will translate into different CPU times in the end. Note that in the best case, the first element in the unsorted list could be the element searched for. Therefore, only one step is needed. On average, and given a completely random input, $n/2$ steps will be needed to locate an element.

In complexity theory, the so-called 'Big-O' notation or, more specifically, the Bachmann-Landau notation is used. This notation describes the limiting behaviour of a function when the argument tends towards a particular value or infinity, usually in terms of simpler functions. Big-O notation simplifies the running times in order to concentrate on their growth rates. This requires the suppression of multiplicative constants and lower order terms. Therefore, running time expressed with the Big-O notation is said to be the asymptotical running time, i.e. as the input size goes to infinity.

Continuing with the search algorithm example, we would say that the best case running time is $O(1)$ or constant time, the average running time would be $O(n/2)$ or linear time and the worst case running time $O(n)$ or linear time also. Note that, for very large input sizes (big value of n), the large time needed is only divided by a constant.

Normally, running times are measured for the average running time, and for simplicity the O notation is used. As a result, our basic searching algorithm has a running time complexity of $O(n)$ or linear running time. Notice that this means that n basic operations are needed. However, at least three several basic operations are performed at each step j, $j = \{1, 2, \ldots, n\}$: reading the j-th element from the unsorted list and comparing this element with the desired element. If the comparison is positive, end the search; if not, increase the counter and repeat for the next element in the list. However, for large values of n, $O(3n)$ is equal to $O(n)$.

Let us picture now a different algorithm. The algorithm just adds a constant value to each element of a square matrix of size $n \times n$. This algorithm will have one outer

Table 7.2 Total CPU times for two types of computers for applying the Quicksort algorithm of $O(n \log n)$ computational complexity

n	$n \log n$	Computer time (slow)	Computer time (fast)
1,000	3,000	3.00 μs	0.60 ps
10,000	40,000	40.00 μs	8.00 ps
100,000	500,000	0.50 ms	100.00 ps
1,000,000	6,000,000	6.00 ms	1.20 ns
10,000,000	70,000,000	70.00 ms	14.00 ns
100,000,000	800,000,000	0.80 s	160.00 ns
1,000,000,000	9,000,000,000	9.00 s	1.80 μs

Slow computer running at 1 GHz. Fast computer running at 5 PHz

loop to traverse all rows (or columns) and an inner loop to traverse columns (or rows) with one basic addition operation. The running time of the algorithm will be $O(n^2)$ or quadratic time. In general, many algorithms have polynomial running times like for example $O(n^3)$ or $O(n^4)$. Note that something like $O(10n^4 + 3n^3 + n^2 + n)$ is not correct under the Big-O notation and must be written as $O(n^4)$. Recall that for large n values, the dominating term is the n^4, not the remaining polynomial.

There are many different categories of running times. For example, the most efficient comparison sort algorithm known, Quicksort, has an average running time of $O(n \log n)$ or linearithmic time, in particular, or polynomial time, in general. In the previous single machine problem example, the brute-force approach needed $O(n!)$ steps, something referred to as factorial time, in particular, or exponential time, in general.

Not all polynomial running time algorithms are equal. After all, an algorithm with a polynomial running time of $O(n^{10})$ is surely going to be slower than another with $O(n^3)$. Well, this is obviously true, but recall the previous single machine example depicted in Table 7.1. Exponential running times of $O(n!)$ quickly run out of hand for extremely small values of n. However, $O(n^{10})$ with $n = 20$ needs 248 h on the 'slow' computer of the example, whereas the brute-force approach needed 77.09 years. Furthermore, the same algorithm would need 2.05 ms on the fast computer of the example (compared to 8.11 min of the other algorithm). In other words, in a polynomial running time algorithm, a computer that is 5,000,000 times faster reduces the real CPU times linearly, i.e. 2.05 ms is 5,000,000 times faster than 248 h. This is not the case for exponential running time algorithms. Table 7.2 shows the CPU times that the two types of computers would need for very big n values of the Quicksort algorithm.

As can be seen, sorting one billion items with this algorithm would require in the worst case barely 9 s in the slow computer and an almost instantaneous 1.8 μs in the fast computer.

7.3.3 Model Complexity

In the previous subsection, we have discussed the complexity of algorithms when applied to a specific scheduling model. Informally speaking, scheduling models for which a polynomial running time algorithm is known to exist are 'easy', since such algorithms are able to produce, on average, an optimal solution in a computational time that at least does not grow 'too much' with the instance size. In contrast, scheduling models that do not have a polynomial running time algorithms are 'hard'. These scheduling models are said to belong to the NP-hard class of computational problems. Determining whether a model belongs to the NP-hard class requires a mathematical proof and it is based on a conjecture (not demonstrated until today) that the NP-hard class exists on their own.

Although very approximate, the previous definitions serve to classify scheduling models according to their complexity. This is useful as, quite often, one algorithm for a scheduling model can be applied (in a straightforward manner of with some modifications) to another scheduling model. The simplest case are the weighted and unweighted version of many criteria. For instance, it is clear that $\alpha|\beta|\sum C_j$ is a special case of $\alpha|\beta|\sum w_j C_j$ and therefore, *any* algorithm for the latter model could be used for the former. We can then say that $\alpha|\beta|\sum C_j$ *reduces* to $\alpha|\beta|\sum w_j C_j$, and it is denoted as:

$$\alpha|\beta|\sum C_j \propto \alpha|\beta|\sum w_j C_j$$

Using these reduction ideas, a hierarchy of complexity of (some) scheduling models can be established depending on the layout, on the criterion and on the constraints. Some of these reductions are rather straighforward. For instance, it is clear that $1||\sum C_j$ reduces to $F_m||\sum C_j$.

Other reductions may require some (moderate) explanation. For instance, if we have an optimal algorithm for the $\alpha|\beta|\max L_j$ model, then we can construct a special case of this model for which all $d_j = 0$. For this instance, $L_j = C_j$; therefore $\max L_j = \max C_j$. Consequently, having an algorithm for solving the $\alpha|\beta|\max L_j$ model implies that it can solve (with the same computational complexity) the $\alpha|\beta|\max C_j$ model. Thus, $\alpha|\beta|\max C_j \propto \alpha|\beta|\max L_j$.

Finally, some other reductions require more thorough proofs. In Figs. 7.1, 7.2 and 7.3, the complexity hierarchy (also named *reduction tree*) is given for most models with respect to different layouts, constraints and criteria, respectively. It is perhaps interesting to note that adding constraints to a model does not necessarily turn it into less or more complex.

The reductions also work the other way round, i.e. if a model lower in the complexity hierarchy is of a given complexity, then clearly the models to which this former model is reduced are at least of the same complexity. In a sort of tautology, if a scheduling model is difficult, then more complex models are more difficult.

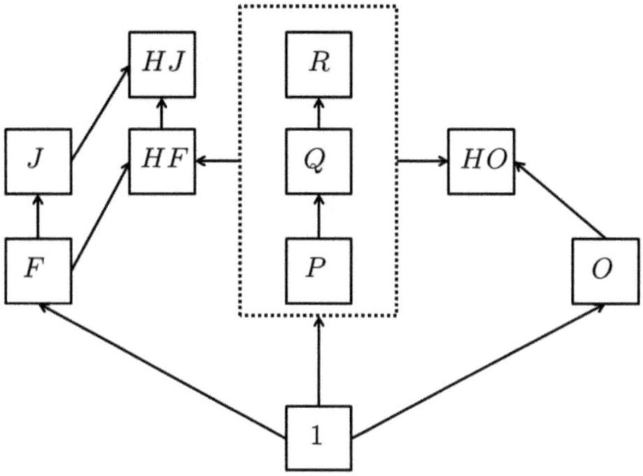

Fig. 7.1 Complexity hierarchy with respect to the layouts

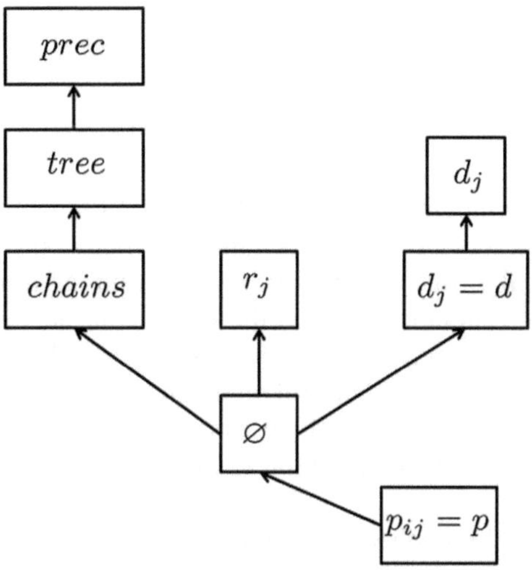

Fig. 7.2 Complexity hierarchy with respect to the constraints

Unfortunately, most scheduling problems belong to the NP-hard class. Only the simplest cases, and most of them are single machine layouts, do not belong to the NP-hard class. The question is: What to do then? The answer is in the very heart of all the scientific efforts that the scientific scheduling community has been doing over the past decades. If the problem belongs to the NP-hard class all that is known is that an optimum solution for large-sized instances is unlikely to be viable in acceptable, i.e. polynomial time. However, for practical scheduling there are several alternatives to handle this computational complexity. These are discussed in the next section.

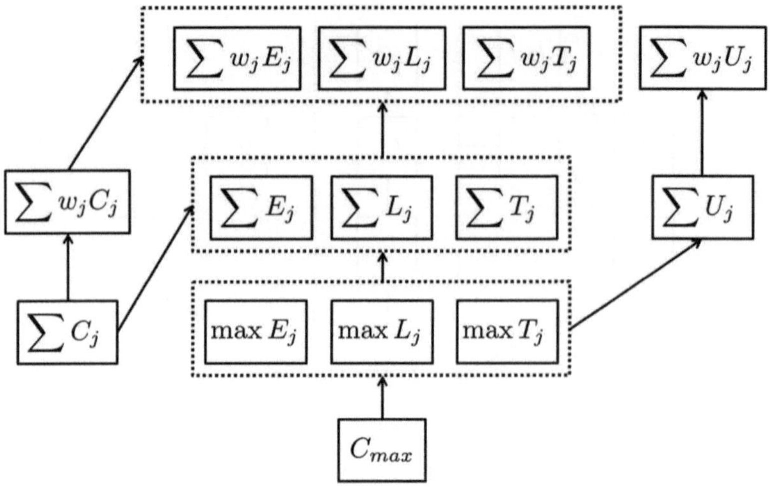

Fig. 7.3 Complexity hierarchy with respect to the criteria

7.3.4 Handling Computational Complexity

When faced with a computationally complex model (which happens most of the cases), several alternatives are open, depending on whether the model complexity is acceptable in practical terms, or not. Note that the running time is instance dependent, and that the Big-O notation neglects the size of constant factors or exponents in the running time and these may have practical importance. Therefore, an exact exponential running time algorithm might still return an optimal solution in an acceptable time for moderately sized problems. If these problems are of interest to the practical application, then the computational complexity is (at least from a practical viewpoint) acceptable.

If the computational complexity is not acceptable, then again two options (not necessarily alternative) are available:

1. Try to reduce the complexity of the model and/or that of the associate algorithms to an acceptable level, or
2. Avoid the use of algorithms to solve the model.

With respect to the first option, recall that two types of complexity in manufacturing scheduling were introduced in Sect. 2.3.2, according to the discussion in Sect. 1.5: A computational complexity reduced to a formal sphere, and a real-world complexity. While the reduction of the latter was discussed in Sect. 6.3, we will address here the reduction of the computational complexity. A general framework for complexity reduction with respect to model and/or method can be seen from Fig. 7.4.

Our point of origin is situation 1 in Fig. 7.4, in which a (original) model is shown in combination with a (original) exact algorithm. If the complexity of this combination of model and algorithm is not acceptable, the required complexity reduction

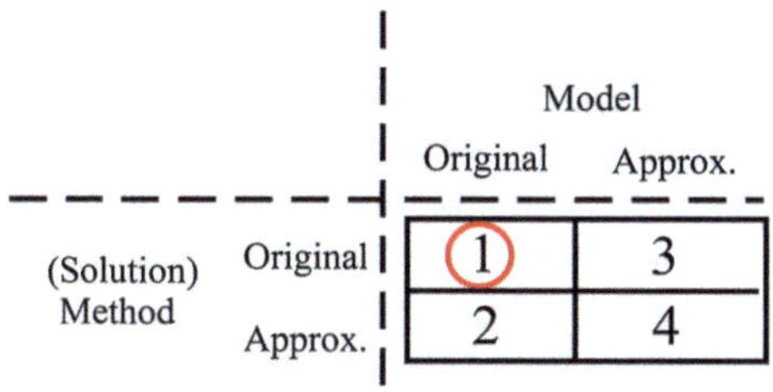

Fig. 7.4 Complexity reduction and approximate methods

might take place into two directions. On one hand, the same original model might be approximately solved with an approximate method. This case is represented by situation 2 in Fig. 7.4. For example, a given model can be solved by an algorithm returning a solution without guarantee of optimality. On the other hand, the model might be simplified and solved with the same algorithm which is available for the original model (situation 3 in Fig. 7.4). For example, jobs of the original model might be clustered into job groups which are to be executed consecutively within their group and the exact algorithm for the original model is applied only to the job group model which is hopefully, significantly less complex than the original one. Finally, both approaches can be combined (situation 4 in Fig. 7.4).[1]

Note that, after having solved the model- and/or method-simplified setting, the solution derived for this simplification has to be adjusted to the requirements of a solution for the original model. Despite of possible ambiguities, we will follow this intuitive line of thinking in our discussion of selected ideas for approximate methods which, accordingly, are seen as a simplification with respect to model and/or method in combination with a subsequent transfer of the solution of the simplified setting to the original model.

With respect to the second option—avoiding the use of algorithms to solve the model—it may seem a bit unrealistic, but yet it has been (and still is) one of the favourite options for the scheduler. While algorithms are designed to obtain a schedule, it may be that one can schedule jobs without a 'true' schedule. Instead, some rules can be given to assign on-the-fly jobs to machines as soon as they enter the shop floor. Such a rule is called a *dispatching rule*, and the set of rules that one can use to avoid deriving a schedule is called *scheduling policy*. More formally, a scheduling policy consists of a set of principles usually based in 'common sense' or 'rule of thumb' that are used to generate a schedule. However, note that a schedule itself is not explicitly given, although it can be obtained from this set of principles. For instance, a reasonable scheduling policy may be to process first those jobs whose due date is closest, in order to honour the commitment with the customer. Clearly, given a known set of jobs with known due dates, this policy can be translated into a

[1] It should be noted that this separation of model-oriented and of method-oriented simplification approaches might appear rather intuitive. In fact, it is somewhat artificial and the unique separation of simplification approaches into these two categories might be problematic. Ambiguities with respect to this separation might occur.

schedule. Quite clearly as well, this policy can lead to disastrous results, as it does not take into account the different processing times of the jobs, and therefore jobs with short processing times but relatively close due dates would be prioritised in front of jobs with further due dates, but also high processing times and therefore less slack.

The two aspects mentioned before summarise the main features of using scheduling policies as a scheduling method, i.e.: their simplicity and their myopic nature. The first feature is the reason why scheduling policies were and still are so popular in practice, since we will see in this chapter that developing optimal or even good schedules is, in general, far from being an easy task. In contrast, the second feature exposes their great disadvantage in front of other methods. There is no point in trying to make the principles more sophisticated so they can take into account most possible situations. Every rule has an exception and except for few cases, the performance of the dispatching rules is rather poor. Even worse, for most models the performance of the dispatching rule heavily depends on the problem instance, so they are not very reliable.

The most employed scheduling policies are described in detail in Sect. 7.4.

7.4 Scheduling Policies

Industrial activity dates back to the dawn of the industrial revolution. The need for scheduling is also very old and can be traced back to the onset of the production line at the beginning of the twentieth century. Obviously, at that time there were no computers and things were done manually. Even today, a large percentage of companies rely on human schedulers or on simple spreadsheets and rules of thumb to schedule jobs in the shop.

Manual schedulers do lots of iterations over production plans that evolve over time from production wishlists to more detailed production needs. Schedules are continually revised, changed, adapted and patched on a daily, even hourly basis. Most manual methods work in a forward fashion, i.e. jobs are launched to the shop and sequenced at machines to be started as soon as those machines are ready. However, there is seldom any accuracy when determining when jobs will be completed. Priority choices are made at some decision points, subject to many judgmental exceptions and special cases. While this might sound chaotic, we have to remind that it has worked and still works for countless firms. Human schedulers have a number of advantages like the ability to react quickly to changes in the production floor, common sense, experience, etc. However, complex production systems are often out of the possible reach of human beings. The intricacies and dynamics of several hundred jobs moving inside a shop is beyond the cognitive capability of even the best scheduling expert.

There are several ways for dealing with this situation. Sometimes, changing the whole production system to eliminate the need for scheduling is the best approach. If a production line has become too complex to handle since it is processing many different products, a possibility is to reduce the number of products or to break down that line into several different lines. Ideally, one line for each product and the scheduling problem is mostly gone or at least greatly reduced.

Manual scheduling is often based on the so-called dispatching rules. As the name implies, these rules basically launch or dispatch tasks to machines according to usually a simple calculation that returns an index or relative importance for pending tasks. Dispatching rules are also known as priority rules.

Dispatching rules are classical methods and therefore cannot be easily attributed to a single author and their origins are hard to trace. Basically, a dispatching rule maintains a list of eligible tasks or pending tasks. These tasks are entire jobs in a single machine layout or the tasks that make up a full job in a flow shop or job shop layout. Tasks enter the eligible set whenever it is actually possible to start processing them, i.e. when the preceding task is finished, when the release date has elapsed, etc. Let us denote the list of eligible tasks as $\wp$. Similarly, $|\wp|$ denotes the number of tasks in the eligible list.

In a nutshell, general dispatching rules carry out a simple calculation for all tasks in $\wp$ and dispatch the tasks according to the result of this calculation, sometimes referred to as priority. Dispatching rules can be generally classified in two dimensions:

1. Local/Global rules. This refers to the data considered in the priority index calculation. Some rules only consider the data of each task in $\wp$ separately. These are called local dispatching rules. Rules that consider the data of more than one task in $\wp$ simultaneously or other additional information not just related to the task itself are referred to as global dispatching rules.
2. Static/Dynamic rules. The result of the dispatching rule depends on the time at which it is applied. Static dispatching rules always return the same priority index, regardless of the state of the schedule or the list $\wp$. On the contrary, dynamic dispatching rules depend on the instant of time t at which they are calculated and hence on the information resulting from the partial schedule derived up to time t.

7.4.1 Basic Dispatching Rules

- First Come First Served (FCFS). The oldest pending task is scheduled. It is equivalent as ordering the tasks in $\wp$ in a First In First Out (FIFO) fashion. This rule is the simplest one and also seems the fairest when dealing with (human) customers as jobs that arrived first are also processed first. Obviously, this rule is too simple and does not work well in complex settings. FCFS is a local and static rule. In the presence of release dates r_j, this rule can be easily modified to Earliest Release Date first (ERD) which sorts jobs in ascending order of their release dates.
- Last Come First Served (LCFS). It is the opposite of FCFS. The most recent task is scheduled. It is equivalent to Last In First Out (LIFO). LCFS is a local and static rule.
- Random Order. Pending tasks are sequenced in a random order. While this might seem naive, we have a guarantee that no specific objective will be either favoured or penalised in a consistent way. Random Order is often used as a reference solution in simulation studies when comparing different dispatching rules.

- Shortest Processing Time first (SPT). The task in the pending list $\wp$ with the shortest processing time has the highest priority. More specifically, among m machines, indexed by i, the task k that satisfies the following expression has the highest priority:

$$p_{ik} = \min_{j=1}^{|\wp|} \{p_{ij}\}$$

  SPT is a local and static rule. Most interestingly, SPT results in an optimal solution for the problem $1||\sum C_j$. SPT just requires the pending job list to be ordered. Furthermore, this can be done once as it is a static rule. This results in a computational complexity for SPT of $O(n \log n)$.
- Longest Processing Time first (LPT). Contrary to SPT, LPT gives higher priority to the pending task with the longest processing time, i.e. the task k satisfying the following expression is given the highest priority:

$$p_{ik} = \max_{j=1}^{|\wp|} \{p_{ij}\}$$

  LPT is a local and static rule. It also has a computational complexity of $O(n \log n)$.
- Earliest Due Date first (EDD). This rule considers the due dates of the pending tasks in $\wp$. The highest priority is given to the task with the smallest due date or deadline, i.e.:

$$d_k = \min_{j=1}^{|\wp|} \{d_j\} \vee \bar{d}_k = \min_{j=1}^{|\wp|} \{\bar{d}_j\}$$

  EDD is also a local and static rule. EDD gives the optimum solution for the problems $1||T_{\max}$ and $1||L_{\max}$. It has a computational complexity of $O(n \log n)$. EDD is sometimes referred to as Jackson's rule. EDD has some important drawbacks for other objectives different than $L_{\max}$. Take for example, the $\sum T_j$ or $\sum_{j=1}^{n} w_j T_j$ criteria. It is very easy to get a very bad result if there is a job with a large p_j and an early d_j. That job would be sequenced first by EDD, even if doing so will delay the start of many other jobs, possibly forcing them to finish later than their due dates even if the first job is impossible to finish on time if $d_j < p_j$. In order to solve this issue, a dynamic version of the EDD is commonly used, this is described in the following.
- Minimum Slack first (MS). The slack with respect to the due date is the amount of time that a task still has before it would be tardy. This slack decreases as more and more jobs are scheduled and the current time t increases. Therefore, from the pending jobs list $\wp$, and at any given time t, a job k with the minimum following calculation is selected:

$$k = \operatorname*{argmin}_{j=1}^{|\wp|} \Big\{ \max\{d_j - p_j - t, 0\} \Big\}$$

Therefore, a job with the earliest due date but short processing time is not necessarily sequenced first. MS is an example of a dynamic dispatching rule, as it has to be recalculated each time a job is ready to be dispatched. Assuming that there are n jobs and that after completing a job, the pending job list $\wp$ is examined again, the computational complexity of MS can be stated as $O(n^2 \log n)$. While considerably slower than the more simple EDD, with modern computers there is no actual difference in running times between EDD and MS even for thousands of jobs.

- Largest Number of Successors (LNS). In problems with precedence relationships among jobs, it is interesting to process early jobs that 'unlock' several other jobs once finished. These jobs correspond to those that have many successors. In order to work correctly, this rule has to be combined with others that further give priority to some specific objective, such as due dates or completion times. LNS is a local and static rule. However, this last classification is a bit blurry depending on how LNS is implemented. If we only consider the immediate successors, then LNS is a local rule. However, if all successors are considered (i.e. the successors of the successors and so on), then LNS could be considered as a global rule.
- Shortest Setup Time first (SST). This rule helps in problems where setup times are a concern. Basically, from all the pending jobs in the $\wp$ list, the job with the highest priority is the one with the shortest setup time. SST is a local and static rule.
- Least Flexible Job first (LFJ). In problems with machine eligibility, it might be of use to sequence jobs that have very specific machine needs, in order to avoid scheduling them later where those specific machines might be already busy with previously scheduled jobs. LFJ is a local and static rule.
- Shortest Remaining Work Next Machine first (SRWNM). This is another example of a global rule. In flow or shop layouts, when assigning jobs to a specific machine, we might be interested to assign a job to a current machine looking ahead in the information of that job for subsequent machines. For example, two jobs to be processed on a given machine might have similar requirements in that machine and similar due dates so one could myopically think that it does not matter which job to assign to that specific machine. However, the next machine to be used for those jobs (picture that the next machine for those two jobs is not the same) could give us more information. From those two machines, we could look into the one that has the least remaining work left. Giving more work to that machine is crucial so to avoid imminent idleness. Note that there are many possibilities as regards global dispatching rules and SRWNM is just an example.

Let us give an example of the EDD rule for the $1||\max L_j$ problem with six jobs. The processing times and due dates of these jobs are given in Table 7.3.

EDD just needs to sort the jobs in ascending order of d_j, this requires the application of the Quicksort algorithm which has a running time complexity of $O(n \log n)$. For the example, the sorted list, and EDD solution is $\pi_{EDD} = (1, 2, 4, 3, 5, 6)$. As we can see, the completion times are $C_1 = p_1 = 6$, $C_2 = C_1 + p_2 = 6 + 3 = 9$,

Table 7.3 Processing times (p_j) for the single machine maximum lateness minimisation problem example

j	1	2	3	4	5	6
p_j	6	3	9	7	8	2
d_j	3	7	12	10	20	29

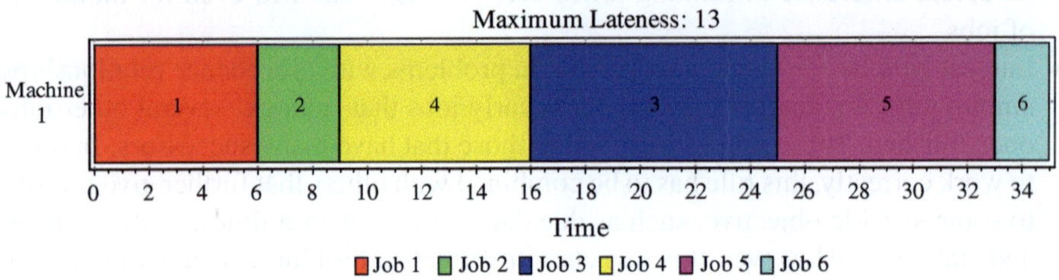

Fig. 7.5 Gantt chart with the EDD solution or the single machine maximum lateness minimisation problem example

$C_4 = C_2 + p_4 = 9 + 7 = 16$, $C_3 = C_4 + p_3 = 16 + 9 = 25$, $C_5 = C_3 + p_5 = 25 + 8 = 33$ and finally, $C_6 = C_5 + p_6 = 33 + 2 = 35$. As a result, the lateness of the jobs are: $L_1 = C_1 - d_1 = 6 - 3 = 3$, $L_2 = 9 - 7 = 2$, $L_3 = 25 - 12 = 13$, $L_4 = 16 - 10 = 6$, $L_5 = 33 - 20 = 13$ and $L_6 = 35 - 29 = 6$. Therefore, the maximum lateness $L_{\max} = L_3 = L_5 = 13$ which is optimal for this example. The resulting Gantt chart is pictured in Fig. 7.5.

There are variants for dispatching rules that consider weights. For example, the Weighted Shortest Processing Time first (WSPT) is the weighted variant of the SPT where the task k that satisfies the following expression is given the highest priority:

$$k = \operatorname*{argmax}_{j=1}^{|\wp|} \left\{ \frac{w_j}{p_{ij}} \right\}$$

Basically, all previous dispatching rules can be modified in order to consider weights in a straightforward way.

7.4.2 Advanced Dispatching Rules

The dispatching rules in the previous section are useful mainly for sequencing jobs. In problems with parallel machines or where there are hybrid layouts as hybrid flow or job shops, there is an additional decision to be made about to which machine each job must be assigned to. There are some basic dispatching rules for this as well:

- First Available Machine (FAM). Assigns the job to the first eligible machine available. This is the machine with the minimum liberation time from its last scheduled job, or lowest availability date for the machine if no job is scheduled at the machine yet.

- Earliest Starting Time (EST). Chooses the machine that is able to start job j at the earliest time. Note that only in simple layouts with no constraints this EST rule coincides with FAM. For example, availability of the job, lags, overlaps and a possible setup time can be considered. EST rule measures when the job will be able to start at the machine, not only when the machine is free from previous jobs.
- Earliest Completion Time (ECT). Takes the eligible machine capable of completing job j at the earliest possible time. Thus the difference with the previous rule is that this rule includes processing (possibly machine-dependent) times and other constraints.

Note than the EST and ECT rules need more calculations than the FAM as all pending jobs in $\wp$ have to be tested for earliest starting time and earliest completion times in all possible eligible machines in the hybrid environments. However, in some cases, this added CPU time might prove crucial for a good machine assignment decision.

Similarly, in recent years, some other more advanced machine assignment rules have been proposed. Some of them are discussed below:

- Earliest Preparation Next Stage (EPNS). The machine able to prepare the job at the earliest time for the next stage to be visited is chosen. It is a hybrid layout adaptation of the SRWNM rule. Many possible special situations can be considered like time lags between the current and the next stage, transportation constraints, etc. The rule uses more information about the continuation of the job, without directly focusing on the machines in the next stage. This is a global machine assignment rule.
- Earliest Completion Next Stage (ECNS). The availability of machines in the next stage to be visited and the corresponding processing times are considered as well. This rule also considers all the eligible machines at the next stage to make a decision. ECNS is also a global machine assignment rule.
- Forbidden Machine (FM). Excludes machine l^* that is able to finish job k at the earliest possible time, where job k is the next job in the sequence immediately following job j. ECT is applied to the remaining eligible machines for job j. This rule tries to minimise the excessive greediness of the ECT and similar rules. The fastest machine for a given stage might create problems in downstream stages. This rule is expected to obtain better results for later jobs, as it reserves some capacity in the machine able to finish the next job earliest. As with the previous rules, this is also a global or a 'look ahead' rule.
- Next Job Same Machine (NJSM). The assumption is made that two consecutive jobs in the sequence j and k are assigned to the same machine. The machine chosen for assignment is the one also able to finish jobs j and k at the earliest time although only job j is assigned. This is a look-ahead rule that also considers the next job in the sequence. This rule is especially useful if setups are relatively large, as not the setup for job j is considered, but also the setup between jobs j and k.

- Sum Completion Times (SCT). Completion times of job j and job k are calculated for all eligible machine combinations. The machine chosen is the one where the sum of both completion times is the smallest. This rule is similar to NJSM, but without the assumption that job k is assigned to the same machine.

Note that extremely complex dispatching rules or machine assignment rules can be devised. However, whether these more complicated rules perform better or not is even today a matter of research. For example, Urlings et al. (2010) demonstrated that the more complex machine assignment rules are often not the best ones. The added CPU time needed for the more complex assignments did not pay off when compared to faster and simpler rules. Let us not forget that the main benefit of dispatching rules is their simplicity and speed; therefore, going in the opposite direction might cancel out these benefits.

7.4.3 Combination of Dispatching Rules

Sometimes dispatching rules are combined. For example, in hybrid environments we could have a EDD-FAM rule that would possibly solve the job scheduling and machine assignment interrelated problems. In some other scenarios, specially when the optimisation of more advanced objectives is desired, composite dispatching rules are of interest. One well-known example is the Apparent Tardiness Cost (ATC) dispatching rule, proposed by Vepsalainen and Morton (1987), also explained in detail in Pinedo (2012). This rule was proposed for the job shop with total weighted tardiness objective but has been widely applied also to the single machine total weighted tardiness problem or $1||\sum w_j T_j$. Recall that this problem was shown to be NP-hard by Du and Leung (1990). Note that in this single machine problem, some of the earlier studied dispatching rules provide excellent results. For example, the rule WSPT will be optimal if all due dates are zero or if release dates are all equal to a large value greater than $\sum_{j=1}^{n} p_j$. EDD and MS rules will also result in optimal or close to optimal schedules if the due dates are easy to satisfy. By easy we refer to well spread out due dates (not clustered around a point in time) and loose (due dates much larger than processing times). ATC tries to combine the good characteristics of both WSPT and MS. ATC is a dynamic rule that is applied each time the single machine is freed from the last assigned job. Therefore, if there are n jobs, ATC is firstly applied for the n jobs and one job, the one with the highest priority according to ATC is scheduled. Then, when the machine is free, the rule is calculated again for all remaining $n-1$ jobs, then $n-2$ and so on until only one job remains, which is directly appended to the schedule. In total, ATC is calculated $n \cdot (n+1)/2-1$ times, which results in a running time of $O(n^2)$.
The priority index is calculated as follows:

$$I_j(t) = \frac{w_j}{p_j} \cdot e^{\left(\frac{\max\{d_j-p_j-t,0\}}{K\overline{p}}\right)}$$

Note that $I_j(t)$ has two main components. The first one corresponds to the WSPT rule and the second is the MS rule. $\overline{p}$ is just the average of the processing times on the single machine or $\overline{p} = \frac{\sum_{j=1}^{n} p_j}{n}$. K is referred to as the look-ahead parameter and has to be set in a particular way. For the simple machine problems, the due date tightness factor τ and the range of due dates R are employed as follows:

$$\tau = 1 - \frac{\sum_{j=1}^{n} d_j}{n \cdot C_{\max}}$$

$$R = \frac{d_{\max} - d_{\min}}{C_{\max}}$$

where $d_{\max}$ and $d_{\min}$ are the minimum and maximum due dates, respectively. Note that in the single machine case, the makespan is trivially calculated as $C_{\max} = \sum_{j=1}^{n} p_j$.

From τ and R there are several ways of calculating K. In any case, it can be seen that the larger the K value, the lower the quotient of the exponent of the e in the $I_j(t)$ priority index, which translates in a lower participation of the 'MS part' in the ATC rule. In the extreme, if that quotient is 0, and knowing that $e^0 = 1$, the $I_j(t)$ priority index is reduced to the WSPT rule.

The main drawback of dispatching rules is their intrinsic myopic nature. Decisions are too static and too near-sighted for a good overall performance. Furthermore, dispatching rules tend to be tailored for specific objectives and favour some performance measures over others. However, they are general in the sense that they can be applied with little modifications to most machine layouts and constraints. Considering that advanced approaches are very slowly gaining terrain in real factories and are still not widespread, it is easy to understand that dispatching rules are still nowadays used intensively at companies.

7.5 Scheduling Algorithms

As already discussed in Sect. 7.2.1, it is customary to classify algorithms into exact and approximate. Regarding exact algorithms, there is a guarantee that no other schedule performs better than the one obtained with respect to the objective sought. Such guarantee does not exist for approximate algorithms and their performance is established upon experience. Nevertheless, in Sect. 7.6.2.1 we will discuss some cases where the maximum deviation of the schedule obtained by an approximate algorithm with respect to the optimal solution can be obtained.

Exact and approximate algorithms are discussed in the next subsections.

7.5.1 Exact Algorithms

Broadly speaking, there are two types of exact algorithms for scheduling problems:

- Exact constructive algorithms. These algorithms exploit some properties of the specific model in order to construct a solution which is guaranteed to be optimal. Constructive exact procedures will be discussed in detail in Sect. 8.2, but a hint on its logic is given here using the problem of scheduling a number of jobs on one machine with the goal of minimising the total completion time or flowtime. According to the notation introduced in Chap. 3, this problem can be modelled as $1|d_j|\sum C_j$. For a given sequence $\pi = (\pi_1, \ldots, \pi_n)$, the value of the objective function can be expressed as $\sum_j C_j(\pi) = n \cdot p_{\pi_1} + (n-1) \cdot p_{\pi_2} + \cdots + 2p_{\pi_{(n-1)}} + p_{\pi_n}$, where p_{π_k} is the processing time of job π_k in the sequence π. It is easy to see then (and can be formally proved) that the best value of the objective function is obtained scheduling first these jobs with lowest processing times. Therefore, it is possible to construct a solution which is optimal. Since the procedure requires to sort the jobs in ascending order of their processing times, its computational complexity is given by $O(n \log n)$.[2]
  Although such constructive procedures cover only a very limited spectrum of rather simple problems if they are computational manageable, they can be used as sub-modules of more complex approaches for the approximate solution of more complex scheduling problems. For example, if one machine or stage can be identified as a distinctive bottleneck of the real-world problem under consideration, an adequate solution procedure for a single machine might yield an initial solution for this bottleneck machine. Afterwards, this solution is complemented and possibly adjusted by a schedule for the operations before and behind this bottleneck machine/level.
- Enumerative algorithms. These algorithms implicitly or explicitly guarantee to evaluate all possible solutions of the model. It includes complete enumeration as well as branch and bound, cutting plane, or branch and cut approaches (as well as other implicit enumeration approaches, often based on some decision tree consideration)—if these procedures are executed comprehensively (which might be problematic because of time reasons). The main types of enumerative algorithms for manufacturing scheduling are discussed in Sects. 8.3.1–8.3.3.
  As already mentioned, enumeration approaches offer the comfort and safety of exact optimal solutions at the expense of computational effort that, in many cases, grows exponentially with the problem size making large problem sizes almost intractable for solution.

---

[2] The thorough reader should have recognised that such procedure is identical to the SPT rule discussed in Sect. 7.4.

7.5.2 Approximate Algorithms

Approximate methods in optimisation can be defined as procedures yielding a solution to the optimisation problem under consideration which is supposed to be of adequate or good quality. However, this assumed quality cannot be formally proven in advance but is simply based on experience. Of course, such methods will only be applied if either no exact procedure is available or the application of the latter is too expensive and/or time-consuming.

There is a wide spectrum of approximate methods for almost every NP-hard manufacturing scheduling problem. And as wide are also the ideas and concepts behind them, so even a classification of these methods can be done under different approaches. In this introduction, we will distinguish between *heuristics*, which are approximate methods specifically tailored for a particular decision problem, and more generic procedures known as *metaheuristics*. Obviously, this classification is not unambiguous, as many heuristics originally designed for a specific problem can be successfully applied outside the original scope. Additionally, we find it useful to differentiate between *constructive heuristics* and *improvement heuristics*. These types are discussed in the next sections.

7.5.2.1 Constructive Heuristics

Constructive heuristics in manufacturing scheduling generate a hopefully feasible and good sequence/schedule from scratch, i.e. exclusively based on the input data of the problem, without referring to another solution to the problem under consideration. These approaches are sometimes used as stand-alone solution approaches, i.e. their solution is implemented as is—or at least taken as the result from the formal optimisation procedure and as basis for the real-world solution to be implemented. Within more sophisticated (heuristic) optimisation procedures, constructive heuristics are often used as the starting point for improvement approaches (see next section).

Constructive approaches use rules for schedule construction which have been shown or are at least supposed to return good or acceptable schedules. This rating of their performance may simply arise from the thumb or the experience of the decision makers, may these experiences stem from other (probably similar) scheduling problems and/or from more or less sophisticated results from optimisation in general.

Many constructive approaches are of (iterative) *greedy* or *myopic* type, i.e. out of the remaining operations/tasks to be included to a so far generated partial schedule, the next operation is chosen as the one which is contributing best to the objective function under consideration. Interaction with the other previously scheduled operations is exclusively regarded with respect to the partial schedule derived so far. Interaction with other non-scheduled operations is ignored. Myopic here refers to the fact that only the very next operation to be scheduled is considered with respect to its interaction with the schedule derived so far. (Of course, the expression 'myopic' might also be used for a limited scope of the manufacturing setting considered; e.g.

the restriction of the consideration to a single machine or job. However, this refers more to the decomposition and coordination aspects mentioned in Sect. 6.3 and is basically not addressed here.) It should be mentioned that the myopia in this type of approaches is, as a start, addressing operations as scheduling objects. Since operations are combining jobs and machines/stages, the prioritisation of operations might be with respect to both jobs (on a given machine or group of machines) or machines (for given jobs or group of jobs) or even a combination of both.

The strictness of pure greedy approaches can by reduced by simultaneously considering the inclusion of more than one operation into a sequence/schedule and/or by not only appending an operation to a given partial sequence but to consider more than one position within this partial schedule for insertion of one or more new operations.

Summarising, constructive heuristics in manufacturing scheduling usually work iteratively: First, they sort the objects (operations, jobs etc.) to be scheduled statically or dynamically and, second, they choose one or more of these up to now not scheduled objects and insert them at one or several potential slots in the so far determined sequences. Third, one or more of the best insertions are kept and represent the sequence/schedule of the so far considered objects for the next iteration. This procedure is continued until all objects have been scheduled.

Finally, a short remark is given with respect to so-called ties in the iterative process of generating a solution: Both, in constructive as well as in improvement approaches referred to in the next section, it might happen that the criterion for the choice of one or more new scheduling objects to be inserted in the partial schedule derived so far does not yield a unique selection. For example, if one out of several remaining operations should be chosen to be scheduled next, the criterion might give more than one proposal for the next best insertion. (Also more than one 'best' position in the sequence could appear for only one object to be inserted.) This ambiguity is called a tie. The scheduling literature does not provide many hints on how to deal with such a situation. Apart from the simple suggestion to consider all of these alternatives separately (which might result in exponential computational effort), only few hints are given. For example, as an adaptation from multi-criteria optimisation, it is proposed to use a secondary criterion which then has to be determined carefully in addition.

7.5.2.2 Improvement Heuristics

In contrast to the constructive approaches, improvement heuristics try to derive a better solution based on one or more solutions found so far. 'Better' here in the wide majority of scientific contributions refers to the improvement of the objective function value of solutions but might also refer to the reduction of infeasibility, or both. (In many problem settings for manufacturing scheduling (in-) feasibility is not seen as the main field of study. However, especially in more complex real-world settings, reaching feasibility might be a challenge as well.)

To set up improvement approaches, a single modification rule or a set of modification rules which is intended to yield better solutions has to be provided as well as a stopping criterion to finish the improvement process. This holds for both, exact and heuristic approaches. However, the stopping criterion for exact approaches is implicitly or explicitly the achievement of the optimal solution while the stopping criterion for a heuristic improvement approach will not and cannot include this optimality condition.

A core problem in enumerative improvement of most combinatorial optimisation problems—including most in manufacturing scheduling—is their non-convexity. Consequently, especially looking for improvements in the surrounding of a given solution (so-called local search procedures) will often lead to local optima while the intended determination of global optima will not be possible by these heuristic approaches (see, e.g. Groetschel and Lovasz 1993). Here we distinguish two different approaches, which are shown below.

Limited enumeration
Limited enumeration as a heuristic for combinatorial optimisation problems here basically comprises enumerative approaches where the limitation results from a direct or indirect limitation of the number of solutions considered directly or indirectly to a number below the overall number of solutions. Consequently, these approaches do not evaluate (explicitly or implicitly) all solutions and therefore cannot guarantee the achievement of an optimal solution.

The limitation might result from:

- a direct upper bound on the number of solutions considered,
- an indirect upper bound on the number of solutions considered, e.g. given by a bound on the computation time or other resources,
- a limitation to the quality of the solution to be derived, e.g. by some upper bound on the deviation from the optimal objective function value which has to be fulfilled by the final solution from the approach (this requires an additional procedure for determination of tight upper bounds),
- another type of limitation of the number of solutions evaluated, e.g. by some 'distance' constraint which means that only solutions in the surrounding of a given solution are explored as in local search procedures.

Of course, these limitation procedures can also be applied in combination.

Local Search
Local search approaches try to explore the neighbourhood of one or more given solutions. Defining local search procedures requires:

- the definition of a neighbourhood of a solution,
- a rule for the sequence of the solutions in the neighbourhood to be explored
- a rule which of the solutions examined should be kept, i.e. for the next iteration of the neighbourhood exploration,
- a rule whether the whole neighbourhood should be explored or only a limited part of it, e.g. only until a first better solution has been detected, as well as
- a stopping criterion to finish the search procedure (see also Domschke et al. 1997).

Neighbourhoods of sequences and schedules can be determined in many different ways: Exchanging two or more consecutive or non-consecutive jobs or operations of a given solution are standard ways to define neighbourhoods of this given solution. More sophisticated and more problem-specific definitions of neighbourhoods are available for many problem settings.

Within the local search process it has to be guaranteed as far as possible that a solution already examined is not considered again (so-called cycling) to avoid superfluous computational effort.

There are two main kinds of accepting solutions from a neighbourhood. In the 'permanent improvement' approach, new solutions are only accepted if the objective function value of the new solutions is better or at least not worse than the best one so far. Obviously, this procedure might quickly run into a suboptimal solution. 'Allowance of intermediate worsening' approaches temporarily also accept solutions for further consideration which might even be (slightly) worse than the current best solution. These approaches might avoid the process to get stuck too quickly in a suboptimum and have been shown to work rather well when applied to scheduling problems or combinatorial optimisation problems in general. These latter class of approaches may be further subdivided into those with deterministic (but maybe varying, i.e. usually iteratively reduced) maximum degree of worsening or with stochastic limitation of this degree of worsening.

Local search approaches usually stop after a certain number of solutions explored, after a pre-determined time limit and/or when no improvement has been generated for a certain amount of time or a certain number of solutions explored.

7.5.2.3 Metaheuristics

Because of the non-convexity of many combinatorial optimisation problems and its consequence of easily getting stuck in local suboptima when applying constructive heuristics in combination with deterministic 'permanent improvement' neighbourhood search approaches, during the last decades several approaches have been developed which try to overcome or at least reduce this problem. These approaches—collectively known as metaheuristics—systematically try to expand the search space by allowing temporal and stochastically worsening of solutions to be considered for the next neighbourhood search and/or to adapt principles from phenomena of nature to explore also rather different solutions which are excluded by pure neighbourhood search.

Because manufacturing scheduling problems represent a prominent group within the class of combinatorial optimisation problems, scheduling problems have been intensively analysed with respect to the performance of metaheuristic approaches. The basic principles for controlling of search procedures from metaheuristics, however, are applicable for a multiplicity of problems and neighbourhood definitions. There are many classes of metaheuristics, including simulated annealing, tabu search, genetic algorithms, ant colony algorithms, particle swarm algorithms, etc. Metaheuristics for manufacturing scheduling are discussed in Sect. 9.4.

7.6 Assessing Scheduling Methods

To give some hints for a performance evaluation of a scheduling method, let us first recall the decision-making process already discussed in Chap. 2 (Sect. 1.5), and particularly Fig. 1.2 where the flow of the decision-making process was presented. Evaluating the performance of the process sketched out above may be accomplished (at least) in two ways:

- The 'formal adequacy' can be checked. Then, the performance of the method on the formal level is restricted to the link between the formal model and the formal solution. This task is exclusively dedicated to the formal sphere and it will be discussed in Sect. 7.6.2.
- The 'real-world adequacy' can be evaluated. Evaluating the contribution of the process to the solution of the real-world problem needs to have a broader perspective, since it requires to evaluate the overall process from Fig. 1.2. This type of adequacy is discussed in Sect. 7.6.1.

Evaluating real-world adequacy is usually case specific to a high degree (i.e. strongly dependent on the particular manufacturing scheduling decision problem to be addressed) and will include many soft aspects. Therefore, classifying this type of evaluation will probably remain on a rather abstract level. However, the evaluation of formal adequacy, as is or as a sub-component of real-world adequacy is rather advanced, since restricting to the formal sphere offers a perspective which can be seen more or less neutral with respect to and independent of the real-world background of the problem. In this sequel, we will address both types.

7.6.1 Real-World Adequacy

From a real-world application point of view, the overall process of manufacturing scheduling is successful if the solution finally to be implemented fulfils the requirements of the decision maker and/or of the real-world problem for what it is intended for. Therefore, the evaluation refers to the overall process from identifying the real-world problem via its simplification to a formal model (discussed in Chap. 6) and the solution of this model to the already mentioned transfer and implementation of the solution in the real-world setting. This evaluation requirement is obvious on one hand and depends in main parts more or less completely on the specific situation on the other hand.

Given the already mentioned case-specific nature of this type of adequacy, only few non case-specific aspects can be addressed. These are more or less valid for all real-world decision problems which are solved in a way as sketched out in Fig. 1.2. More specific statements could be possible by referring to the specific case under consideration.

The aspects addressed here are the following:

- Designing of the decision process on one hand and executing the process on the other may be seen separately but interrelated and maybe iteratively connected with each other. In particular, time budgets for the overall process as well as individually for the design and the execution phase of this process have to be determined and accepted.
- Related to the first aspect, as part of the decision-making process the time to set-up and to execute the process (not only of the formal part) might be decisive for the success of the decision, i.e. most probably there will be a trade-off between accuracy of the process and of the solution on one hand and the time needed to execute the process on the other hand.
- The decision maker or the real-world participants will often evaluate the quality of the solution itself—but not the decision process as a whole. Retracing their (dis-) satisfaction and deriving modifications of the process is task of those who are responsible for the design and the execution of the decision-making process itself.
- As mentioned in many references, robustness, flexibility, stability and nervousness of solutions are relevant aspects for analysing the solution quality, maybe limited to the formal sphere alone but of course more important with respect to the real-world problem and its solution. With respect to scheduling, these expressions include sensitivity of the solution to changes in face of parameter uncertainty and disruptive events. Also the consequences caused by upcoming new information over time, maybe before the plan is implemented or while it has already partly been executed, have to be anticipated in terms of robustness etc. Schedule robustness gives a measure for the effect of how much disruptions would degrade the performance of the system as it executes the schedule. Flexibility covers the ability of a system or a solution to adapt itself in the best way possible to unforeseen changes. Scheduling stability measures the number of revisions or changes that a schedule undergoes during execution, may it be caused by internal or external reasons. Scheduling nervousness (already introduced in Sect. 1.2) is defined as the amount or the share of significant changes in the execution phase of the schedule. All four expressions are closely related to each other and, depending on the problem under consideration might be precisely defined and/or measured in many different ways.
- Partly already belonging to the aspects to be discussed next with respect to formal adequacy, it is part of the analysis of real-world adequacy whether and how good the interface between the formal sphere and the real-world sphere fits to the solution of the underlying (real-world) decision problem.

7.6.2 Formal Adequacy

As already mentioned, the check of formal adequacy refers exclusively to the formal sphere of a decision problem as sketched out in Fig. 1.2. If we suppose that a given formal problem in connection with one or several algorithms yields one or more solutions, formal adequacy then refers exclusively to the evaluation of the quality of the solution(s) within this formal framework. The issue of formally evaluating algorithms becomes particularly acute in the case of approximate algorithms, since we already know that, for most scheduling decision-making problems, we will obtain models that cannot be solved in an exact manner, at least not in reasonable time.

Since some algorithms—and in particular metaheuristics—depend on some parameters, the term 'different algorithms' also include comparing two versions (i.e. with different parameters) of the same algorithm.

The evaluation should refer to (at least) two indicators, e.g. objective function value and computation time, which results in a multi-dimensional evaluation of the algorithms. Usually, there will be a trade-off between these two dimensions which has to be interpreted as in standard multi-criteria evaluations. Without loss of generality, we will usually refer to the first indicator, i.e: the objective function value.

With these two performance indicators in mind, two possibilities are usually employed for the evaluation of algorithms:

- Benchmark performance. This often means to give an upper bound for the deviation of the algorithm's solution from the optimal objective function value. For example, the algorithm can be shown to yield a solution that is at most x % worse than the optimal solution's objective function value. However, results of this type are not as often to be derived. With respect to the computation time, an upper bound for the effort related to the accomplishment of the algorithm and depending on the problem instance length (e.g. measured by the number of jobs or operations to be scheduled) might be given. This upper bound analysis for the computational effort is provided by the computational complexity, which has been addressed in Sect. 7.3.
- Comparing two or more algorithms among each other. In this case, two different approaches (which will be discussed in the next sections) can be employed:

    1. A formal proof can be given showing one algorithm to perform better than another one, always or for some characteristics of the instances of the model.
    2. A simulation study (also many times referred in the literature as to *computational experiment*) tries to derive some dominance results between the algorithms.

7.6.2.1 Formal Proofs

A formal proof could be given, e.g. for the trivial situation that an algorithm is a refinement of another one, in the sense that it examines explicitly or implicitly all solutions of the non-refined algorithm. Then, obviously, the refined algorithm will never be worse than the non-refined one. However, conclusions of above type mostly are either trivial or non-available. An alternative is to compare the performance of several algorithms by means of *worst-case analysis*.

Let us formalise the main concepts behind worst-case analysis: Assume that, for a given problem instance l, the objective function value for the algorithm under consideration is $OFV_a(l)$ and the respective objective function value of the reference algorithm is $OFV_{\text{ref}}(l)$. A worst-case analysis results usually expresses that $OFV_a(l)$ and $OFV_{\text{ref}}(l)$ (or the respective optimal objective function values) are linked by some function as

$OFV_a(l) \leq f(OFV_{\text{ref}}(l))$ or $OFV_a(l) \leq f(OFV_{\text{ref}}(l), m, n, ...)$ for instance l
or
$OFV_a^* \leq f(OFV_{\text{ref}}^*)$ or $OFV_a^* \leq f(OFV_{\text{ref}}^*, m, n, ...)$ for the respective optimal objective function values.

An example for the latter is given for the 2-machine flow shop problem with makespan objective and limited intermediate storage of b jobs between the two machines (Papadimitriou and Kanellakis 1980). These authors compare the (polynomially solvable) approach for the respective no wait problem (obviously not needing any buffer between the two machines) yielding makespan $C^*_{\max,NW}$ with the (optimal) solution to the buffer-constrained problem $C^*_{\max,b}$, i.e. they apply the generation of the no-wait solution as an approximate method to the buffer-constrained problem and use an exact enumerative approach for the buffer-constrained problem as reference approach. They prove that

$$C^*_{\max,NW} \leq \frac{2b+1}{b+1} C^*_{\max,b}$$

holds. Obviously, therefore the optimal solution to the no-wait problem, used as heuristic solution to the buffer-constrained problem, is not worse than twice the optimal solution for the buffer-constrained problem.

Although these analytical results are smart, their contribution to the evaluation of an approximate approach might be regarded restrainedly, due to the following reasons:

- The results are only valid for the worst case. To what extent this worst case is relevant for the problem solution has to be evaluated separately.
- The quality of worst-case conclusions might be somewhat disillusioning. In the above example, from a mathematical point of view, the result is definitely interesting. To what extent a limitation of the worst case deviation from the optimal solution by 100 % provides valuable information for a possibly underlying real-world situation or even only for the formal approximate solution approach via the no-wait problem might be discussed controversially.
- Analytical results of the above nature are only available for a few and rather simple problem settings.

7.6.2.2 Simulation Study

Given the limitations of formal proofs, simulation studies using a set of sample instances for the evaluation of an approximate method are the most widespread method for the evaluation of the performance of several algorithms. There is no universally accepted approach for conducting such simulation studies, but it can be summarised into two steps:

1. Testbed generation. A certain number of instances of the model are either generated specifically for the model, or adapted from those employed for other models. This set of sample instances are usually denoted as *testbed*. Note that this adaptation might take place twofold: Either the adaptation refers only to the method of generating test instances while the instances themselves are generated anew, or the instances are taken directly from the references or from some public databases.
2. Algorithms analysis. The algorithms to be compared are run with every single problem instance, and some metrics/analyses are employed to establish the relative performance of each algorithm on the testbed. If, according to the metrics/analysis, one algorithm is found to outperform the other on the testbed, then it is assumed that the latter would, in general, perform better than the former.

It is clear that the correctness of each step greatly affects to the goal, which is to establish, in general, the relative performance of different algorithms. Next we give some remarks on each one of these steps.

Regarding testbed generation, the following consideration should be followed:

- The size of the problem instances (e.g. indicated by the number of jobs and/or machines) should be in line with real-world problem dimensions and/or the dimensions of problem instances used in comparable studies. This includes also the different combinations of number of jobs and number of machines, number of operations, etc.
- The number of problem instances should be sufficient to derive useful results, i.e. this number should reflect the real-world problem setting and/or the statistical significance of the conclusions to be obtained in the next step. Note that, if the algorithms under comparison are deterministic, a single run of each algorithm on every instance will suffice in step 2. However, if the algorithms under comparison are of stochastic nature, then one option is to design a testbed large enough to average the performance of these algorithms over all the testbed.
- The main parameters of the model (such as processing times, set up times, release dates, etc.) should be chosen adequately with respect to the intended goal. For instance, highly diverse processing times (e.g. randomly drawn from a uniform distribution between 1 and 99, as is often proposed) might yield complicate problem instances and therefore might give indications for some kind of robustness of an algorithm relative to the heterogeneity of processing times. However, as compared with real-world problem settings, more or less significant correlations of processing times with respect to different jobs on a specific machine/stage and/or with respect to different operations of a specific job on different machines might be related closer to real-world applications—and might yield different conclusions with respect to the (relative) performance of the algorithms considered. Note that a hypothesis often addressed with respect to correlations in processing times is that the relative performance differences of algorithms decrease if the correlation of processing times increases. However, this is by no means a conclusion valid for

all scheduling models. A loophole from this problem with respect to real-world applications is to base the generation of test instances on observations from the specific real-world case.

- When extending/adapting an existing testbed to different models, the additional features of the model under consideration, e.g. due dates, availability assumptions, precedence constraints, have to be carefully determined and embedded into the test instances. For instance, taking literature-based standard testbeds for permutation flow shop models with makespan objective and using these instances for problems with due dates and tardiness objective requires a careful determination of the due dates. However, setting the due dates too tight on one hand, results in not reachable due dates with the consequence that approaches which intend to maximise throughput will be adequate while very loose due dates on the other hand will always enable to reach these due dates and eases the optimisation problem from the opposite side. Although this observation may seem trivial, it is sometimes ignored or at least not explicitly regarded in some references, i.e. the discussion of the adequacy of adapted testbeds is often neglected or at least not dealt with satisfactorily.

Regarding testbed analysis, once the objective function values for each algorithm on each instance have been obtained, some metrics can be derived for analysis. Although one option would be to employ the objective function values as a performance indicator, it is usually preferred to employ the *Relative Percentage Deviation* (RPD) metric for each algorithm and instance. RPD of an instance l ($l = 1, \ldots, L$) is usually defined as:

$$RPD(l) = RPD_{a,\text{ref}}(l) = \frac{OFV_a(l) - OFV_{\text{ref}}(l)}{OFV_{\text{ref}}(l)}.$$

Recall from Sect. 7.6.2.1 that $OFV_a(l)$ is the objective function value obtained by algorithm l when applied to instance l, and $OFV_{\text{ref}}(l)$ the respective objective function value of the reference algorithm.

Considering the whole set of L test instances the *Average Relative Percentage Deviation* of a algorithm l is defined as

$$ARPD = ARPD_{a,\text{ref}} = \frac{1}{L} \sum_{l=1}^{L} \frac{OFV_a(l) - OFV_{\text{ref}}(l)}{OFV_{\text{ref}}(l)}.$$

In many references, particularly older ones, the algorithms under consideration are exclusively evaluated by just comparing the *ARPD* values in a more or less ‘folkloristic’ manner, i.e. by verbally interpreting similarities and/or differences. In addition, but only sometimes standard deviation or variance values of *ARPD* are addressed in the verbal discussion of the performance of the algorithms. Note that *ARPD* refers to an average performance, while usually scheduling algorithms do not have the same ‘performance profile’ over the whole benchmark, i.e. one algorithm might be better for some type of instances than another algorithm and vice versa.

Recently, more and more statistical test procedures become standard to evaluate the relative performance of approximate methods. We explicitly approve such statistical approaches because they give at least some statistical reasoning for the relative performance of the approaches under consideration.

To apply statistical test procedures, first a hypothesis has to be formulated. Afterwards, an adequate test procedure has to be chosen and the number of problem instances for a certain level of significance of the evaluation of the hypothesis has to be determined or, vice versa, for a given number of instances a level of significance can be derived.

For small sizes of problem instances, especially with respect to the number of jobs, $OFV_{\text{ref}}(l)$ might be the value of the optimal solution (obtained e.g. by means of an enumeration approach). Then *ARPD* gives the average deviation from the optimal objective function value. For larger problem sizes, an algorithm cannot be compared to the optimal solution but only to another algorithm or to (the best value of) a set of algorithms. An alternative is to use some bound value for the optimal objective function value as a reference point.

For those instances for which the optimal solution can be determined, a hypothesis concerning the mean values may be formulated as follows:

$$\text{Hypothesis: } ARPD_{am,\text{ref}} \leq \alpha \text{ or equivalently}$$

i.e. subject to a certain level of significance, the solution value of the approximate method will not deviate more than α % from the optimal objective function value.

For larger problem sizes, a similar hypothesis can be stated as well. However, since in this case the optimal objective function value is not available for comparison, the deviation from the best approach among the ones considered can be covered by respective hypotheses.

In addition and more common for larger problem sizes, comparing two different algorithms $a1$ and $a2$, an appropriate hypothesis can be stated as

$$\begin{aligned} \text{Hypothesis: } & ARPD_{a1,\text{ref}} \leq ARPD_{a2,\text{ref}} \text{ or equivalently} \\ & DIST_{a2,a1} := ARPD_{a2,\text{ref}} - ARPD_{a1,\text{ref}} \geq 0, \end{aligned}$$

The reference then can be defined as the best solution of $a1$ and $a2$ for every instance, i.e. $OFV_{\text{ref}} = \min(OFV_{a1}, OFV_{a2})$. The confirmation of this hypothesis (or the rejection of its opposite) would state that $a1$ performs better than $a2$, relative to the reference approach and subject to a certain level of significance. (Remark: Sometimes choosing random solutions as one approximate method and doing a respective statistical comparison analysis proves or refutes the assertion that a proposed method is at least competitive with a random choice of a solution. However, in most—but not all!—cases of methods proposed, the result of this comparison of hypotheses should be obvious.)

Having precisely defined the hypothesis to be tested, the toolbox of statistical test procedures yields adequate approaches to decide whether to accept or to reject the hypothesis. We will not describe these test procedures in detail but refer the reader to numerous textbooks on statistical testing (see Sect. 7.7 on further readings).

Instead, we give some hints for the application of test procedures when evaluating (approximate) methods in manufacturing scheduling.

If simply one approximate method is compared with another one, paired t-test procedures can be applied. t-test approaches can be used if $DIST_{a2,a1}$ follows a normal distribution. This requirement is often supposed to be fulfilled without explicit proof. If normal distribution of $DIST_{a2,a1}$ is questionable, an approximate Gauss test approach (parametric test) or Wilcoxon's signed ranks test (non-parametric test) are more appropriate since they do not require any assumption with respect to the type of the distribution of $DIST_{a2,a1}$.

If more than two heuristics are to be compared with respect to their performance, ANalysis Of VAriance (ANOVA) can be used to accept or to reject the hypothesis that all algorithms l under consideration perform equally with respect to their $ARPD_{a,l,\text{ref}}$-values. If this hypothesis is rejected (what is desired to determine a best method or at least significant differences between (some of) the methods under consideration) it has to be regarded that the identification of better or best approaches among the approaches under consideration is based on the (pairwise) comparison of these approaches on the same testbed. This, however, requires statistical corrections when applying test procedures to more than one pair of approaches. Among others, the most popular of these correction approaches are by Tukey, Tukey-Kramer, Bonferroni and Holm-Bonferroni. Again, we will not discuss this aspect in detail, but just mention that the correction procedure refers to some adjusted interpretation of the p-values when determining the level of significance of the evaluation of the respective hypotheses.

Additionally, we point out that applying ANOVA demands for the fulfilment of three requirements, i.e.

1. Randomness and independence
2. Normality: sample values are from a normal distribution
3. Homogeneity of variance/homoscedasticity: variances of every indicator are the same.

Before applying ANOVA to hypotheses testing, these requirements have to be checked. If these requirements are not fulfilled, the Kruskal-Wallis rank test might be applied.

Finally, we point out that most of the statistical test procedures mentioned here and the respective discussions concerning the relative performance of the approximate scheduling approaches explicitly or implicitly refer to the so-called α error (type 1 error), i.e. how large is the risk to reject a hypothesis although it is true. The β error (type 2 error), i.e. the risk to accept a hypothesis although it is false, is usually not considered when statistically evaluating the performance of scheduling approaches.

Concluding this section, we would like to point out that also other more or less different statistical approaches can yield additional insight in the performance of approximate methods. For example, referring to the complete distribution of objective function values for a problem and comparing this distribution with the solutions generated by an approximate method may yield additional insight into the approximate method's performance.

7.6.2.3 Further Remarks on the Evaluation of Algorithms

Summarising the previous section, partly also from a real-world perspective, we point to the following aspects:

1. We recall once again that a properly defined and executed statistical analysis of the performance of algorithms for manufacturing scheduling models should be standard when evaluating these methods. However, we have the impression that this has not been accomplished so far.
2. Another aspect, also relevant from a real-world application point of view is the (non-) necessity of a certain level of solution quality. Especially because of arguments referring to the uncertainty of many problem data, a medium-size (whatever that means) deviation from a best solution might be accepted within the formal problem since the real-world implementation will differ from the solution derived for the formal problem more or less significantly anyway. Therefore, pre-determination of non-zero levels of acceptance (with respect to optimisation error and/or degree of infeasibility) might be a setting even in the formal sphere of the optimisation process. However, an inevitable but also rather trivial requirement for the solution of the formal problem is its ability to be transformed to an adequate solution of the real-world problem or at least to deliver some valuable additional information to such a solution.
3. For an externally (by the decision maker) given level of significance for the acceptance or the rejection of a hypothesis of the above type, the test procedures will give a minimum number of problem instances to be generated and calculated (or vice versa: for a given number of instances a level of significance can be derived).
4. If several (i.e. more than one) hypotheses shall be accepted or rejected using the same testbed, i.e. more than two approaches are to be compared, the respective comparisons are not statistically independent anymore. Therefore, in this case either the level of significance decreases (if the number of instances is not increased) or the number of instances has to be increased (to maintain a pre-specified level of significance). In statistical analyses of the performance of approximate methods for manufacturing scheduling (where often more than two methods are compared), this aspect is ignored rather often.
5. If a thorough statistical analysis of the performance of scheduling approaches is performed, both types of errors in statistical test theory, i.e. α error (type 1 error) as well as β error (type 2 error), should be addressed.
6. Finally, not only the quality of a given algorithm for a given formal manufacturing scheduling problem has to be considered. Also the effort for the process of setting up the model itself as well as of the solution algorithm might be limited.

7.7 Conclusions and Further Readings

This chapter has presented a brief overview on many possible scheduling methods. We have initially discussed the complexity of scheduling problems from a simple approach avoiding as much as possible the technicalities of computational complexity. The conclusion from this is that most scheduling problems are computationally complex and have an astronomical number of possible solutions. We have also pointed out the relative importance of obtaining the exact optimum solution for a scheduling problem, given that what we are actually solving is a model from the studied reality and the approximations and simplifications made to that model might perfectly offset the gains of a precisely obtained optimal solution.

The main scheduling methods have been introduced in this chapter as well. Each method has its own advantages and disadvantages. In a nutshell, dispatching rules are easy to understand, to implement and to calculate. However, their results are often poor except for some simplistic problems and are easily surpassed by most scheduling algorithms. Exact approaches provide the optimum solution but are only viable for simple problems and for small instances as well. Unless some very specific problem is being studied for which a very effective exact optimisation approach exists, they are not advisable to be used in complex production shops where hundreds, potentially thousands of jobs are to be scheduled. Heuristics are often tailored for some specific problems and objectives but they usually provide much better solutions when compared to dispatching rules. With today's computing power, heuristics are practically as fast as dispatching rules.

The final part of the chapter is devoted to discussing how to assess scheduling methods, both from formal and real-world points of view. From a formal viewpoint, the predominant approach for approximate methods are the simulation studies, for which no standard procedure yet exists, although we give some insight into the basics of these analyses.

Interested readers wishing to delve into the world of computational complexity should start with the books of Garey and Johnson (1979), Papadimitriou (1979) or Arora and Barak (2009), just among the many more specialised texts. Studies about computational complexity for scheduling problems appeared in Garey et al. (1976), Rinnooy Kan (1976), Błazewicz et al. (1983), Lawler et al. (1993) and many others. The book of Brucker (2007) and the one of Brucker and Knust (2006) contain a large deal of complexity results for scheduling problems. The same authors run a website with complexity results for scheduling problems at http://www.mathematik.uni-osnabrueck.de/research/OR/class/. This website is very detailed and contains lots of up-to-date information.

Many dispatching rules and heuristics are covered by the works of Panwalkar and Iskander (1977), Blackstone et al. (1982), Haupt (1998), Rajendran and Holthaus (1999) or Jayamohan and Rajendran (2000). Some advanced dispatching rules (including many global machine assignment rules) can be found in Urlings et al. (2010).

A classical text for heuristics is the book of Morton and Pentico (1993). For a much more detailed overview on exact algorithms, dispatching rules and heuristic algorithms, the reader may consult the general books on scheduling: Conway et al. (1967), Baker (1974), French (1982), Błazewicz et al. (2002), Brucker (2007), Pinedo (2012), Baker and Trietsch (2009) and Pinedo (2009). Regarding the discussion of the simulation studies, the reader might want to have a look at Bamberg and Baur (1998), Berenson et al. (2006), Montgomery (2012), particularly for covering the technical/statistical issues. Finally, an interesting simulation study regarding correlations in processing times and the relative performance of algorithms is Watson et al. (2002).

References

Arora, S. and Barak, B. (2009). *Computational Complexity: A Modern Approach.* Cambridge University Press, Cambridge.

Baker, K. R. (1974). *Introduction to Sequencing and Scheduling.* John Wiley & Sons, New York.

Baker, K. R. and Trietsch, D. (2009). *Principles of Sequencing and Scheduling.* Wiley, New York.

Bamberg, G. and Baur, F. (1998). *Statistik.* Oldenbourg, Muenchen, Wien.

Berenson, M. L., Levine, D. M., and Krehbiel, T. C. (2006). *Basic business statistics: Concepts and applications.* Pearson Education, Upper Saddle River, NJ.

Blackstone, Jr, J. H., Phillips, D. T., and Hogg, G. L. (1982). A state-of-the-art survey of dispatching rules for manufactuing job shop operations. *International Journal of Production Research,* 20(1):27–45.

Błazewicz, J., Ecker, K. H., Pesch, E., Schmidt, G., and Węglarz, J. (2002). *Scheduling Computer and Manufacturing Processes.* Springer-Verlag, Berlin, second edition.

Błazewicz, J., Lenstra, J. K., and Rinnooy Kan, A. H. G. (1983). Scheduling Subject to Constraints: Classification and Complexity. *Discrete Applied Mathematics,* 5:11–24.

Brucker, P. (2007). *Scheduling Algorithms.* Springer, New York, fifth edition.

Brucker, P. and Knust, S., editors (2006). *Complex Scheduling.* Springer-Verlag, Berlin.

Conway, R. W., Maxwell, W. L., and Miller, L. W. (1967). *Theory of Scheduling.* Dover Publications, New York. Unabridged publication from the 1967 original edition published by Addison-Wesley.

Domschke, W., Scholl, A., and Voss, S. (1997). *Produktionsplanung: Ablauforganisatorische Aspekte.* Springer, Berlin. 2nd, revised and upgraded edition.

Du, J. and Leung, J. Y. T. (1990). Minimising total tardiness on one machine is NP-hard. *Mathematics of Operations Research,* 15(3):483–495.

French, S. (1982). *Sequencing and Scheduling: An Introduction to the Mathematics of the Job-Shop.* Ellis Horwood Limited, Chichester.

Garey, M. R. and Johnson, D. S. (1979). *Computers and Intractability. A Guide to the Theory of NP-Completeness.* W. H. Freeman and Company, New York.

Garey, M. R., Johnson, D. S., and Sethi, R. (1976). The Complexity of Flowshop and Jobshop Scheduling. *Mathematics of Operations Research,* 1(2):117–129.

Groetschel, M. and Lovasz, L. (1993). *Geometric algorithms and combinatorial optimization.* Springer, Berlin [a.o.]. 2th, corrected edition.

Haupt, R. (1989). A survey or priority rule-based scheduling. *OR Spectrum,* 11(1):3–16.

Jayamohan, M. S. and Rajendran, C. (2000). New dispatching rules for shop scheduling: a step forward. *International Journal of Production Research,* 38(3):563–586.

Lawler, E. L., Lenstra, J. K., and Rinnooy Kan, A. H. G. (1993). Sequencing and Scheduling: Algorithms and Complexity. In Graves, S. C., Rinnooy Kan, A. H. G., and Zipkin, P. H., editors,

Logistics of Production and Inventory, volume 4 of Handbooks in Operations Research and Management Science, Amsterdam. Elsevier Science Publishers, B. V.
Montgomery, D. C. (2012). *Design and Analysis of Experiments.* Wiley; 8 edition.
Morton, T. E. and Pentico, D. W. (1993). *Heuristic Scheduling Systems With Applications to Production Systems and Project Management.* Wiley Series in Engineering & Technology Management. John Wiley & Sons, Hoboken.
Panwalkar, S. and Iskander, W. (1977). A survey of scheduling rules. *Operations Research,* 25(1):47–55.
Papadimitriou, C. H. (1993). *Computational Complexity.* Addison-Wesley, Reading.
Papadimitriou, C. H. and Kanellakis, P. C. (1980). Flowshop scheduling with limited temporary-storage. *Journal of the ACM,* 27(3):533–549.
Pinedo, M. (2009). *Planning and Scheduling in Manufacturing and Services.* Springer, New York, second edition.
Pinedo, M. L. (2012). *Scheduling: Theory, Algorithms, and Systems.* Springer, New York, fourth edition.
Rajendran, C. and Holthaus, O. (1999). A comparative study of dispatching rules in dynamic flowshops and jobshops. *European Journal of Operational Research,* 116(1):156–170.
Rinnooy Kan, A. H. G. (1976). *Machine Scheduling Problems: Classification, Complexity and Computations.* Martinus Nijhoff, The Hague.
Urlings, T., Ruiz, R., and Sivrikaya-Şerifoğlu, F. (2010). Genetic algorithms with different representation schemes for complex hybrid flexible flow line problems. *International Journal of Metaheuristics,* 1(1):30–54.
Vepsalainen, A. P. J. and Morton, T. E. (1987). Priority rules and lead time estimation for job shop scheduling with weighted tardiness costs. *Management Science,* 33(8):1036–1047.
Watson, J.-P., Barbulescu, L., Whitley, L., and Howe, A. (2002). Contrasting structured and random permutation flow-shop scheduling problems: Search-space topology and algorithm performance. *INFORMS Journal on Computing,* 14:98–123.

Chapter 8
Exact Algorithms

8.1 Introduction

In the previous chapter we discussed that most scheduling models are hard to be solved optimally and that there is little hope in expecting the optimal solution for complex problems arising in real production settings. However, some simple problems might arise as subproblems or surrogate settings in complex shops. Additionally, exact algorithms help in closely and deeply understanding scheduling models and therefore, they have an intrinsic utility. Moreover, some exact algorithms do exist for some special cases of scheduling problems and are worth of study. Finally, we have already mentioned that some of the exact procedures can be employed as approximate methods (e.g. by terminating them before optimality is reached), and can serve to inspire the construction of approximate methods for more complex problems.

After the introduction to exact methods for scheduling models carried out in Sect. 7.5.1, this chapter deepens in these methods, and we discuss both exact and constructive algorithms, and some widely employed exact algorithms, i.e. branch and bound approaches as well as dynamic programming procedures.

More specifically, in this chapter we

- discuss the most important exact constructive algorithms and their limitations (Sect. 8.2),
- present exact algorithms based on integer programming (Sect. 8.3.1) and on problem-specific branch and bound approaches (Sect. 8.3.2) and
- introduce the main concepts of dynamic programming employed in manufacturing scheduling (Sect. 8.3.3).

J. M. Framinan et al., *Manufacturing Scheduling Systems*,
DOI: 10.1007/978-1-4471-6272-8_8, © Springer-Verlag London 2014

8.2 Exact Constructive Algorithms

Recall from Sect. 7.5.1 that exact constructive algorithms attempt to exploit specific properties of the scheduling model in order to construct a solution which is guaranteed to be optimal.

There are some cases for which finding exact constructive algorithms is, if not trivial, rather straightforward. For instance, for the single-machine model with makespan objective (model $1||C_{\max}$) every schedule yields the same makespan if the operations are assigned left shifted to the timescale. Therefore every solution is optimal. For the 1-machine model with total completion time objective (model $1||\sum C_j$), we have already discussed in Sect. 7.5.1 that sorting the jobs according to the shortest processing time-rule (*SPT*-rule) yields the optimal solution. This sorting can be performed with computational effort which is limited by $O(n \cdot \log n)$. Similarly, the 1-machine problem with due dates and the objective of minimising maximum lateness (model $1||\max L_j$) can be solved easily by sorting the jobs according to the earliest due date rule (*EDD*-rule).

For other models, deriving exact constructive algorithms is not trivial or straightforward. Indeed, apart from some single machine layouts and some rather simplistic two-machine settings, there are not many other known exact algorithms. However, exact constructive algorithms may play an important role in the development of approximate algorithms, as we will discuss in Chap. 9. Therefore, it is worth studying in the next subsections some of the most famous.

8.2.1 Johnson's Algorithm

Probably the most famous case of an exact constructive algorithm is the well-known *Johnson's rule* (Johnson 1954) for solving the 2-machine permutation flow shop model with makespan criterion or $F2|prmu|C_{\max}$. Note than in Conway et al. (1967), as well as in many other texts, it is actually demonstrated that the permutation schedules suffice for obtaining the optimum makespan in 2-machine flow shops. Johnson's rule determines that a given job j should precede in the sequence another job k if the following expression is satisfied:

$$\min\{p_{1j}, p_{2k}\} < \min\{p_{1k}, p_{2j}\}$$

In this way, jobs with a short processing time on the first machine go first in the sequence (this in turn minimises the idle time in the second machine). Jobs with a short processing time in the second machine are processed last also to avoid idle times in the second machine. Note that in a 2-machine flow shop without constraints, there are no idle times on the first machine. Therefore, minimising makespan is equivalent to minimising idle times on the second machine. Johnson's algorithm is further described Algorithm 1.

Table 8.1 Processing times for a six job and two machine flow shop example to be solved with Johnson's rule

Machine (i)	Job (j)					
	1	2	3	4	5	6
1	8	4	10	5	3	12
2	12	1	6	5	2	8

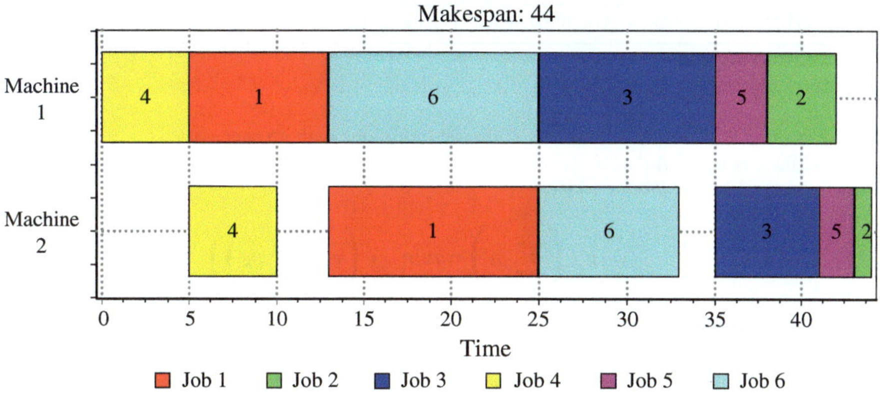

Fig. 8.1 Gantt chart with the resulting sequence in the Johnson's rule example

Algorithm 1: Johnson's Algorithm

Input: Instance data
Output: Optimal sequence Π
begin
  Let $\Pi = \varnothing$, $J = \{1, \dots, n\}$;
  Let $J_1 = \{j \in J / p_{1,j} \leq p_{2,j}\}$;
  Let $J_2 = \{j \in J / p_{1,j} > p_{2,j}\}$;
  Construct Π_1 sorting jobs in J_1 by increasing values of $p_{1,j}$;
  Construct Π_2 sorting jobs in J_2 by decreasing values of $p_{2,j}$;
  $\Pi = (\Pi_1 | \Pi_2)$;
  return Π
end

Let us apply Johnson's algorithm to an instance comprised of six jobs whose processing times on the two machines are given in Table 8.1.

According to Johnson's rule, $J_1 = 1, 4$ and $J_2 = 2, 3, 5, 6$. Consequently, $\Pi_1 = (4, 1)$ and $\Pi_2 = (6, 3, 5, 2)$. Therefore, $\Pi = (4, 1, 6, 3, 5, 2)$ with $C^*_{max} = 44$. This optimal schedule is represented by a Gantt chart in Fig. 8.1.

Johnson's rule can be implemented with a computational complexity of $O(n \log n)$. This means that even for thousands of jobs, Johnson's rule can obtain the optimum solution in a very short amount of time for 2-machine flow shop layouts and makespan objective. Most regrettably though, adding almost any constraint and/or changing the objective or increasing the number of machines, results in Johnson's rule is not being able to return the optimum solution.

8.2.2 *Lawler's Algorithm*

This algorithm was shown optimal by Lawler (1973) for the model $1|prec|\gamma$, being γ any regular objective function with max-form. Let $\max g_j$ be a regular max-form function. The algorithm is shown in Algorithm 2.

Algorithm 2: Lawler's Algorithm for $1|prec|\max g_j$ Lawler (1973)

Input: Instance data
Output: Optimal sequence Π
begin
  Let $\Pi = \varnothing$, $J = \{1, \ldots, n\}$ and J' the set of all jobs with no successors.;
  while J *is* not $\varnothing$ **do**
    Let j^* be the job for which

$$g_{j^*}\left(\sum_{k\in J} p_k\right) = min_{j\in J'}\left(g_j\left(\sum_{k\in J} p_k\right)\right)$$

    Add j^* to Π;
    Delete j^* from J;
    Modify J' to represent the new set of schedulable jobs;
  return Π
end

8.2.3 *Moore's Algorithm*

The problem $1||\sum U_j$ is solved optimally by the algorithm presented in Algorithm 3 by Moore (1968).

Algorithm 3: Moore's Algorithm for $1||\sum U_j$ Moore (1968)

Input: Instance data
Output: Optimal sequence Π
begin
  Let $\Pi = \varnothing$, $J = \{1, \ldots, n\}$ verifying $d_1 \leq d_2 \leq \ldots d_n$;
  Let $\Pi = (1, \ldots, n)$;
  Let $Tardy = \varnothing$;
  while $\exists l \in \Pi$ *such as* $C_l > d_l$ **do**
   L L
  et k be such that $C_k > d_k$ and $\forall i < k,\ C_i \leq d_i$;
  Let j be such that $j \leq k$ and $p_j = \max 1 \leq i \leq k p_i$;
  $\Pi = \Pi - \{j\}$; $Tardy = Tardy \cup \{j\}$; **return** Π
end

8.3 Enumerative Algorithms

Enumerative algorithms implicitly or explicitly guarantee to evaluate all possible solutions of the model. These algorithms include a large number of different approaches, and discussing all of them will be outside the scope of this book. Instead we will focus in the next subsections on the three types of solution methods that are more popular in manufacturing scheduling namely Integer Programming (Sect. 8.3.1), Branch and Bound (Sect. 8.3.2), and Dynamic Programming (Sect. 8.3.3). Note that, from a pure technical viewpoint, Integer Programming could be seen as a particular type of (generic) branch and bound. However, given the specific characteristics of Integer Programming, we prefer to discuss it in a separate section.

8.3.1 Integer Programming (IP)

Many manufacturing scheduling models can be formulated as integer or binary (linear or nonlinear) programs, therefore the whole variety of branch and bound methods in binary or integer programming can be applied. This method would consist basically of two steps:

1. Formulation of the manufacturing scheduling model under consideration in terms of MILP (Mixed Integer Linear Programming). This formulation will usually either directly derive a solution or refer to some standard intermediate model which can be transferred to an integer program in a standard way, e.g. by using models from graph theory (see, e.g. Pinedo 2012; Tseng et al. 2004 for examples of such formulations).
2. Applying IP software to the MILP model generated in the previous step. Here, smart parameter adjustments within the IP software, e.g. error bounds, branching strategies, etc., have to be executed. Examples of software dealing with MILP models are the solver of Microsoft Excel, LINGO from Lindo Systems (http://www.lindo.com/), CPLEX from IBM (http://www-01.ibm.com/software/integration/optimisation/cplex/) and many others, including the Gurobi solver (http://www.gurobi.com/).

Since the last step is more or less exclusively referring to the standardised formal problem and to the software used, we will not refer to this here. Instead, we will sketch out some model formulations which cover the first step. More specifically, we will present in the next sections MILP formulation of several models in order to illustrate how to capture the processing constraints and formulate the objectives. It is to note that a smart formulation of the MILP model might reduce the computational effort significantly.

8.3.1.1 Parallel Machine Scheduling Models with Makespan Objective

In this section we consider two parallel machines scheduling models. Recall from Sect. 3.2.3.2 that in this layout there is a set N of n independent jobs that have to be processed on a set M of m machines arranged in parallel. The following (rather standard) assumptions are adopted:

- Each job j, $j = 1, \ldots, n$ has to be processed by exactly one out of the m parallel machines.
- No machine can process more than one job at the same time. Furthermore, once the processing of a job by a given machine has started, it has to continue until completion (e.g. no preemption is allowed).
- The processing time of a job is a known, finite and fixed positive number. In general, machine i ($i = 1, \ldots, m$) has a different speed p_{ij} when processing each job j, i.e. machines i will be occupied by p_{ij} units of time when processing job j.

For now, we will consider that there are no further restrictions (later we will introduce sequence-dependent set-up times in this model), so this model can be denoted as $R||C_{\max}$. Note that this is, in reality, an assignment problem since the processing order of the jobs assigned to a given machine do not alter the maximum completion time at that machine. Therefore, there are m^n possible solutions to the problem after all possible assignments. The $R||C_{\max}$ has been shown to be NP-Hard, since the special case with identical machines (referred to as $P||C_{\max}$) was already demonstrated by Garey and Johnson (1979) to belong to that class. Even the 2-machine version ($P2||C_{\max}$) is already NP-Hard according to Lenstra et al. (1977).

As we can see, even the two identical parallel machines case is a hard problem to solve. Furthermore, Lenstra et al. (1990) showed that no polynomial time algorithm exists for the general $R||C_{\max}$ problem with a better worst-case ratio approximation than 3/2.

A MILP model for $R||C_{\max}$ would contain the following variables:

$$x_{ij} = \begin{cases} 1, & \text{if job } j \text{ is assigned to machine } i \\ 0, & \text{otherwise} \end{cases}$$

$$C_{\max} = \text{Maximum completion time or makespan}$$

The objective function is the makespan minimisation:

$$\min \; C_{\max} \tag{8.1}$$

And the constraints:

$$\sum_{i=1}^{m} x_{ij} = 1 \quad \forall j \in N \tag{8.2}$$

$$\sum_{j=1}^{n} p_{ij} \cdot x_{ij} \leq C_{\max} \quad \forall i \in M \tag{8.3}$$

$$x_{ij} \in \{0, 1\} \quad \forall j \in N, \forall i \in M \tag{8.4}$$

Note that the constraint of type (8.2) ensures that each job is assigned to exactly one machine. The second set of constraints just states that the makespan is equal or greater than the sum of the processing times of the jobs assigned to each machine. This is repeated for all machines since in this problem, the makespan is given by the machine that finishes all its assigned jobs at the latest time. The last set of constraints just defines the nature of the variables.

The previous model was thoroughly tested by Fanjul-Peyró and Ruiz (2010) using IBM CPLEX 11.0 and the results were very encouraging. From a complete set of 1,400 instances, and with a maximum CPU time limit of 2 h, CPLEX 11.0 was able to solve instances of up to 1000 jobs and 50 machines in many cases, with optimality rates reaching 88 % in some cases and maximum-observed optimality gaps for the unsolved instances of 8.66 % and average-observed gaps usually below 2 %. As one can see, although the MILP model does not result in optimal solutions in all cases, the results are very good in most scenarios. This enforces the already commented idea that although most scheduling problems are indeed NP-hard, sometimes usable solutions for sizeable instances are possible.

Let us add sequence-dependent set-up times to the previous setting. As the parallel machines are unrelated, we assume that there is a set-up time matrix per machine. Therefore, S_{ijk} denotes the machine-based sequence-dependent set-up time on machine $i, i \in M$, when processing job $k, k \in N$, after having processed job $j, j \in N$.

The so-resulting model can be denoted as $R|S_{sd}|C_{\max}$. One could think that adding sequence-dependent setup times to the problem is not a big deal. However, the implications of the setups are far reaching as far as the model goes. Notice that when adding setups, this is no longer just an assignment problem. Depending on how the different jobs are sequenced among those assigned to a machine, the completion time of that machine will differ. As a result, the sequence has to be determined by the model as well, not just the assignment. The new MILP model involves the following decision variables:

$$X_{ijk} = \begin{cases} 1, & \text{if job } j \text{ precedes job } k \text{ on machine } i \\ 0, & \text{otherwise} \end{cases}$$

$$C_{ij} = \text{Completion time of job } j \text{ at machine } i$$

$$C_{\max} = \text{Maximum completion time}$$

As one can see, there are $n \cdot (n-1) \cdot m$ binary variables, which quickly grow to unacceptable numbers in medium-sized problems. Note that X_{ijj} is not defined. However, the total number grows because some dummy jobs are needed as well. Below are the details.

The objective function is:

$$\min C_{\max} \tag{8.5}$$

And the constraints are:

$$\sum_{i\in M}\sum_{\substack{j\in\{0\}\cup\{N\}\\ j\neq k}} X_{ijk} = 1, \qquad \forall k \in N \tag{8.6}$$

$$\sum_{i\in M}\sum_{\substack{k\in N\\ j\neq k}} X_{ijk} \leq 1, \qquad \forall j \in N \tag{8.7}$$

$$\sum_{k\in N} X_{i0k} \leq 1, \qquad \forall i \in M \tag{8.8}$$

$$\sum_{\substack{h\in\{0\}\cup\{N\}\\ h\neq k, h\neq j}} X_{ihj} \geq X_{ijk}, \qquad \forall j,k \in N, j\neq k, \forall i \in M \tag{8.9}$$

$$C_{ik} + V(1 - X_{ijk}) \geq C_{ij} + S_{ijk} + p_{ik}, \quad \forall j \in \{0\}\cup\{N\}, \forall k \in N, j\neq k, \forall i \in M \tag{8.10}$$

$$C_{i0} = 0, \qquad \forall i \in M \tag{8.11}$$

$$C_{ij} \geq 0, \qquad \forall j \in N,\ \forall i \in M \tag{8.12}$$

$$C_{ij} \leq C_{\max}, \qquad j \in N,\ i \in M \tag{8.13}$$

$$X_{ijk} \in \{0,1\}, \qquad \forall j \in \{0\}\cup\{N\}, \forall k \in N, j\neq k, \forall i \in M \tag{8.14}$$

Constraint set (8.6) ensures that every job is assigned to exactly one machine and has exactly one predecessor. Notice the usage of dummy jobs 0 as $X_{i0k}, i \in M, k \in N$. With constraint set (8.7) we set the number of successors of every job to a maximum of one (a job could be the last one on a machine). Set (8.8) limits the number of successors of the dummy jobs to a maximum of one on each machine. With set (8.9), we ensure that jobs are properly linked on their machine. if a given job j is processed on a given machine i, a predecessor h must exist on the same machine. Constraint set (8.10) is to control the completion times of the jobs at the machines. Basically, if a job k is assigned to machine i after job j (i.e. $X_{ijk} = 1$), its completion time C_{ik} must be greater than the completion time of j, C_{ij} plus the setup time between j and k and the processing time of k. If $X_{ijk} = 0$, then the big constant V renders the constraint redundant. Sets (8.11) and (8.12) define completion times as 0 for dummy jobs and non-negative for regular jobs, respectively. Set (8.13) defines the makespan. Finally, set (8.14) defines the binary variables.

We can see how the sequence-dependent set-up times version of the unrelated parallel machines problem is significantly larger and more complex. As a matter of fact, and as shown in Vallada and Ruiz (2010), this model is quite unsolvable. The previous model, when tested also with CPLEX 11.0 and under comparable settings, resulted in optimum solutions for problems of up to 8 jobs and 5 machines in all cases. However, only some problems of 10 jobs and 12 jobs could be solved to optimality. We can see that these are extremely small problems.

8.3.1.2 Flow shop Scheduling Models with Makespan Objective

Several MILP formulations of flow shop or job shop problems (especially with makespan objective) are due to Liao and You (1992); Manne (1960); Wagner (1959). In this section, we will sketch out three different MILP formulations for elementary models for the permutation flow shop with makespan objective, i.e. the models by Manne (1960); Wagner (1959) and by Wilson (1989). We will then analyse the different formulations in order to highlight their implications in terms of their applicability beyond small-sized problems.

Manne's model (adapted for permutation flow shop problems as in Tseng et al. (2004):

The original version of the model by Manne is suited to solve the general job shop problem. However, for reasons of simplification and comparability to other approaches, we present the model in its permutation flow shop version which can be found (Table 8.2), e.g. in Tseng et al. (2004).

Suppose a standard permutation flow shop problem setting with makespan objective. The model is formulated as follows:

Table 8.2 Manne model: Notation

Indices	
i	Machine index, $i = 1, \dots, m$
j	Job index, $j = 1, \dots, n$
Variables	
$D_{j,j'}$	Binary variable with $D_{j,j'} = 1$ if job j is scheduled before (not necessarily immediately before) job j', $D_{j,j'} = 0$ otherwise, $j < j'$
C_{ij}	Completion time of job's j operation on machine i
$C_{\max}$	Finishing time of the last operation on machine m
Parameters	
p_{ij}	Processing time of job j on machine i
M	Large number

Model:

$$\min \quad C_{\max} \tag{8.15}$$

s. t.

$$C_{1j} \geq p_{1j} \qquad j = 1, \ldots, n \tag{8.16a}$$

$$C_{ij} - C_{i-1,j} \geq p_{ij} \qquad i = 2, \ldots, m; \;\; j = 1, \ldots, n \tag{8.16b}$$

$$C_{ij} - C_{ij'} + M \cdot D_{j,j'} \geq p_{ij} \qquad i = 1, \ldots, m; \;\; j, j' = 1, \ldots, n, \;\; j < j' \tag{8.17a}$$

$$C_{ij} - C_{ij'} + M(D_{j,j'} - 1) \leq -p_{ij'} \qquad i = 1, \ldots, m; \;\; j, j' = 1, \ldots, n, \;\; j < j' \tag{8.17b}$$

$$C_{\max} \geq C_{mj} \qquad j = 1, \ldots, n \tag{8.18}$$

This model is more or less straight forward. Equation (8.15) represents the objective function of makespan minimisation. Equations (8.16a, 8.16b) give the job availability constraints, i.e. they guarantee that an operation (i, j) of a job j on machine i cannot be finished before its preceding operation $(i - 1, j)$ of the same job j on the preceding machine $i - 1$ is finished (no such operation on machine 1 in (8.16a)) and its processing time p_{ij} has passed. Equations (8.17a, 8.17b) give the machine availability constraints. In case of job j precedes job j' in a solution/permutation, constraint (8.17a) becomes a 'true' constraint while if job j' precedes job j, constraint (8.17b) is the relevant constraint while the respective other constraint becomes irrelevant in the respective situation. Constraints (8.18) give the lower limit of the makespan as the completion time of every single job j.

This model includes $0.5n(n - 1)$ binary variables $D_{j,j'}$ and $n \cdot m$ continuous variables C_{ij} plus one continuous variable $C_{\max}$, i.e. in total $n \cdot m + 1$ continuous variables. The number of constraints is nm (8.16a, 8.16b) plus $2mn(n - 1)/2$ (8.17a, 8.17b) plus n (8.18), i.e. in total $n(mn + 1)$ constraints.

One characteristic of this adapted Manne model is that it basically only implicitly constructs the permutation by the jobs' precedence relation imposed by the binary variables $D_{j,j'}$. In contrast, the following models of Wilson and Wagner explicitly define permutations as solutions by an assignment problem structure.

Wilson model:

By including an assignment problem the Wilson model avoids the dichotomic constraints of the Manne model which cover both cases of job precedence for all pairs of jobs, i.e. job j_1 preceding job j_2 and vice versa (Table 8.3).

Model:

$$\min \quad C_{\max} = S_{m[n]} + \sum_{j=1}^{n} p_{mj} \cdot Z_{j[n]} \tag{8.19}$$

s. t.

$$\sum_{j=1}^{n} Z_{j[j']} = 1 \qquad j' = 1, \ldots, n \tag{8.20a}$$

Table 8.3 Wilson model: Notation

Indices	
i	Machine index, $i = 1, \ldots, m$
j	Job index, $j = 1, \ldots, n$
$[j']$	Position index, $j' = 1, \ldots, n$. This index indicates the j'-th position of a job permutation
Variables	
$Z_{j,[j']}$	Binary variable with $Z_{j,[j']} = 1$ if job j is assigned to position j' in the sequence/permutation, $Z_{j,[j']} = 0$ otherwise
$S_{i,[j']}$	Starting time of the operation of the job in position $[j']$ on machine i
$C_{\max}$	Finishing time of the last operation on machine m
Parameters	
p_{ij}	Processing time of job j on machine i

$$\sum_{j'=1}^{n} Z_{j[j']} = 1 \qquad j = 1, \ldots, n \tag{8.20b}$$

$$S_{1[1]} = 0 \tag{8.21a}$$

$$S_{1[j']} + \sum_{j=1}^{n} p_{1j} Z_{j[j']} = S_{1,[j'+1]} \qquad j' = 1, \ldots, n-1 \tag{8.21b}$$

$$S_{i[1]} + \sum_{j=1}^{n} p_{ij} Z_{j[1]} = S_{i+1,[1]} \qquad i = 1, \ldots, m-1 \tag{8.21c}$$

$$S_{i[j']} + \sum_{j=1}^{n} p_{ij} Z_{j[j']} \leq S_{i+1,[j']} \qquad i = 1, \ldots, m-1; \quad j' = 2, \ldots, n \tag{8.22}$$

$$S_{i[j']} + \sum_{j=1}^{n} p_{ij} Z_{j[j']} \leq S_{i,[j'+1]} \qquad i = 2, \ldots, m; \quad j' = 1, \ldots, n-1 \tag{8.23}$$

The objective function, i.e. the makespan, is given by (8.19) as the sum of the starting time of the last operation on machine m plus its processing timc. Observe that the sum contains only one 'true' summand since only one job is assigned to the last position of the permutation and therefore all other binary variables in this sum are 0. Constraints (8.20a, 8.20b) represent the classical assignment constraints, i.e. every position of the permutation $[j']$ has to be assigned to exactly one job j (8.20a) and every job has to be assigned to exactly one position in the permutation (8.20b). Equations (8.21a, 8.21b, 8.21c) give the jobs' initialisation constraints on machine 1 (8.21b) and the machine initialisation constraints for the job in position [1]

Table 8.4 Wagner model: Notation

Indices	
i	Machine index, $i = 1, \ldots, m$
j	Job index, $j = 1, \ldots, n$
$[j']$	Position index, $j' = 1, \ldots, n$. This index indicates the j'-th position of a job permutation
Variables	
$Z_{j,[j']}$	Binary variable with $Z_{j,[j']} = 1$ if job j is assigned to position j' in the sequence/permutation, $Z_{j,[j']} = 0$ otherwise
$X_{i,[j']}$	Idle time on machine i before the operation of the job in position $[j']$ on machine i starts
$Y_{i,[j']}$	Idle time on machine i after the operation of the job in position $[j']$ on machine i is finished
Parameters	
p_{ij}	Processing time of job j on machine i

(8.21c). Equation (8.22) represent the job availability constraints while (8.23) give the machine availability constraints.

This model includes n^2 binary variables $Z_{j,[j']}$ (being n of them non-zero in every solution) and nm continuous variables $S_{i,[j']}$ plus one continuous variable $C_{\max}$, i.e. in total $nm + 1$ continuous variables. The number of constraints is $2n$ (8.20a, 8.20b) plus $(n + m - 1)$ (8.21a, 8.21b and 8.21c) plus $2(m - 1)(n - 1)$ (8.22 and 8.23), i.e. in total $2mn + n - m + 1$ constraints.

Wagner model (modified as in Tseng et al. 2004):
The Wagner model is probably the oldest MILP model for permutation flow shop scheduling. It combines the job availability constraints and the machine availability constraints of the Wilson model into one type of constraints (Table 8.4).

Model:

$$\min \quad C_{\max} = \sum_{j=1}^{n} p_{mj} + \sum_{j'=1}^{n} X_{m[j']} \tag{8.24}$$

s. t.

$$\sum_{j=1}^{n} Z_{j[j']} = 1 \qquad j' = 1, \ldots, n \tag{8.25a}$$

$$\sum_{j'=1}^{n} Z_{j[j']} = 1 \qquad j = 1, \ldots, n \tag{8.25b}$$

$$\sum_{j=1}^{n} p_{ij} Z_{j[j'+1]} - \sum_{j=1}^{n} p_{i+1,j} Z_{j[j']} + X_{i[j'+1]} - X_{i+1,[j'+1]} + Y_{i[j'+1]} - Y_{i[j']} = 0$$

$$i = 1, \ldots, m-1; \quad j' = 1, \ldots, n-1 \qquad (8.26a)$$

$$\sum_{j=1}^{n} p_{ij} Z_{j[1]} + X_{i[1]} - X_{i+1,[1]} + Y_{i[1]} = 0 \qquad i = 1, \ldots, m-1 \qquad (8.26b)$$

The objective function, i.e. the makespan, is given by (8.24) as the sum of the processing times of all jobs on the last machine (which is constant and decision irrelevant) and the idle time on the last machine. (This means nothing but that the minimisation of makespan is equivalent to minimising idle time on the last machine m.) Constraints (8.25a, 8.25b) represent the classical assignment constraints, as in the Wilson model above. Equations (8.26a, 8.26b) links the jobs on positions $[j']$ and $[j'+1]$ on all machines, (8.26a) couples jobs 1 through n (or jobs [1] through $[n]$, respectively) while (8.26a) gives the initialisation for job 1. Equation (8.26b) indicates that

- the waiting time on machine i before the first job in the sequence [1] will start on i plus
- the processing time of [1] on i, i.e. $p_{i[1]}$, plus
- this job's waiting time after it is finished on i

are equal to this first job's waiting time on the next machine $i+1$. This can be seen easily by rearranging (8.26b) so that $X_{i+1,[1]}$ is isolated on one side of the equation.

Rearranging (8.26a) accordingly yields the expanded expression for the finishing time of job's $[j'+1]$ operation on machine $i+1$ as follows:

$$Y_{i[j']} + \sum_{j=1}^{n} p_{i+1,j} Z_{j[j']} + X_{i+1,[j'+1]} = X_{i[j'+1]} + \sum_{j=1}^{n} p_{ij} Z_{j[j'+1]} + Y_{i[j'+1]}$$

This formula can be interpreted by means of Fig. 8.2. The figure, for reasons of clarity, supposes that the underlying permutation is the identical permutation, i.e. $j = [j]$ for all $j = 1, \ldots, n$. The equation can be deducted from the time interval which is marked by the curly brace in Fig. 8.2, i.e. the situation after job j's operation on machine i has been finished. (It should be mentioned that in many cases several of the X and Y variables sketched out in Fig. 8.2 will be zero for an efficient (with respect to makespan) schedule).

This modification of Wagner's model includes n^2 binary variables $Z_{j,[j']}$ (being n of them non-zero in every solution) and $2nm - (n-1)$ continuous variables $X_{i,[j']}$ and $Y_{i,[j']}$. The number of constraints is $2n$ (8.25a, 8.25b) plus $n(m-1)$ (8.26a, 8.26b), i.e. in total $n(m+1)$ constraints.

As already mentioned, we will not discuss the B&B optimisation approaches with respect to these 'pure' MILP formulations of the problem under consideration. However, we give a table that compares the different dimensions of these three models since these model dimensions might be one crucial aspect for the solvability of the models.

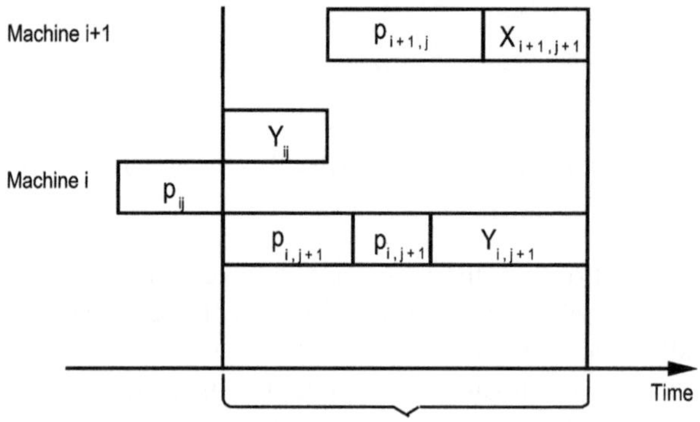

Fig. 8.2 Time calculation for the adapted Wagner MILP model for permutation flow shops relative to operation of job j on machine i

As can be seen from Table 8.5, the number of binary variables is the smallest in the Manne model. However, since the binary variables' influence in the Wilson model and the Wagner model is of assignment model type, additional features of this problem type might be advantageous for the solution of the problem. Therefore, the lower number of binary variables in the Manne model may not pay off in the solution process. The number of continuous variables can be neglected for the evaluation of the expected computational effort. However, the number of constraints might influence the performance of the model solution. With respect to this aspect, the Wagner formulation is best as compared with the other formulations. For more detailed numerical comparison of these models see Tseng et al. (2004). Their numerical results indicate that the Wagner formulation performs best, the Wilson formulation is slightly worse and the Manne formulation performs worst—at least for exactly solvable small problem instances. This demonstrates that the same (formal) problem might be solved by different (formal) models and the performance of the model in terms of computational effort to derive the same optimal solution might differ significantly from model to model.

8.3.2 Problem-Oriented B&B Approaches

Branch and bound approaches are one of the most wide-spread types of approaches to solve combinatorial optimisation problems exactly or, if prematurely stopped, in an approximate manner. We will not discuss their mathematical, formal aspects in detail but refer the interested reader to the relevant literature cited as further readings in Sect. 8.4.

The problem-oriented approaches to branch and bound procedures refer to the underlying manufacturing scheduling problem itself. For these approaches we present some aspects of how the design of the four components that could be

Table 8.5 Model sizes with respect to number of machines m and number of jobs n (see also Tseng et al. 2004)

	Manne	Wilson	Wagner
Continuous variables	$nm+1$	nm	$2nm-(n-1)$
Binary variables	$0.5n(n-1)$	n^2	n^2
Constraints	$n(mn+1)$	$2mn+n-m+1$	$n(m+1)$

Table 8.6 Processing times (p_{ij}) for a four jobs and three machines flow shop scheduling problem

Machine (i)	Job (j)			
	1	2	3	4
1	6	9	2	6
2	8	4	2	7
3	7	4	3	8

customised, i.e. solution/node definition, branching definition, bound computation and branching strategy.

To exemplify some basic ideas, we will refer to the simple permutation flow shop scheduling problem with makespan objective and a given instance with four jobs and three machines. The processing times can be seen from Table 8.6.

A complete decision tree for this problem is given by Fig. 8.3. (We will refer to the numbers in the nodes of this figure further down.)

(1) *Node definition*: The nodes coincide with a subset of the overall solution set. In Fig. 8.3 these subsets are defined by a sequence S^+ of jobs that have already been scheduled, i.e. the subset is defined by all sequences which have the same initial subsequence S^+. Example, the node 23** indicates all sequences/permutations of the $F|prmu|C_{\max}$ problem instance under consideration which start with subsequence 23 while jobs one and four have not yet been scheduled. These jobs constitute the set of non-scheduled jobs in node 23** called S^-. (It should be mentioned that S^+ is a

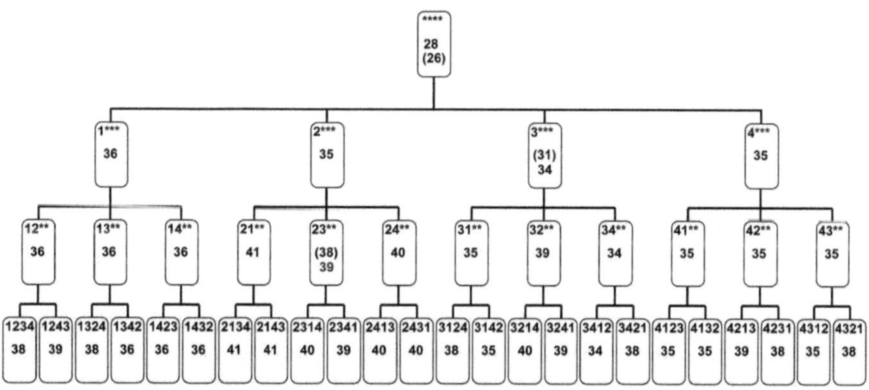

Fig. 8.3 Complete decision tree of the $F|prmu|C_{\max}$ example instance

(well-ordered) vector of jobs while S^- is an unsorted set of remaining jobs which have not yet been scheduled. The final nodes of the decision tree in Fig. 8.3 represent the complete sequences. Looking only at the bottom level and evaluating all solutions is the well-known complete enumeration).

However, even in this very basic example, definition of nodes can be done different than it is done in the standard way depicted in Fig. 8.3. For example, the sequences must not be constructed forward from the first job to the last one but can also be generated backward from the last job to the first one. Especially in those types of problems where it does not make a difference whether a sequence is constructed from the first operation of the first job on its first machine to the last operation of the last job on its last machine or vice versa (which is obviously the case for the simple flow shop problem in our example), exact as well as approximate/heuristic procedures can be applied 'in both directions'. Since the (approximate) procedures will often give two different sequences by the application of both, the forward and the backward approach, the best of them can be chosen while the effort is only doubled. Exact procedures, such as the branch and bound procedure discussed here, might yield different optimal solutions (if more than one exists) and/or may result in different computational effort for the same solution if a forward or a backward approach is used. This aspect of reversibility is, e.g., discussed by Pinedo (2012). It was applied to buffer-constrained flow shop problems by Leisten (1990). However, although the effort of this combined forward/backward approach is (coarsely) only twice as high as the application of one of the single direction approaches and it might result in some improvements of the single direction approaches, this double direction approach is not very popular so far.

Many other approaches can be considered here as well. Potts (1980) describes a simultaneous forward/backward approach in a branch and bound context and Leisten (1990) applies this idea to heuristics for buffer-constrained flow shops. Every intermediate approach could be used as well. However, forward, backward, forward/backward or a combination of these approaches are those being discussed in the literature. From an application point of view, scheduling problems with a distinct bottleneck stage could focus on planning this bottleneck first and then apply backward and forward strategies based on the schedule for this bottleneck stage.

(2) *Branching definition*: Within the branching definition, the transition from one (predecessor) node to its successor(s) has to be defined. Having already scheduled a subset of jobs (or operations respectively) which results in a solution subvector S^+, one (or more) of the remaining jobs/operations from the currently unscheduled objects in the set S^- have to be selected to generate the new solution subvector(s) S^+ on the next level of the decision tree. In our permutation flow shop example with forward construction of solutions, a currently unscheduled job $j^* \in S^-$ is selected, appended to the sequence of S^+ and deleted from S^-. This defines a node on the next level. Doing this for all jobs $j^* \in S^-$ in a given node S^+ defines the whole set of successor nodes of the node under consideration. In our example, e.g. the successor nodes of 23** are 231* and 234*. However, since on the third level only one job remains to be scheduled, 231* can be identified with 2314 and 234* with 2341.

(3) *Bound computation*: Bound computation is one of the most crucial items of a branch and bound approach. It substantially determines the quality of the approach, i.e. the ability to cut off subtrees of the overall decision tree before they are refined to the final leaves of complete solutions. In manufacturing scheduling (with a minimisation objective function), bounds are calculated to give a lower limit on the best solution/schedule which can be derived from this node (and all its successors). If this lower bound indicates that the best solution from this node and its successors is worse (or at least not better) than the objective function of another, already known solution, this node can be 'cut off' and must not be considered further. And obviously: The earlier a subtree can be cut off (i.e. the better the bounds are), the less is the computational effort.

(If only one optimal solution is to be determined, then 'not better' suffices as criterion. If all optimal solutions are to be determined, then 'worse' is required to cut off a subtree. In manufacturing scheduling problems, especially those with processing times with moderate variance or variability, this distinction might result in significant different computational effort since the number of optimal solutions might increase significantly with a decrease in variance/variability of processing times).

In manufacturing scheduling problems in every node, lower bounds in principle based on the objective function contribution of the subsolution in S^+ try to estimate the contribution to the objective function of the remaining jobs/operations in S^- to the final solution without computing all complete solutions (which can be generated departing from S^+).

Sophisticated bounds for many branch and bound approaches are available in the literature. To structure or even only to describe this variety of bounds would go far beyond the scope of this book. Therefore, we will only present two common bounds to the $F|prmu|C_{\max}$ problem, i.e. the above example, to demonstrate a basic approach to bound computation in manufacturing scheduling. These bounds focus on the remaining operations on the different machines (machine-based bound) or on the jobs still to be scheduled (job-based bound). See Ignall and Schrage (1965) for an early description of these bounds.

1. Machine based bound.
   If $C(S^+, i)$ is defined as the completion time of the subsequence S^+ on machine i, i.e. the time when the last operation scheduled so far on machine i is completed, then a bound for the makespan of the overall schedule can be determined as follows:

$$\text{MBB}(S^+) = \max_{i=1,\dots,m} \left\{ C(S^+, i) + \sum_{j \in S^-} p_{ij} + \min_{j \in S^-} \sum_{i'=i+1}^{m} p_{i'j} \right\}$$

   The second summand of the formula guarantees that every job not scheduled so far is processed on machine i (without idle time on i) while the third summand indicates that, after finishing all jobs on machine i, at least the job with the minimum remaining processing times has to be processed on all subsequent machines

$i+1$ through m (while ignoring potential waiting time). Since this is a lower bound for the makespan of the overall schedule for all machines, the maximum of this indicator over all machines gives a valid lower bound for all schedules having the same initial subsequence S^+.

2. Job based bound. A job-based bound can be determined as follows:

$$\text{JBB}(S^+) = \min_{j \in S^-} \left\{ \max_{i=1,\ldots,m} \left\{ C(S^+, i) + \sum_{i'=i}^{m} p_{i'j} + \sum_{j' \in S^-; j' \neq j} \min\left\{p_{ij'}, p_{mj'}\right\} \right\} \right\}$$

The second summand of the formula indicates that job $j \in S^-$ is to be processed on machines i through m and the third summand says that every other job $j' \in S^-$ not scheduled so far is either to be processed before job j, i.e. its processing time on machine i takes place before job j starts on machine i, or it is processed after job j, i.e. its processing time on machine m takes place after job j is finished on machine m. The minimum of both processing times (of every job $j' \neq j$) on machines i and m therefore can be included in a lower bound for the overall makespan. This can be summarised over all remaining jobs $j' \neq j$ since this holds for all other jobs j'. And since this conclusion holds also for every single machine i, the maximum over all machines can be taken. However, this conclusion holds only, if job j is scheduled immediately after S^+. Since all jobs $j \in S^-$ might immediately follow S^+, the minimum over all remaining jobs gives a valid lower bound for the makespan of all schedules starting with S^+.
The first summand in the job-based bound derived above can be improved as follows: Since the calculation of the summands refers to the assumption that job j is scheduled immediately after S^+, job's j 'true' starting time on machine i can be determined by temporarily scheduling this job after S^+, i.e. by determining $C\big((S^+, j), i\big)$ which includes possible idle time of machine i if job j has not been finished on machine $i-1$ yet. The earliest starting time of job j on machine i is then $min\Big(C(S^+, i), C\big((S^+, j), i-1\big)\Big)$ where $C\big((S^+, j), i-1\big)$ can be determined recursively from machine 1 through m by placing job j immediately after S^+. (It should be mentioned that the above considerations require the permutation property of the solution of the problem under consideration).

In Fig. 8.3, for every node the bound values are shown within the node. If there is only one bound value, both values are the same. If two values are shown (only in ****, 3***, 23**), the first value gives the machine-based bound, the second the job-based bound while the value in brackets is the worse of both.

Summarising with respect to bound calculation approaches it should be mentioned that the basic ideas of bound computation schemes refer to two aspects.

1. What is the minimum additional contribution to the objective function value of the subsequence identified with the node under consideration when considering all remaining jobs/operations to be scheduled?
2. Use a calculation scheme that considers basically the manufacturing stages (which might be advantageous especially if a clear bottleneck stage can be identified) or use a job wise calculation scheme (especially if clear 'bottleneck jobs'/'bottleneck operations' occur)—or combine both approaches.

(4) *Branching strategy*: Finally, the branching procedure has to be determined. As in all branch and bound approaches, also in those exclusively determined by the formal IP problem, depth-first or breadth-first strategies or combinations of both are known and used. It is well known that depth-first search procedures generate a feasible solution rather quickly while the number of nodes to be explored until one optimal solution has been determined might be rather large while in breadth-first search approaches the opposite holds. We will not discuss these approaches further here.

Referring to Fig. 8.3, the breadth-first search approach requires the analysis of 9 nodes while the depth-first search approach analyses 20 nodes before the optimal solution 3412 with makespan of 34 is determined (and its optimality is confirmed). We mention that the same approaches applied to the reverse problem as described above requires the analysis of 14 nodes for the first-depth search approach while 12 are required for the breadth-first search approach. (The interested reader might calculate the bound values to confirm these numbers).

Here, we referred more or less exclusively to the simple permutation flow shop setting. Every other (combinatorial) manufacturing scheduling problem may be formulated and solved by branch and bound techniques as well. The more general flow shop problem (without permutation constraint), job shop problems and/or other objective functions as well as many more settings have been intensively discussed. Nevertheless, the four components of B&B approaches, i.e. node definition, branching definition, bound computation and branching strategy, in their problem-specific setting have to be executed explicitly.

The advantage of branch and bound approaches as compared with simple, complete enumeration approaches arises if a larger subset (possibly on an early solution set separation level) can be identified to not contain an optimal solution. In this case, the further consideration of this subset including the complete calculation of every single solution in this subset can be omitted. However, to cover the worst case, for every problem setting and every definition of a branch and bound approach, usually a problem instance can be constructed so that the respective branch and bound approach has to finally calculate every single discrete solution. Therefore, the intended advantage of reducing the calculation effort cannot be guaranteed for any branch and bound approach in general. However, computational experience shows that usually a significant reduction of this computational effort can be expected.

There are numerous alternative and modified branch and bound approaches to combinatorial optimisation problems in general and to scheduling problems as (formal) applications of these approaches. Branch and cut (Crowder and Padberg

1980) or branch and price methods are widely discussed in more mathematically oriented references on combinatorial optimisation. We will not discuss these approaches here since they more or less exclusively refer to the above mentioned IP-based B&B approaches and emphasize on formal aspects of these problems. However, since these approaches have proven to be numerically rather efficient, they might become relevant also for the problem-oriented B&B approaches: If these newer formal approaches would significantly outperform numerically the problem-oriented approaches, one might consider to emphasize on the mapping of manufacturing scheduling problems into formal IP problems (and solve them with the advanced formal IP techniques) instead of developing and refining more problem-specific approaches to manufacturing scheduling.

Other approaches try to identify good feasible solutions before a B&B procedure is started, e.g. by sophisticated constructive and/or improvement heuristics. This usually reduces the error bound in a B&B procedure in the very beginning (as compared with the simple upper bound for the objective function value which might be derived from a first-depth search approach), giving reason for stopping the B&B procedure rather quickly if the remaining maximum error is below a pre-specified limit.

Another, heuristic approach based on B&B is beam search where, within a node under consideration only a limited number of subsequent nodes, i.e. the most promising ones, are taken into account for the B&B procedure while the other nodes are excluded from further consideration. Fixing the number of nodes to be explored further and determination of the respective selection criterion are decisive for the quality of these approaches. Beam search approaches are usually combined with breadth-first B&B approaches to limit the number of open nodes to be stored simultaneously.

8.3.3 Dynamic Programming

Roughly spoken, dynamic programming in mathematical optimisation is a classical technique applicable if

- the overall decision can be separated into partial decisions which (including their objective function contribution) can be assigned to points/stages of a continuous or a discrete scale
- while the optimal (partial) decision sequence *up to* a stage depends completely on the unvarying optimal (partial) decision sequence *up to* the stage in front of this stage under consideration *and* the decision *in* this stage under consideration.

Under these conditions, dynamic programming approaches can be applied which use the recursive optimisation approach induced by the well-known Bellman principle/the Bellman recursion equation. For a discrete scale, this setting allows the application of the discrete backward or forward recursive optimisation scheme of dynamic programming stage by stage. It means that an optimal solution up to a stage consists of the optimal combination of a decision *on* this stage in combination with the best decision *up to* the previous stage. In fact, this limits the effort, e.g. as compared with complete enumeration, by avoiding every solution on every stage to be computed from the beginning stage (may it be the first stage if forward calculation is used or the last stage if backward calculation is executed) onwards. The (optimal sub-) decisions (or subdecision sequences) up to the previous stage remain unchanged and only the decision of the current stage has to be linked with exactly these previous (sub-) decisions to get the best decisions up to the stage under consideration.

While in many dynamic programming applications the scale is representing time and the decisions might be taken at certain points in time (discrete dynamic programming approach) or continuously (continuous dynamic programming approach), in manufacturing scheduling mostly discrete approaches are used and the discrete stages often represent the number of jobs/operations scheduled so far. On every stage, the different settings which might occur on this stage (i.e. the states on the stage) have to be identified. For a simple 1-machine scheduling problem with n jobs and forward recursion scheme, a stage might be defined as the number of jobs already scheduled so far (e.g. the stage index $j'-1$ indicates that $j'-1$ jobs have been scheduled already) while the maximum set of states on this stage consists of the $(j'-1)!$ permutations of $j'-1$ jobs. Moving to stage j' then means taking one by one every remaining job out of the $\big(n-(j'-1)\big)$ so far non-scheduled jobs, constructing the expanded (sub-) sequences of length j' and looking for the best combinations for the specific job added.

We explain the approach by a most simple example problem which tries to find the optimal solution for a one-machine problem with four jobs and flowtime objective as well as no further 'specialities'. (As is well known, this simple problem can be easily solved to optimum by using the shortest processing time rule. However, for the sake of simplicity of the demonstration of the approach, we stick to this trivial example and give hints on extensions later on.)

Suppose four jobs to be scheduled having processing times 6, 2, 4, and 3, respectively. The objective function is to minimise flowtime, i.e. the sum of completion times of every job, $\sum_j C_j$. The stages are defined by the number of jobs j' being scheduled up to this stage while the possible states on the stage are characterised by all subsequences of length j' having a certain job k on the last position j'. We use a forward recursive approach. Then, the state transition from a sequence $S_{j'-1}$ on stage $j'-1$ to a sequence including the so far non-scheduled job k in position j' of $S_{j'}$ is $S_{j'-1} \rightarrow S_{j'} = \big(S_{j'-1}, k\big)$. The Bellman recursion equation per state on stage j' is

$$C_{\sum}\left(S_{j'} \text{ s.t. job } k \text{ on position } j' \text{ in } S_{j'}\right)$$

$$= \begin{cases} p_k & \text{for } j = 1 \\ \min_{S_{j'-1},\, k \notin S_{j'-1}} \left(C_{\sum}(S_{j'-1}) + \left(\sum_{j' \in S_{j'-1}} p_{j'} + p_k\right)\right) & \text{for } j = 2, \ldots, n \end{cases}$$

Figure 8.4 shows the computational steps. In Fig. 8.4a, the states as the partial job sequences with a certain last job on each stage are mentioned first. Then the possible realisations of these subsequences are mentioned. In italics letters, the flowtime values of every single possible subsequence are given while a '*' gives $C_{\sum}(S_{j'})$,

(a)

Stage 1	Stage 2			Stage 3			Stage 4		
1 6 *	x1	21	10 *	xy1	421	22	xyz1	2431	42 * ***
		31	14		231	22		2341	44
		41	12		241	20 *			
2 2 *	x2	12	14	xy2	-	∞	xyz2	-	∞
		32	10						
		42	8 *						
3 4 *	x3	13	14	xy3	213	24	xyz3	2413	44 *
		23	8 *		423	20			
		43	10		243	18 *			
4 3 *	x4	14	15	xy4	214	23	xyz4	-	∞
		24	7 *		234	19 *			
		34	11						

(b)

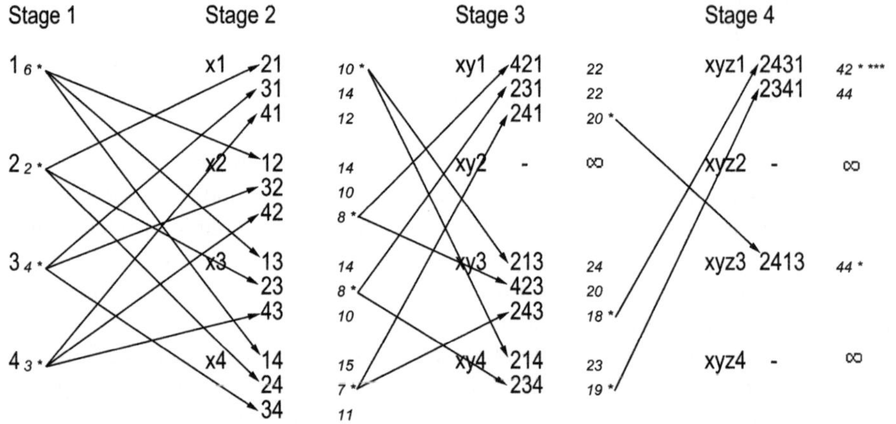

Fig. 8.4 Dynamic programming approach for the example instance of the one-machine flowtime minimisation problem, **a** Results per stage, **b** Additionally indicating the calculation paths

i.e. the optimal subsequence with minimum recursion value and an additional '***' indicates the overall optimal solution (on stage 4). Figure 8.4b shows the same values but additionally gives the relevant state transitions considered in every step. Note that an infinite value (e.g. in xy2) indicates that there is no feasible sequence left to place job 2 in stage/position 3 since job 2 is already scheduled in positions 1 or 2 in all (!) optimal subsequences up to stage 2. Analogously, the same holds for xyz2 and xyz4.

As already mentioned, the example problem presented above can be easily solved by just sorting the jobs according to the shortest processing time rule. For this type of problem, a dynamic programming approach therefore is superfluous. However, if the objective function is more complicate, the solution is not as easy and a dynamic programming approach might become adequate. Early contributions concerning this

Stage 1		Stage 2			Stage 3			Stage 4		
1	21 *	x1	21	47	xy1	321	58 *	xyz1	3421	95 *
			31	30 *		231	69		3241	97
			41	49		341	60			
2	17 *	x2	12	46	xy2	312	57 *	xyz2	-	∞
			32	26 *		342	57 *			
			42	46						
3	7 *	x3	13	45	xy3	-	∞	xyz3	-	∞
			23	37 *						
			43	45						
4	21 *	x4	14	50	xy4	314	61	xyz4	3214	98
			24	37		324	58 *		3124	95 *
			34	30 *		234	69			

Fig. 8.5 Dynamic programming calculation for the three machine, four jobs problem instance from Sect. 8.3.2 for the permutation flow shop problem with flowtime objective

matter are by Held and Karp (1962); Rinnooy Kan et al. (1975) or Schrage and Baker (1978). Instead of the pure linearity of the influence of jobs' completion times with respect to the flowtime objective, initially constant and later linear influence on the objective function as, e.g. in tardiness problems result in problems which cannot be solved by polynomial approaches. Therefore, the above-mentioned papers and the dynamic programming approaches proposed therein allow rather general objective functions, being monotonicity (not necessarily strict monotonicity) in time of jobs objective function contribution the more or less only relevant requirement for the application of these approaches. Also other specific constraints, e.g. precedence relations between the operations, can be advantageously handled by dynamic programming approaches.

We will not deepen the discussion on dynamic programming in manufacturing scheduling but give only some additional remarks on the application of these approaches in manufacturing scheduling.

As in all dynamic programming approaches, the adequate definition of stages and their states as well as the stage separability as mentioned above are the crucial aspects to setup a dynamic programming procedure for a scheduling problem under consideration.

State separability may be guaranteed by an adequate separation of the set of jobs or operations into operations/jobs scheduled so far and not scheduled so far.

The separability with respect to objective functions is often fulfilled for additive (or averaging) objective functions as (mean) flowtime or (mean) tardiness. In contrast, separability for maximin objective functions cannot be expected in general. Therefore, e.g. makespan or maximum tardiness problems are often not suited to be solved by means of dynamic programming approaches.

However, to demonstrate problems which might occur with respect to separability, we refer to the standard permutation flow shop with flowtime objective. We suppose the three machine, four jobs problem instance given as a B&B example in Sect. 8.3.2. But now we suppose flowtime to be minimised. Performing the dynamic programming approach as in the above example yields the results as in Fig. 8.5.

As can be seen, the dynamic programming approach yields two 'optimal' permutations, i.e. 3421 and 3124, with a flowtime of 95. However, the true optimal solution for this problem instance is 3412 with a flowtime of 94. This solution is not generated by the dynamic programming approach because the initial partial sequence 341 of the overall optimal solution is excluded from further consideration on stage 3 since 321 gives a better solution for the state xy1. Therefore, even in this simple example, the separability requirement of dynamic programming is not fulfilled and dynamic programming cannot be applied to this kind of problem—at least not with this definition of stages.

Summarising, dynamic programming (in manufacturing scheduling) can be classified as a limited enumeration approach such as B&B approaches. In contrast to B&B approaches, dynamic programming requires certain additional and specific structural features of the manufacturing scheduling problem considered, i.e. separability. If these requirements are fulfilled, dynamic programming will be usually competitive to respective B&B methods. However, on one hand dynamic programming is also a non-polynomial enumerative approach which is, in general, only applicable for small sizes of problem instances if the optimal solution is to be determined. On the other hand, as in B&B, dynamic programming can be stopped prematurely and/or combined with other heuristic approaches.

8.4 Conclusions and Further Readings

As discussed in the previous chapter, exactness in solving manufacturing scheduling problems is closely related to the formal definition and simplification of these scheduling problems and can only be interpreted relative to these formal problems. With respect to the underlying real-world problems, because of their complexity, 'exact' (i.e. optimal) solutions can never be expected.

Referring to this notion of exact approaches to (formal) manufacturing scheduling problems, only few and rather simple problems can be solved exactly by constructive methods, i.e. by polynomial effort. However, since manufacturing scheduling problems represent a subset of combinatorial optimisation problems the more complex formal problems, in principle, can be solved by enumerative approaches such as complete enumeration, branch and bound or dynamic programming. In the worst case, these approaches require non-polynomial, exponential effort relative to the problem size. Therefore, these procedures, on one hand have to be designed carefully and sophisticated. On the other hand, since the 'pure' exact procedures will usually have to be stopped prematurely, at least for larger problem instances, these approaches should and can successfully be combined with advanced heuristic approaches. It should be mentioned that there are numerous practical cases where the problem under consideration is of non-polynomial type. However, since the problem size of these cases often remains moderate, these instances can be solved to optimality anyway.

Regarding further readings, early scheduling books such as Conway et al. (1967); French (1982) cover exact procedures for many scheduling problems, while a key book on integer programming and branch and bound procedures is the one by Nemhauser and Wolsey (1988). The paper of Johnson is one of the first that appeared in the scheduling literature and some authors, like the already mentioned Conway et al. (1967), attribute the extensive use of the makespan criterion (probably this is overrated) to this initial seminal paper. This has been criticised by other authors, like Gupta and Dudek (1971) who stated that makespan in not realistic. A detailed description of Lawler's and Moore's algorithms (among many others) can be found in several scheduling textbook, such as for instance T'Kindt and Billaut (2006).

References

Conway, R. W., Maxwell, W. L., and Miller, L. W. (1967). *Theory of Scheduling*. Dover Publications, New York. Unabridged publication from the 1967 original edition published by Addison-Wesley.

Crowder, H. and Padberg, M. W. (1980). Solving large-scale symmetric travelling salesman problems to optimality. *Management Science*, 26(5):495–509.

Fanjul-Peyró, L. and Ruiz, R. (2010). Iterated greedy local search methods for unrelated parallel machine scheduling. *European Journal of Operational Research*, 207(1):55–69.

French, S. (1982). *Sequencing and Scheduling: An Introduction to the Mathematics of the Job-Shop*. Ellis Horwood Limited, Chichester.

Garey, M. R. and Johnson, D. S. (1979). *Computers and Intractability. A Guide to the Theory of NP-Completeness*. W. H. Freeman and Company, New York.

Gupta, J. N. D. and Dudek, R. A. (1971). Optimality Criteria for Flowshop Schedules. *IIE Transactions*, 3(3):199–205.

Held, M. and Karp, R. M. (1962). A dynamic programming approach to sequencing problems. *Journal of the Society for Industrial and Applied Mathematics*, 10(1):196–210.

Ignall, E. and Schrage, L. E. (1965). Application of branch- and bound technique to some flow shop problems. *Operations research*, 13(3):400–412.

Johnson, S. M. (1954). Optimal two- and three-stage production schedules with setup times included. *Naval Research Logistics Quarterly*, 1:61–68.

Lawler, E. L. (1973). Optimal Sequencing of a Single Machine Subject to Precedence Constraints. *Management Science*, 19(5):544–546.

Leisten, R. (1990). Flowshop sequencing problems with limited buffer storage. *International Journal of Production Research*, 28(11):2085.

Lenstra, J. K., Rinnooy Kan, A. H. G., and Brucker, P. (1977). Complexity of machine scheduling problems. *Annals of Discrete Mathematics*, 1:343–362.

Lenstra, J. K., Shmoys, D. B., and Tardos, E. (1990). Approximation algorithms for scheduling unrelated parallel machines. *Mathematical Programming*, 46(3):259–271.

Liao, C.-J. and You, C.-T. (1992). An improved formulation for the job-shop scheduling problem. *Journal of the Operational Research Society*, 43(11):1047–1054.

Manne, A. S. (1960). On the job-shop scheduling problem. *Operations Research*, 8(2):219–223.

Moore J. (1968). An *n* job, one machine sequencing algorithm for minimizing the number of late jobs. *Management Science*, 15 (1):102–109.

Nemhauser, G. L. and Wolsey, L. A. (1988). *Integer and combinatorial optimization*. Wiley, New York.

Pinedo, M. L. (2012). *Scheduling: Theory, Algorithms, and Systems*. Springer, New York, fourth edition.

Potts, C. N. (1980). *An adaptive branching rule for the permutation flow-shop problem*. Elsevier, Amsterdam.

Rinnooy Kan, A., Lageweg, B., and Lenstra, J. (1975). Minimizing total costs in one-machine scheduling. *Operations Research*, 23(5):908–927.

Schrage, L. E. and Baker, K. R. (1978). Dynamic programming solution of sequencing problems with precedence constraints. *Operations Research*, 26(3):444–449.

T'Kindt, V. and Billaut, J.-C. (2006). *Multicriteria Scheduling: Theory, Models and Algorithms*. Springer, New York, second edition.

Tseng, F. T., Stafford, Jr, E. F., and Gupta, J. N. D. (2004). An empirical analysis of integer programming formulations for the permutation flowshopn. *OMEGA, The International Journal of Management Science*, 32(4):285–293.

Vallada, E. and Ruiz, R. (2010). Genetic algorithms for the unrelated parallel machine scheduling problem with sequence dependent setup times. *European Journal of Operational Research*. In review.

Wagner, H. M. (1959). An integer linear-programming model for machine scheduling. *Naval Research Logistics Quarterly*, 6(2):131–140.

Wilson, J. M. (1989). Alternative formulations of a flowshopn scheduling problem. *Journal of the Operational Research Society*, 40(4):395–399.

Chapter 9
Approximate Algorithms

9.1 Introduction

In this chapter we deal with approximate algorithms. In principle, one may think that approximate algorithms are not a good option if exact approaches exist and indeed we already discussed that their widespread use is basically justified by the computational complexity inherent for most manufacturing scheduling models. As mentioned in Sect. 7.5.2, approximate algorithms are usually divided into heuristics and metaheuristics, being the main difference among them that the first are specifically tailored for a particular model, while the second constitute more generic procedures. This difference—although not entirely unambiguous—is very important when it comes to describing both approaches, as we intend to do in this chapter: While it is clear that it is possible to discuss the templates of the different metaheuristics that can be employed for manufacturing scheduling models, such thing is not possible regarding heuristics, as they are heavily problem-dependent. Instead we will first illustrate how heuristic algorithms work by presenting perhaps the two most popular heuristics for manufacturing scheduling, i.e. the NEH heuristic (Nawaz et al. 1983) for flowshop scheduling with makespan objective, and the Shifting Bottleneck Heuristic (Adams et al. 1988) for job shop scheduling. Then, in a sort of 'reverse engineering' attempt, we will discuss the main ideas and concepts behind (these and other) heuristics. We believe that this attempt to structure the behaviour of the heuristic may be useful for the reader when faced to the task of designing new heuristics for existing or new scheduling models.

More specifically, in this chapter we

- present two well-known heuristics in order to grasp the general mechanics of heuristic construction (Sect. 9.2),
- give some description of (some) underlying ideas of the approximate methods under consideration (Sect. 9.3),
- present the main concepts behind metaheuristics, and discuss the most widely employed in the manufacturing scheduling context (Sect. 9.4), and
- briefly introduce further approximation techniques (Sect. 9.5).

J. M. Framinan et al., *Manufacturing Scheduling Systems*,
DOI: 10.1007/978-1-4471-6272-8_9,

9.2 Two Sample Heuristics

In this section, two heuristics are presented which have turned out to be rather efficient for the scheduling problems they are created for on one hand. On the other hand, the basic ideas of these heuristics have been the starting point for many adaptations to other heuristics and/or other problem classes.

First, in Sect. 9.2.1, we consider the heuristic by Nawaz et al. (1983) which has been basically the result of a master thesis in the early 1980s and since then is something like a benchmark for the respective type of problems. It deals with the standard permutation flow shop problem with makespan objective. Afterwards, in Sect. 9.2.2, we describe a basic version of the Shifting Bottleneck Heuristic (SBH) which has been published first by Adams et al. (1988). This heuristic is designed for the standard job shop problem with makespan objective.

9.2.1 *The NEH Heuristic*

By far and large, one the best known and highly cited heuristic for the flow shop problem and makespan criterion is the NEH heuristic. This method has been widely recognised as being very high performing, flexible, and efficient. Furthermore, NEH is actively used as a seed sequence for metaheuristic techniques (these will be discussed in the next section). Some examples where NEH is used as a seed sequence are Tabu Search, Simulated Annealing, Genetic Algorithms, Iterated Local Search, Ant Colony Optimisation and many other metaheuristic algorithms.

The NEH heuristic is dedicated to solve the basic permutation flow shop scheduling problem with minimizing makespan as objective function, i.e. $Fm|prmu|C_{max}$. It reduces the exponential effort of complete enumeration of all permutations of job sequences to a polynomial effort by sorting the jobs according to some importance index (see below) and then determines job by job its best position while keeping the relative sequence of the already scheduled jobs unchanged.

Due to the importance of the NEH, we describe it in detail. The NEH procedure is based on the idea that jobs with high sum of processing times on all the machines should be scheduled as early as possible. NEH is explained in Algorithm 1.

Computing P_j has a complexity of $O(nm)$, and sorting n jobs of $O(n \log n)$. Most CPU time is needed in the main loop of the heuristic, where we have a loop of $n-2$ steps and, at each step k, we carry out k insertions of job k in a partial sequence that contains $k-1$ jobs and for each insertion we have to calculate the $C_{\max}$ value of k jobs (including the inserted one). In total, there are $\frac{n(n+1)}{2} - 3$ insertions. Among these, only in the last iteration we have to evaluate n complete sequences. As a result, the computational complexity of the whole method goes to $O(n^3 m)$. This method can be slow for large instances. However, when inserting the k-th job in position, let us say, j, all $C_{i,h}$, $h = j-1, j-2, \ldots, 1$ were already calculated in the previous insertion and we do not need to recalculate these values. A similar approach was

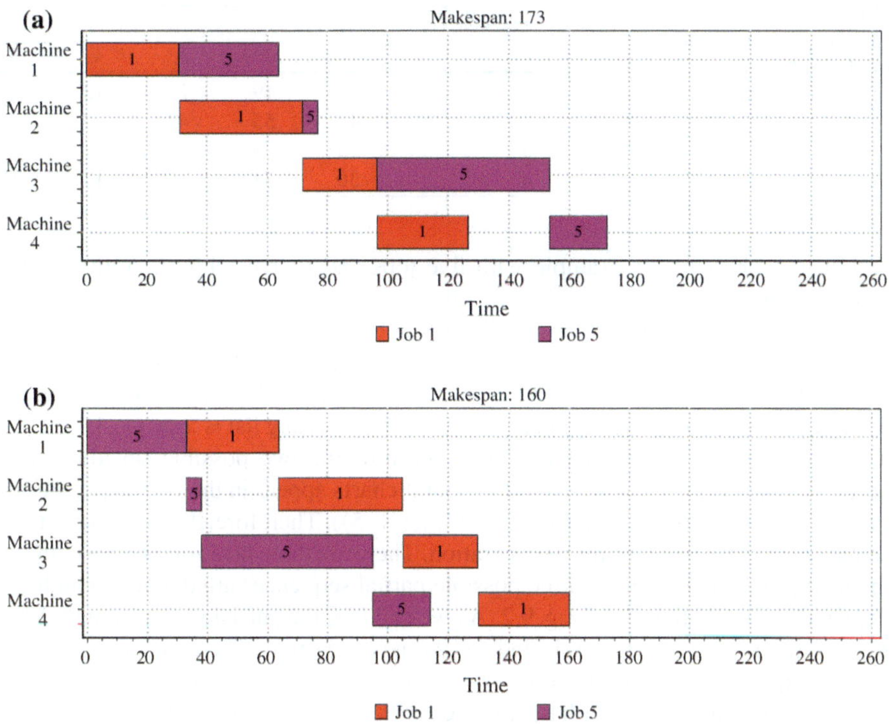

Fig. 9.1 First iteration of the NEH heuristic for the example of Table 9.1. **a** Sequence (1, 5) with a $C_{\max}$ value of 173. **b** Sequence (5, 1) with a $C_{\max}$ value of 160

Algorithm 1: NEH's Algorithm (Nawaz et al. 1983)

Input: Instance data
Output: Sequence Π, Makespan C_{max}
begin
  Let $P_j = \sum_{i=1}^{m} p_{ij}$;
  Let $\Pi = \varnothing$, $J = \{1, \ldots, n\}$ verifying $P_1 \geq P_2 \geq \ldots P_n$;
  Let Π the best permutation between (1, 2) and (2, 1);
  for $k = 3$ *to* n **do**
    Insert job k in the position of Π yielding the lowest makespan value;
  return Π, C_{max}
end

---

followed by Taillard (1990) and as a result the complexity of all insertions in a given step can be calculated in $O(nm)$ by a series of matrix calculations. This reduces the total complexity of NEH to $O(n^2m)$. Such an improved method is many times referred to as NEHT or NEH with Taillard's accelerations.

Table 9.1 Processing times (p_{ij}) for a five jobs and four machines flow shop scheduling problem

Machine (i)	Job (j)				
	1	2	3	4	5
1	31	19	23	13	33
2	41	55	42	22	5
3	25	3	27	14	57
4	30	34	6	13	19

Let us work with an example with five jobs and four machines. Actually, this example was already given in Chap. 3, Table 3.3. However, in order to avoid flipping pages back and forth, the table is replicated in Table 9.1.

We calculate in the first step the total processing times P_j, that result in: $P_1 = 31 + 41 + 25 + 30 = 127$, $P_2 = 111$, $P_3 = 98$, $P_4 = 62$ y $P_5 = 114$. Then we arrange the jobs in decreasing order of P_j and the resulting list is $J = (1, 5, 2, 3, 4)$. The first two jobs are extracted. Let us see how the two possible schedules end up $(1, 5)$ and $(5, 1)$. The corresponding Gantt charts appear in the two subfigures of Fig. 9.1. We can see that $C_{\max}((5, 1)) < C_{\max}((1, 5))$. Therefore, the partial (relative) sequence $(5, 1)$ is fixed for the next iteration. There are three positions where to insert the next job 2 and therefore, three possible partial sequences are derived, which are depicted in the subfigures of Fig. 9.2. As we can see, the partial sequence with least $C_{\max}$ value is $(2, 5, 1)$. It is selected for the next iteration. There are four possible partial sequences resulting from inserting job 3 in the four possible positions. The resulting schedules, along with their makespan values are shown in Fig. 9.3. The last from the previous partial sequences is also the one with the least makespan value, $(2, 5, 1, 3)$, it is selected for the next iteration where job 4 remains to be inserted. There are $n = k$ possible positions where to place this last job 4. These sequences are now full sequences since each one contains n jobs. All of these are shown in Fig. 9.4. In this last step, the first sequence is the one with the minimum $C_{\max}$ value and therefore, it is the result of the application of the NEH procedure to the example: $(4, 2, 5, 1, 3)$. This sequence has a $C_{\max}$ value of 213.

The NEH has been applied to other objectives and even to other layouts. While it has been demonstrated that in other problems, it is no as effective and efficient, it is still competitive.

9.2.2 The Shifting Bottleneck Heuristic

In its original version, the SBH is dedicated to solve the basic job shop scheduling problem with minimizing makespan as objective function, i.e. $Jm|.|C_{max}$. This problem is known to be of the NP-hard class.

The basic idea is as follows: Fix one machine as the bottleneck of the manufacturing system under consideration. Then, for this machine and for every job, it is considered what has to be accomplished *before* the job is processed on this machine (i.e. previous operations of this job on other machines). This results in release times for all jobs on the machine under consideration. Additionally, for this machine and

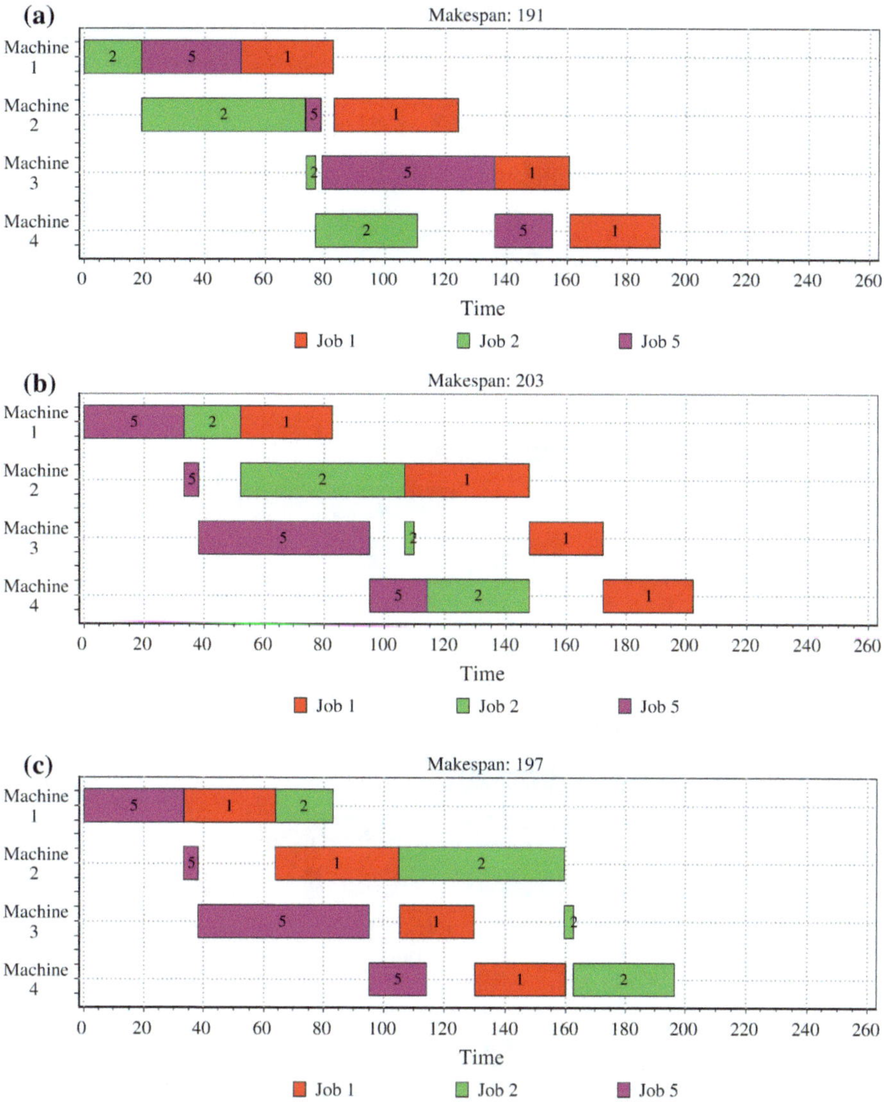

Fig. 9.2 Iteration for $k = 3$ of the NEH heuristic for the example of Table 9.1. **a** Sequence (2, 5, 1) with a $C_{\max}$ value of 191. **b** Sequence (5, 2, 1) with a $C_{\max}$ value of 203. **c** Sequence (5, 1, 2) with a $C_{\max}$ value of 197

for every job, it is considered what has to be accomplished *after* the job is processed on this machine (i.e. subsequent operations of this job on other machines). This results in due dates for all jobs on the machine under consideration. Since the due dates might be exceeded by a current solution, the corresponding $1|r_j, d_j| \max L_j$ problem is (heuristically) solved for the machine under consideration. The resulting

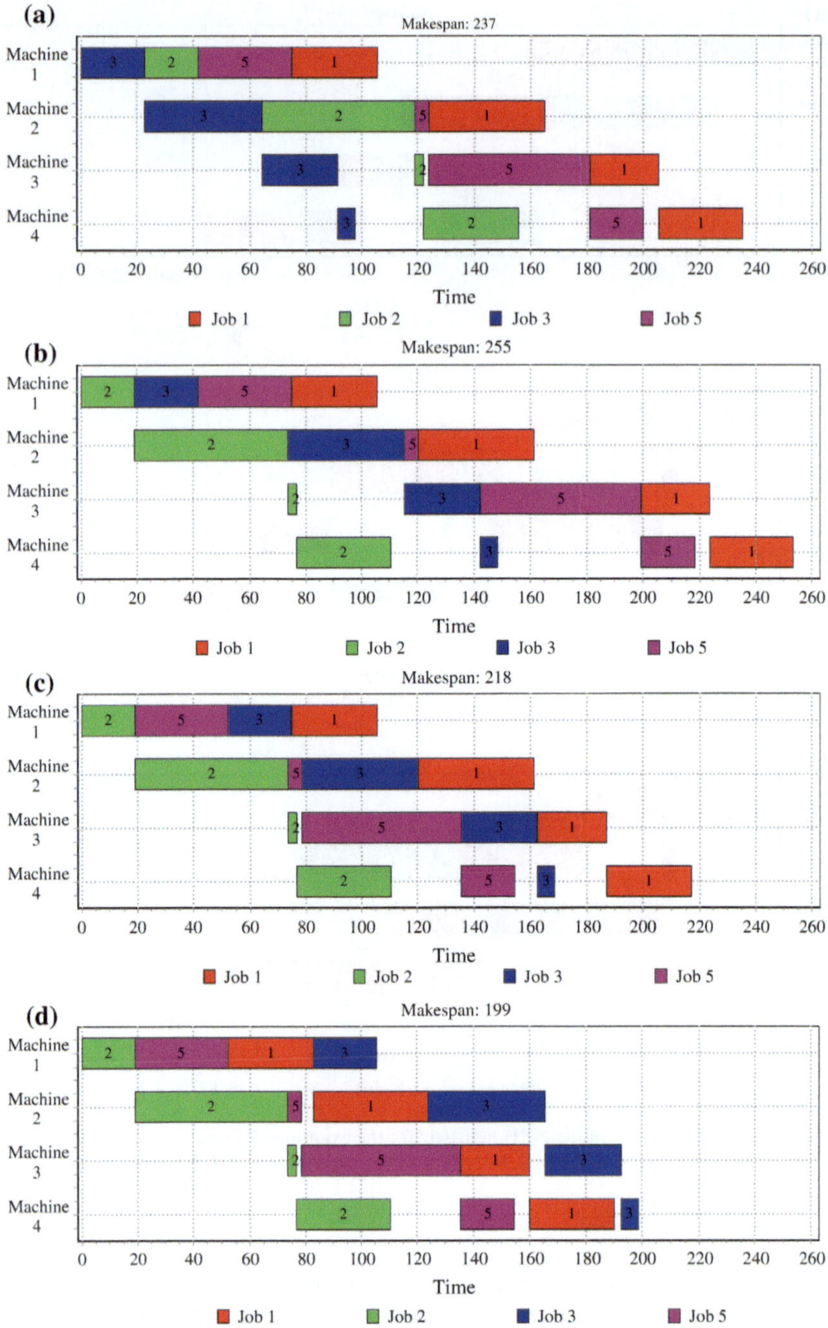

Fig. 9.3 Iteration for $k = 4$ of the NEH heuristic for the example of Table 9.1. **a** Sequence (3, 2, 5, 1) with a $C_{\max}$ value of 237. **b** Sequence (2, 3, 5, 1) with a $C_{\max}$ value of 255. **c** Sequence (2, 5, 3, 1) with a $C_{\max}$ value of 218. **d** Sequence (2, 5, 1, 3) with a $C_{\max}$ value of 199

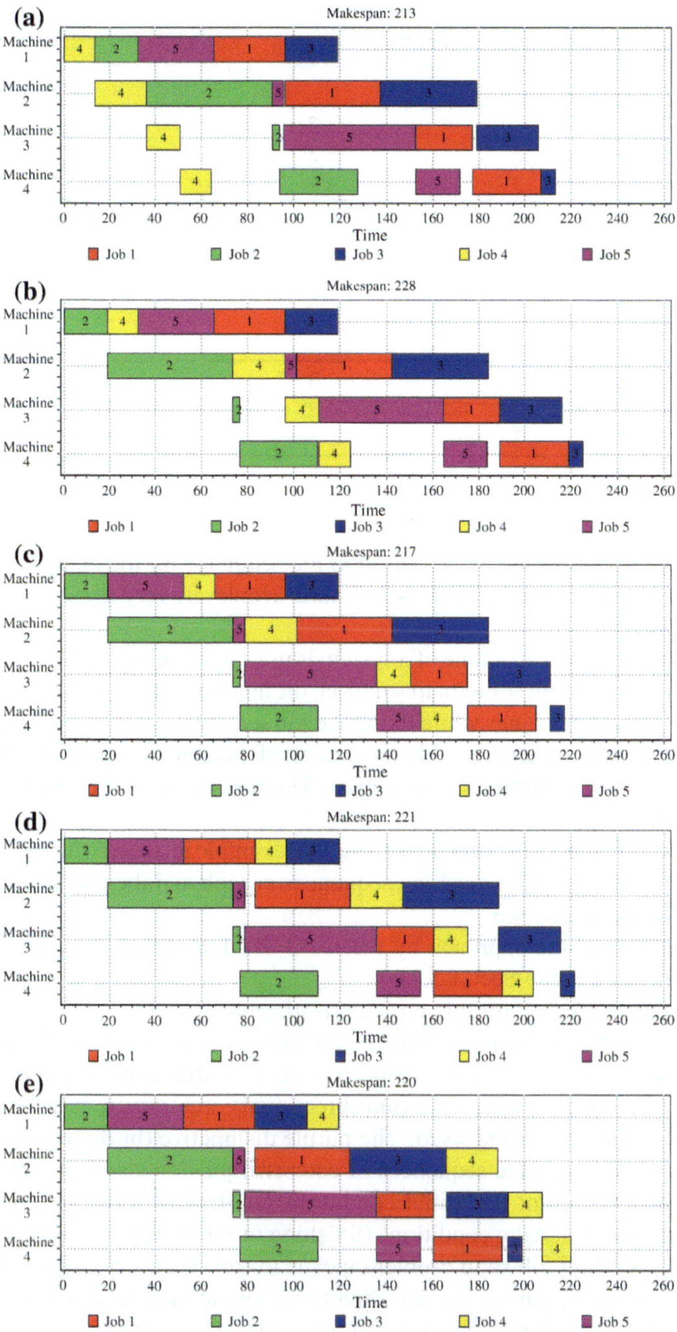

Fig. 9.4 Final iteration ($k = 5 = n$) of the NEH heuristic for the example of Table 9.1. **a** Sequence $(4, 2, 5, 1, 3)$ with a $C_{\max}$ value of 213. **b** Sequence $(2, 4, 5, 1, 3)$ con un $C_{\max}$ value of 228. **c** Sequence $(2, 5, 4, 1, 3)$ con un $C_{\max}$ value of 217. **d** Sequence $(2, 5, 1, 4, 3)$ with a $C_{\max}$ value of 221. **e** Sequence $(2, 5, 1, 3, 4)$ con un $C_{\max}$ value of 220

Table 9.2 Machine sequences (per job)

Job machine	J_1	J_2	J_3	J_4
$M_{[1]}$	1	2	2	3
$M_{[2]}$	2	3	1	1
$M_{[3]}$	3	1	3	–

Table 9.3 Processing times (p_{ij})

Job machine	J_1	J_2	J_3	J_4
M_1	3	3	3	2
M_2	3	2	4	–
M_3	2	3	1	3

job sequence is kept as the SBH's solution job sequence for the machine under consideration. $1|r_j, d_j|\max L_j$ itself is NP-hard. However, very good heuristics exist for this problem.

Since the bottleneck machine of a system is usually not identifiable in advance and might even change from solution to solution, the approach described above is applied successively to all machines while fixing the job sequences on those machines which have already been considered. The sequence of machine consideration is chosen appropriately (see the following example). The procedure of the SBH is described in Algorithm 2.

We exemplify the SBH by means of a simple example instance with three machines and four jobs. Suppose the machine sequences per job and the processing times are given by Tables 9.2 and 9.3.

We demonstrate the approach by means of the disjunctive graph representation of the problem (see also Sect. 1.5.1).

The disjunctive graph for this problem is given by Fig. 9.5.

The nodes in Fig. 9.5 represent the respective operation of J_j on M_i while being the processing times noted in the respective cycle. Initial node A and final node Ω are added. The black conjunctive (uni-directional) arcs represent a job's machine sequence which is given and fixed. The purple disjunctive (bi-directional) arcs represent the possible relative sequences of each two jobs on a specific machine. The dark blue arcs represent the initial conditions (starting of jobs) and the turquoise arcs represent the finishing conditions (finishing of jobs). Minimizing the makespan in the corresponding job shop problem means to transfer every disjunctive arc into a conjunctive one by giving it a uni-directional orientation (and maintaining logic feasibility of the resulting sequence, e.g. no cycles are allowed) so that the longest path from node A to the final node Ω in the network (whose length corresponds to the makespan of the solution derived) is of minimum length.

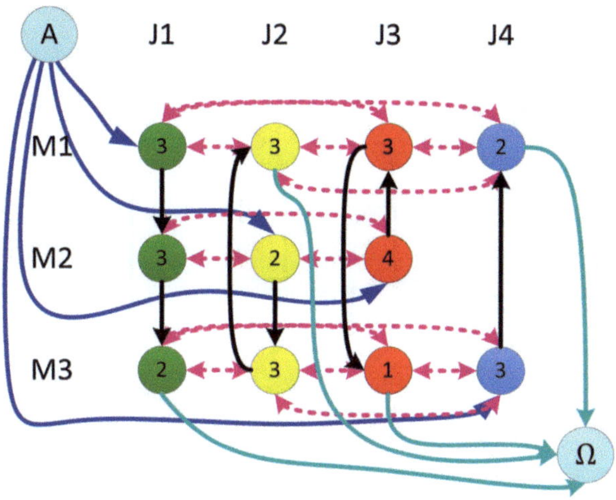

Fig. 9.5 Disjunctive graph of the example for the SBH

Algorithm 2: SBH (Adams et al. 1988)

Input: Instance data
Output: Sequence Π, Makespan C_{max}
begin
Let $UM = \{1, \ldots, m\}$ set of unscheduled machines;
Let $SM = \varnothing$ set of scheduled machines;
Let $\Pi_i = \varnothing, \forall i \in M$;
while *UM is* ***not*** $\varnothing$ **do**
    for *all machines i in UM* **do**
        Compute (regarding the fixing of operations' sequence on the machines in SM and the jobs' machine sequences) r_j and d_j with respect to already scheduled machines *SM* as follows:
            r_j := lower bound for earliest starting time of operation j on machine i;
            d_j := upper bound for the latest finishing time of operation j on machine i;
        Let Π_i the solution of the respective $1|r_j, d_j|\max L_j(i)$ problem
    Let $k \in UM$ be the machine such that $\max L_j(k) = \max i \in UM\_\max L_j(i)\_$(i.e the bottleneck among the remaining unscheduled machines, indicated by the largest maximum lateness);
    Keep Π_k as definitive schedule for machine k (i.e. the job sequence on this machine);
    Insert k in *SM*;
    Remove k from *UM*;
return Π, C_{max}
end

Applying the SBH to this problem instance results in the following.

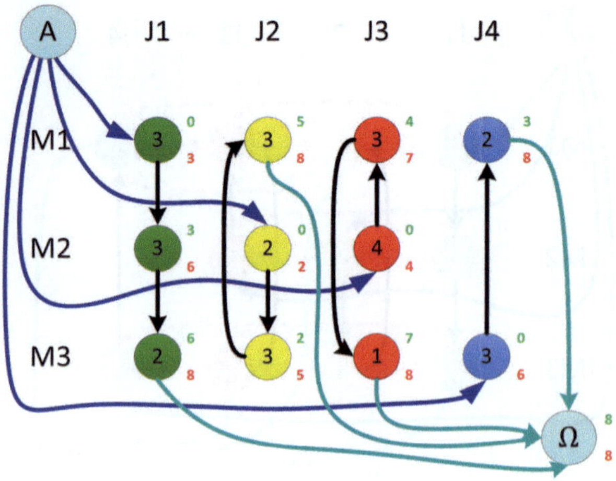

Fig. 9.6 Reduced network in the SBH heuristic (initialization)

Initialization:

Initialize the set of already scheduled machines: $SM = \emptyset$, i.e. the set of unscheduled machines is $UM = (1, 2, 3)$.

Eliminate all disjunctive arcs from the overall problem's disjunctive graph, i.e. the constraints with respect to job sequences on each machine. (In a real-world interpretation, this means that all jobs could be processed in parallel on every machine, assigning every machine (for the time being) an unlimited capacity.)

Determine the longest path in the resulting network (see Fig. 9.6). In Fig. 9.6, the green numbers assigned to every node indicate the earliest starting time of the respective operation while the red numbers represent the latest finishing time of this operation. These time values can be easily computed by the standard forward and backward calculation scheme well-known from time planning in networks.

The C_{max} value of this solution is the maximum of every job's last operation finishing time, i.e.

$C_{max}(SM) = max\ (8, 8, 8, 5) = 8$, as well as the time value at Ω.

The (green) earliest starting time of each operation is now interpreted as the release time r_j of the operation when it is considered on its respective machine. The (red) latest finishing time of each operation accordingly is interpreted as the due date d_j of the operation when it is considered on its respective machine.

Iteration 1: $SM = \emptyset$, $UM = (1, 2, 3)$

Determine min max L_j for all single machine problems incl. release and due dates, i.e. solve $1|r_j, d_j|\max L_j$ for all so far unscheduled machines. As already

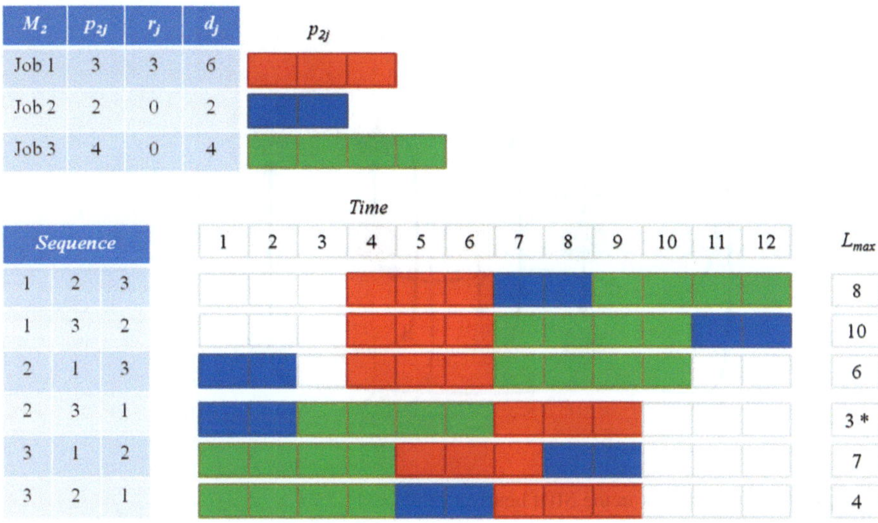

Fig. 9.7 Solving $1|r_j, d_j|\max L_j$ for M_2 in first iteration by complete enumeration

mentioned this problem is NP-hard, but good heuristics exist for it. Here, since there are only $3! = 6$ (on M_2) or $4! = 24$ (on M_1 and M_3) possible solutions for each of these problems, we use complete enumeration to solve the problems exactly:

M_1: min max $L_j = 3$ for job sequence 1432
M_2: min max $L_j = 3$ for job sequence 231
M_3: min max $L_j = 1$ for job sequence 4213

We do not list all possible permutations for all unscheduled machines here. This is left to the reader. Instead, exemplarily we give only the results for M_2 here. Since there are only three operations to be considered on M_2, there are only $3! = 6$ possible sequences to be calculated. The calculation is seen from Fig. 9.7.

We choose the machine with the worst minimum max L_j value and include the disjunctive arcs for the respective solution in Fig. 9.6. The result can be seen from Fig. 9.8. Since there are two machines with the worst max L_j value (namely M_1 and M_2), in the simplest version of the SBH we can choose arbitrarily among those machines. Here we choose M_1 and update $SM = (1)$, $UM = (2, 3)$.

Performing forward and backward calculation of earliest starting times and latest finishing times of every operation yields again the (updated) green and red numbers at every node in Fig. 9.8.

The longest path gives the current 'makespan' (for this relaxed problem) which is the maximum of every job's last operation finishing time in Fig. 9.8, i.e.

$$C_{max}(SM) = max\ (11, 11, 11, 5) = 11, \text{ as well as the time value at } \Omega.$$

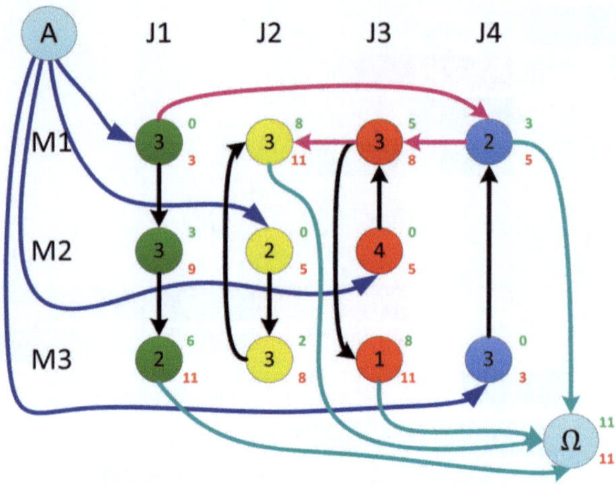

Fig. 9.8 Reduced network in the SBH heuristic ($SM = (1)$, $UM = (2, 3)$)

Iteration 2: $SM = (1)$, $UM = (2, 3)$

Now the procedure is repeated for all remaining unscheduled machines:

Determine min max L_j for all single machine problems incl. release dates and due dates, i.e. solve $1|r_j, d_j|L_{max}$ for all so far unscheduled machines.

M_2: min $L_{max} = 1$ for job sequences 231 and 321
M_3: min $L_{max} = 0$ for job sequences 4213 and 4231.

We choose the machine with the worst minimum max L_j value (i.e. M_2 here) and include for the respective solution the disjunctive arcs in Fig. 9.8. The result can be seen from Fig. 9.9. Here, the machine is uniquely determined. But the respective sequence is not unique (231 and 321 could be chosen). In the simplest version of the SBH we can choose arbitrarily among these sequences. Here we choose 231 on M_2 and update $SM = (1, 2)$, $UM = (3)$.

Performing forward and backward calculation of earliest starting times and latest finishing times of every operation yields the green and red numbers at every node in Fig. 9.9.

The longest path gives the current 'makespan' (for this still relaxed problem) which is the maximum of every job's last operation finishing time in Fig. 9.9, i.e.

$C_{max}(SM) = max\ (12, 12, 12, 6) = 12$, as well as the time value at Ω.

Iteration 3: $SM = (1, 2)$, $UM = (3)$

Determine min max L_j for all single machine problems incl. release dates and due dates, i.e. solve $1|r_j, d_j|L_{max}$ for all so far unscheduled machines.

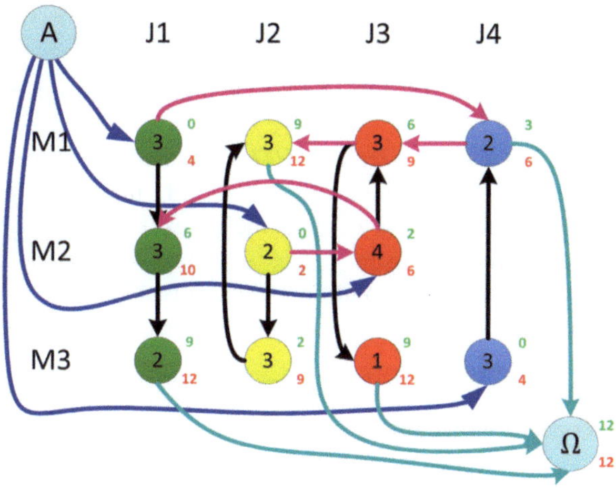

Fig. 9.9 Reduced network in the SBH heuristic ($SM = (1, 2)$, $UM = (3)$)

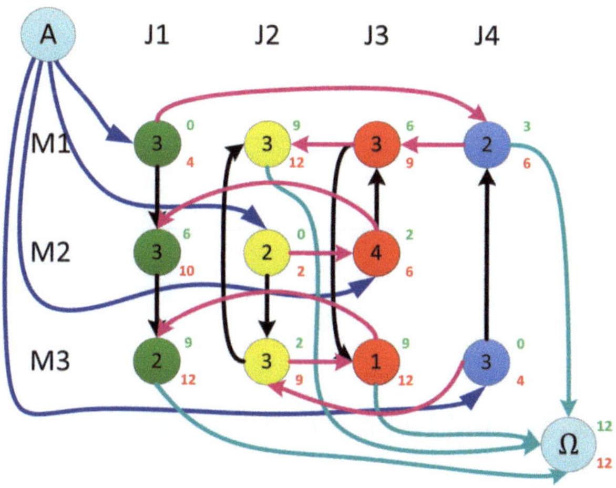

Fig. 9.10 Final network in the SBH heuristic ($SM = (1, 2, 3)$, $UM = \emptyset$)

M_3: min $L_{max} = 0$ for job sequences 4213 and 4231.

The result can be seen from Fig. 9.10. The machine is uniquely determined. But the respective sequence is not unique (4213 and 4231). Here we choose 4231 on M_3 and update $SM = (1, 2, 3)$, $UM = \emptyset$.

Performing forward and backward calculation of earliest starting times and latest finishing times of every operation yields the green and red numbers at every node in Fig. 9.10.

The longest path gives the makespan of the SBH application to the given job shop problem which is the maximum of every job's last operation finishing time in

Fig. 9.10, i.e.

$C_{max}(SM) = max\ (12, 12, 12, 6) = 12$, as well as the time value at Ω.

Since no unscheduled machines remain ($UM = \emptyset$), the SBH procedure is finished. (Note that only by chance the sequence of considering the machines was 1-2-3 in this instance. Usually, the sequence of machine indexes in the iterations must not need to be continuously increasing.)

Finally, we received a complete solution to the original job shop problem instance with makespan minimization which is—in our example instance—also the true optimum solution to this instance.

The solution data from the SBH can be easily deduced from Fig. 9.10.

Job sequences on every machine:

M_1: 1432
M_2: 231
M_3: 4231.

Finishing times of every operation:

$T(i, j)$	J_1	J_2	J_3	J_4
M_1	3	12	9	5
M_2	9	2	6	
M_3	12	6	10	3

Makespan value of this solution: $C_{max} = 12$.

9.3 Some Ideas Behind Heuristics for Manufacturing Scheduling

Once the detailed description of two representative heuristics has been presented, in this section we will outline some of the main ideas behind many heuristics for manufacturing scheduling. We first give some general ideas which are then developed in several subsections. These general ideas are:

- **Relaxation**. A first idea of scheduling procedures is to *relax the model* under consideration until the remaining formal problem becomes tractable. Afterwards the solution of the formal problem is adjusted and modified somehow to a solution of the real-world problem. This approach, as is well-known, is common to almost every model building, not even optimisation-oriented, in almost every real-world based setting.
- **Decomposition**. A second type is to *decompose the problem* under consideration until the remaining partial problems become tractable, to solve these partial

problems afterwards and to *coordinate* the partial solutions finally to receive an integrated solution. This approach is very common to manufacturing scheduling problems, may it be e.g. that scheduling tasks in different plants have to be performed simultaneously, for different manufacturing stages in one shop or for different machines on one stage. These examples imply a machine- or resource-oriented decomposition approach. However, job- or operation-oriented and/or time-oriented decomposition approaches can be developed as well. A main aspect for this approach is, if possible, to identify and treat the core or bottleneck problem (may it be given by a machine, a stage, a job etc.) as a whole and to coordinate the solution of this core problem with its environment.
Decomposition of a scheduling problem might be even pushed so far that on the most detailed levels only *'atoms'* of the problem are scheduled, e.g. jobs waiting in front of a machine are sequenced (and scheduled) by simple or more sophisticated *priority or dispatching rules*. Afterwards, these sequences have to be coordinated to an integrated schedule for the overall scheduling problem under consideration.
- **Adaptation**. Another, very often adapted way for generation of heuristics consists of identifying a smaller or simpler subproblem where a certain approach performs good or even optimal, this *approach is modified according to the more complex situation* under consideration and supposed to yield good results here as well. A prominent and well-known approach belonging to this group is Johnson's algorithm for the 2-machine flow shop problem with makespan objective which has been adapted in many different ways to more general problems, such as the m-machine flow shop problem with makespan objective by the heuristic of Campbell et al. (1970).
With respect to the inclusion of additional resources into a scheduling problem and their temporal exclusion from the respective formal problem, time or capacity *buffers* for jobs/operations and/or resources might give some tolerances to include the limited resources afterwards. (This approach is less concerned with the solution procedure but more with buffers for the input data. In addition, these buffers might not only be included in the model input data but also be implemented after deriving an implementable solution from the formal problem's solution.)
- **Reversibility**. The structure of a schedule as result of some manufacturing scheduling procedure is often *forward-oriented*: Starting from the beginning (e.g. the assignment of the first operation to the first machine) the schedule ends with the final assignment (e.g. the last operation on the last machine). However, using just the opposite, *backward-oriented* perspective (i.e. first assigning the last operation on the last machine and then going backward until the first operation on the first machine has been scheduled) in many settings might give another solution to the decision maker which might even be more in line with the requirement of minimising tied-up capital or Just in Time-delivery.

We want to point out that the ideas and approaches mentioned in this section are by far neither comprehensive nor is the way of their presentation used here mandatory. Many other structuring and other aspects will be found definitely. We restricted ourselves to some main aspects employed in scheduling heuristics. On the other hand, these ideas are not neccessarily confined to the manufacturing scheduling field. Finally, some of these ideas could be incorporated into metaheuristics.

Some specific guidelines can be constructed upon the general ideas outlined above. They start with the consideration of the single-operation as the basic unit of a scheduling model, thus ignoring the possibility of interrupting operations and generating sub-operations thereby. More specifically, we will assume that an operation

- belongs to a job and requires a machine, i.e. it combines a job and a machine,
- possibly requires additional resources,
- cannot be started until certain predecessing operations (usually of the same job) have been finished (or must be finished before its successors can start),
- has to be assigned to one or more (consecutive) time periods,
- ...

Model-based complexity reduction according to the framework discussed in Sect. 7.3.4 therefore implies, implicitly or explicitly,

- reduction of the number of operations to be considered in the reduced model,
- complexity-related modification of some or all operations themselves,
- omission or simplification/relaxation of (resource) constraints,
- ...

With these concepts in view some ways to reduce this complexity appear. Some of them are discussed in the next subsections.

9.3.1 Exclusion of Specific Operations

An obvious but at a first glance not very manageable guideline for the determination of operations which can be excluded in the simplified model is that these operations should not contribute significantly to the quality of the solution derived finally for the original model. However, there are at least three interpretations of this guideline which might be effective and which are used in model simplification.

First, a *machine-oriented interpretation* of this guideline proposes to completely exclude non-relevant machines in the simplified model, i.e. to simplify the model by omitting all operations of all jobs on these machines—and only consider the remaining model with significantly less machines. The choice of the machines to be excluded will often be based on their (non-) bottleneck property in the manufacturing system considered. Bottleneck machines will be preserved in the simplified model while non-bottleneck machines are deleted. Especially, if the number of machines in the simplified model becomes very low (e.g. one or two), the remaining simplified model might be solved with specialised algorithms or heuristics. A typical example of this idea is the SBH already discussed in Sect. 9.2.2. In our context, one of this approach's basic ideas is to identify the bottleneck machine in advance and then confine the simplified model to a one-machine problem which only considers exactly this bottleneck machine.

However, to include this machine's interaction with other machines, on one hand for every job a release time is included which incorporates the time needed for all previous operations of every job or better: the job's expected earliest availability for the bottleneck machine under consideration. On the other hand, a follow-up time is included for every job which represents the time needed to finish all its following operations. (As we have already seen, SBH is much more elaborate. Especially its 'shifting' property is used to consecutively apply the idea iteratively to different machines which, from an application point of view, is very well relevant if there are several bottlenecks or a hierarchy of importance of machines in the system. Several bottlenecks are not as uncommon if the capacities of machines in the manufacturing system under consideration are balanced at least from a mid-term or long-term perspective.)

Second, a *job-oriented interpretation* of the above guideline induces to exclude whole jobs in a simplified model, i.e. to omit all operations belonging to these prespecified jobs. If we remember that the number of jobs in manufacturing scheduling is the main driver of complexity in many solution procedures, reducing the number of jobs in a simplification approach turns out to be a rather attractive simplification strategy from a computational point of view. From an application point of view, job selection might be executed in two ways, i.e. positive selection and negative selection. Positive selection in our context means to choose 'important' jobs to remain in the scheduling model. Importance of jobs might result from their size, from customers' and/or products' importance, from their (expected) processing time variance, etc. Negative selection means to first consider all jobs and successively eliminate unimportant jobs in the simplification process until the number of jobs becomes manageable.

Third, these two approaches to simplification, i.e. job selection and machine selection, might be combined into a *machine/job-oriented interpretation*. And the approach might be even more specific with respect to operations when single operations are included or excluded in the simplification process.

Without going into more details here, it has also to be mentioned that the above outlined simplification approaches might also require some adjustment of the objective function in the simplified model. E.g., using bottleneck ideas might imply to include opportunity cost considerations into the simplified model.

9.3.2 Modification/Aggregation of (Specific) Operations

Somewhat like a generalisation of the approaches presented above in Sect. 9.3.1 is the modification and/or aggregation of operations themselves, i.e. of their processing times. Here as well, machine-oriented and job-oriented approaches can be identified.

Machine-oriented approaches are widespread in operations aggregation. Often processing times of consecutive operations of a job are simply added up straight or a weighted sum of processing times is used, e.g. as follows for a simple flow shop problem:

$$p_{kj}^{\text{agg}} = \sum_{i=\text{lowind}(k)}^{\text{uppind}(k)} g_i p_{ij}.$$

Here the processing time p_{ij} of job j on machine i is weighted using the machine-based weight g_i and added up from machine lowind(k) through machine uppind(k) while these machines define the k-th 'macro' machine in the simplified model by using machine aggregation. Dannenbring (1977) or Park et al. (1984) approximate approaches to multi-machine permutation flow shop problems with makespan objective give examples for a 'true' weighting approach while the procedure of Campbell et al. (1970) applies machine aggregation for the same type of problem by using equal weights $g_i = 1$ for all machines. These approaches simplify, i.e. aggregate the model into a two machine model which can be solved easily to optimum by Johnson's algorithm. This solution, as an approximate approach, is then applied to the original multi-machine problem under consideration.

Job-oriented approaches construct clusters of jobs which are usually interpreted as one 'macro' job in the simplified model. It is supposed that jobs included in one of these macro jobs are then processed consecutively. The processing times of a macro job are usually constructed as

$$p_{il}^{\text{agg}} = \sum_{j=\text{lowind}(l)}^{\text{uppind}(l)} g_j p_{ij}.$$

Here the processing time p_{ij} of job j on machine i is weighted using the job-based weight g_j and added up from job lowind(l) through job uppind(l) while these jobs define the l-th 'macro' job in the simplified model by using job aggregation. It has to be mentioned that this approach requires a prior assignment of jobs to macro jobs. This is usually much less obvious as compared with the machine-aggregation approach since actually the jobs might occur in (almost) every possible sequence (per machine). Therefore, machine-aggregation approaches are often used when the standard assumption of cluster construction is fulfilled, i.e. the objects (jobs) in a cluster are rather homogeneous while objects (jobs) from different clusters differ more or less significantly. This requirement, e.g., often is supposed to be fulfilled if changeover/setup times between jobs of one cluster are negligible while changeover times between jobs of different clusters are of significant size.

Clearly, machine-oriented and job-oriented aggregation approaches might be applied simultaneously. A further generalisation of this machine-based and/or job-based view on aggregation in scheduling problems can be derived from considering the disjunctive graph presentation of the original model and looking for network partitioning and network aggregation based on this problem representation. We will not discuss this further here.

9.3.3 Exclusion of Relations Between Operations

Ignoring constraints of the original model in the simplified one represents another group of simplification approaches. E.g. excluding resource constraints, precedence constraints, the sequence dependency of changeover times, etc. usually simplifies the model significantly. However, to receive a feasible solution to the original model might require sophisticated adjustment procedures with respect to the solution of the simplified model. We will not address these approaches in more detail here.

9.3.4 Modification of Time Scale or Adjustment Policy

Another rather simple but somewhat ambivalent way to reduce the complexity of a model is to reduce the time horizon of the jobs to be considered in the model. By this approach, on one hand probably the number of jobs to be considered is reduced. This reduces the effects of a main complexity driver. On the other hand, excluding later periods from consideration might yield adjustment problems at the end of the scheduling horizon and cause additional feasibility problems.

Instead of reducing the number of jobs to be considered by reducing the time horizon in the model, this number of jobs can also be reduced by performing an add-on scheduling approach instead of starting the procedure from scratch every time a (re-) scheduling is executed. That is instead of scheduling all jobs anew only new jobs are scheduled in an optimisation run while the former jobs' sequence (or only their relative sequence on every machine) remains constant. Basically these approaches adapt the difference known from regenerative and net-change approaches in MRP to manufacturing scheduling. This approach is rather common in dynamic scheduling settings.

9.3.5 Identification of Symmetry Properties

Another (real-world) based idea of constructing an approximate solution is applying an existing approach/method to the inverted or *symmetric* model. The symmetric model inverts the machine sequence of a job's operations, i.e. the last operation (on the job's last machine) is supposed to be the job's first operation in the reverse problem, job's last but one operation is its second operation in the reverse problem and so on. For several scheduling problems it is straightforward that for every problem instance the objective function value of the original model is equal to the objective function value of the symmetric model. The idea behind this 'concept' originates from the consideration that in a network (or in a disjunctive graph) for many problem settings the longest path (e.g., the project duration or the makespan) is the same no matter whether it is calculated from the beginning to the end or vice versa. Therefore, applying a certain approximate approach to the original model or to the symmetric model probably yields two different solutions to the model (where, of course, the solution of the symmetric model has to be reverted again to yield a solution to the original model). However, these two solutions probably will both be of adequate quality and the best one of these two solutions could be finally chosen. Additionally, it should be mentioned that very often when testing approaches using randomly generated test-beds, both, the original instance as well as its symmetric counterpart represent valid problem instances. Therefore, from a numerical testing point of view, there is no difference between the application of an approach to the original or to the symmetric model when using this approach for each of these two instances. Furthermore, the increase in effort by applying an approach also to the symmetric model is nothing but duplication. And by experience, in several problem settings nice improvements might be achieved by using this combined type of approach, i.e. applying a given approach to both, the original as well as the symmetric model.

Symmetry ideas have been described, for instance, in the book of Pinedo (2012). One paper where these ideas are employed for designing a heuristic is Ribas et al. (2010).

9.3.6 Combination of Enumeration and Construction of Solutions

Enumerative constructive methods try to employ enumerative components into the generation of a solution to the scheduling problem. According to Fig. 7.4 in Sect. 7.3.4, these methods therefore refer to a given model and try to simplify the search for a solution, i.e. they usually can be assigned to situation 2 in Fig. 7.4. Simple approximate enumerative constructive methods include

- sequences and accordingly schedules resulting from the simple application of static or dynamic dispatching rules as well as
- every premature stopping of an (exact) explicit or implicit complete enumeration approach.

There is almost no limit to where imagination can take the design of enumerative constructive methods. However, several rather successful methods might be classified according to the following iterative scheme which refers to job-based iteration approaches. Per iteration, these schemes select exactly one job from the set of so far non-sequenced jobs, include this job into the (sub-) sequence(s) resulting from the previous iteration and adjust the sets of sequenced and non-sequenced jobs (as well as the respective (sub-) sequences) as follows:

Suppose that in the beginning of iteration k of the method under consideration the set J of all jobs is partitioned into the set $S(k-1)$ of those jobs which have already been sequenced in iterations 1 through $k-1$ and the remaining jobs are included in set $R(k-1)$. The set of (sub-) sequences sequenced up to iteration $k-1$, called $T(k-1)$, might be of cardinality 1 or higher.

Then one or more or even all jobs from $R(k-1)$ are tentatively and one by one selected according to some selection criterion and integrated into the (sub-)

sequence(s) of $T(k-1)$. The job k^* which yields the best (sub-) sequences is transferred from $R(k-1)$ to $S(k-1)$, i.e.

$$R(k) = R(k-1) - (k^*) \text{ and } S(k) = S(k-1) \cup (k^*).$$

$T(k)$ consists of those (sub-) sequences which have been accepted in iteration k.

Of course, this approach can be expanded in many directions. E.g.,

- some already sequenced jobs can be withdrawn from $S(k)$ and be put back into to $R(k)$,
- several jobs might be assigned to $S(k)$ simultaneously,
- the above approach might be applied to some tree structure, i.e. several jobs can be alternatively assigned to $S(k)$ and each of the resulting alternatives is considered in the following iterations separately,
- ...

All these expansions have to be defined such that the resulting effort for the method remains acceptable, i.e. the computational effort does not grow exponentially but remains polynomially bounded.

9.3.7 Generation of Variants of Efficient Heuristics

Another possibility is to analise the behaviour of existing heuristics and trying to generate new variants by performing a 'reverse engineering' of the heuristic, i.e. decomposing the original heuristic into blocks and generating variants for the blocks. Obviously, the development of this idea is very heuristic-specific, and in order to illustrate it, we take the NEH heuristic already discussed in Sect. 9.2.1. Recall that this heuristic (a) starts from an initial sequence, and (b) construct and evaluates partial sequences in an iterative manner. It has been discussed whether the initial sequence of step (a) or the numerous (partial) sequences considered in step (b) contribute more to the very good performance of NEH. Framinan et al. (2003) show that depending on the objective function other sorting criteria in (a) may be advantageous. However, the numerous sequences considered in (b) contribute as well a lot to the quality of the heuristic.

With respect to generalisations of the NEH approach, several extending links can be identified:

- The sequence of involvement of jobs may be modified. Woo and Yim (1998), e.g. use an ascending ordering of jobs for the permutation flow shop problem with flowtime objective. Also, several different sequences of involvement may be used to generate several schedules and the best of these schedules is then selected as solution to the scheduling problem (see Framinan et al. 2003).
- More than one job can be selected in the iterative loop (b) to be included into the sequence. Again, Woo and Yim (1998) choose every remaining job to be

inserted in every iteration of (b). This approach remains with polynomial complexity and yields good results for the flowtime flow shop problem but is rather time-consuming.

- As already mentioned, ties may occur both in (a) and (b). A sophisticated treatment of ties may improve the approach's performance noticeably. Nevertheless, keeping the computational effort polynomially bounded might be an issue in this case.

9.3.8 Vector Modification

Since manufacturing scheduling models are specific combinatorial optimisation problems and the solutions of these problems can usually be represented as vectors, many improvement approaches are related to vector modification.

First of all, it has to be mentioned that for some models, different schemes of vector representation for the same schedule may exist. Applying the same improvement method to different vector representation schemes might result in different solutions to be checked for improvement and, consequently, also in different improved solutions. We only allude to this aspect here.

Supposing a solution vector, i.e. a feasible schedule is given, many vector manipulation approaches can be designed. These approaches normally withdraw one or more operations (or even whole jobs) from the vector and place it/them at other position(s) in this vector. The approach might be repeated until a certain stopping criterion has been fulfilled. A most simple and sometimes rather efficient vector manipulation scheme is to exchange each two consecutive positions and to (a) look whether the new vector maintains feasibility and (b) yields a better objective function value. This procedure is repeated, e.g., until no further improvement is possible.

More general and supposing that a solution vector is given, vector manipulation approaches consist of the following components:

1. Determine the position(s) of the current vector from where operation(s)/job(s) are withdrawn.
2. Determine the position(s) where this/these withdrawn operation(s)/job(s) are re-inserted.
3. Modify the vector under consideration according to (1) and (2). (Maybe, more than one vector results from this procedure. Step (2) defines the so-called neighbourhood of the vector(s) from (1) which is/are examined by the vector modification approach under consideration.)
4. Check acceptance of the new vector(s) with respect to
   (a) Feasibility
   (b) Acceptance of objective function value.
5. If a pre-specified termination criterion (e.g. maximum number of iterations has been reached, no adequate improvement has been reached,...), then stop.
6. Use the accepted vector(s) from (4) as input for the next iteration and go to (1).

Three additional remarks have to made with respect to this scheme:

1. With respect to 4(b) it has to be mentioned that in its simplest version usually a new vector will be accepted if its objective function value improves the so far best solution. However, because of the pure greedy character of this approach and because of the non-convexity of the underlying combinatorial optimisation problem, this procedure might result in a suboptimal dead end very quickly as has been experienced many times. Therefore, there are approaches which explicitly allow for temporary worsening of the objective function value of intermediate solutions or vectors respectively, just to overcome this quick ending in a suboptimal solution. The extent to which such a temporary worsening of the objective function value is accepted represents another control variable of the enumerative improvement method. Usually, this extent of allowed worsening decreases with the number of iterations of the improvement process. Additionally, the extent of worsening might be determined stochastically, e.g. in combination with a decreasing expected value of accepting this worsening and in combination with a decreasing variance.
2. With respect to (1) and (2) the position(s) of change might be selected deterministically or according to some probabilistic scheme. Random determination of deletion and insertion positions is obviously the most simple of these schemes. However, more sophisticated determination of deletion and insertion locations, e.g. including improvement experience from the last iterations, might be appropriate, may it be deterministically or stochastically.
3. Another aspect which we only allude to here is the requirement that the effort for executing a vector modification approach must remain limited. Therefore, usually a time limit and/or some iteration limit will be set for the procedure. However, even if these two limitation parameters have been determined, (1) and (2) from the above scheme easily might suggest enumeration schemes which can become rather quick rather complex with respect to their computational effort because of the numerous solutions to be considered. Therefore, the limitation of the computational effort has to be kept track of when designing the above procedure, especially with respect to steps (1) and (2).

9.3.9 Modification Approaches Based on Real-World Features

Numerous approaches try to improve a given solution to a manufacturing scheduling problem by analyzing the real-world structure of a given solution. E.g. reducing idle time of machines by bringing forward operations into idle slots of machines might positively contribute to flowtime and makespan while postponing to later idle slots can contribute to reduced earliness. Every plausible consideration with respect to the real-world problem might be checked regarding the modification of solutions. These modifications might be considered myopically, i.e. sequentially for different operations, or jointly for several operations. Note that the diversity of this type of approaches is again enormously.

Another type of approach is to include the human planners' expertise for a given schedule, i.e. to explicitly address the men-machine-interface. Corresponding approaches have turned out to be more or less successful empirically in many settings. Especially, if formerly manually solved manufacturing scheduling problems in practice are automated, on one hand they usually have to be competitive with the former manual processing which then serves as a benchmark with respect to the relevant indicators or objective function(s) for the automated solution. On the other hand, allowing the human planner 'to have a look' at the automated solution might improve even this solution and/or will probably increase acceptance of the automated solution within the real-world manufacturing setting. However, note that these third-type approaches refer to planners' intuition which might be mostly non-quantifiable, including the application of specific methods 'just in time'.

9.4 Metaheuristics

A metaheuristic can be seen as a general algorithmic framework which can be applied to different optimisation problems with relatively few modifications to make them adapted to a specific problem. Metaheuristics are commonly grouped into large classes of methods that need instantiation for a given problem. Although metaheuristics can be seen as a specical type of improvement heuristic, they tend to be more robust and flexible, in the sense that there is a common metaheuristic template for each class of metaheuristic methods that, after instantiation, is able to give reasonable solutions to mostly any scheduling problem. Furthermore, metaheuristics are usually able to obtain much more refined solutions. However, this usually comes at a higher computational cost and complexity than heuristics, although this later fact depends on the level and grade of the instantiation.

Metaheuristics are grouped, as has been stated, in different classes. The earliest methods are population and/or evolutionary based algorithms like genetic algorithms and evolutionary programming. Other classes include simulated annealing, tabu search, ant colony optimisation, particle swarm, estimation of distribution, differential evolution, scatter search, harmony search, honey bees, artificial immune systems and a really long list of other methods. In this section, we first discuss the main concepts in the general framework of metaheuristics, and then present some of the main types of metaheuristics, i.e. simple descent search methods (Sect. 9.4.2), simulated annealing (Sect. 9.4.3), tabu search (Sect. 9.4.4), and genetic algorithms (Sect. 9.4.5). Some other metaheuristics are briefly mentioned in Sect. 9.4.6.

9.4.1 Main Concepts

As a general algorithmic framework, a metaheuristic is composed of a set of concepts that can be used to define or to guide specific heuristic methods that can be applied to a wide set of different problems. These concepts are mainly the following:

- Problem or solution representation.
- Initialisation.
- Neighbourhood definition.
- Neighbourhood search process.
- Acceptance criterion.
- Termination criterion.

These concepts are discussed in detail in the next subsections.

9.4.1.1 Solution Representation

Metaheuristics commonly work with an abstract representation of solutions. Particularly, in scheduling settings, a metaheuristic would often work with a sequence of jobs, rather than with a full schedule. We have already seen that this is also the case with regular construction heuristics. This solution representation abstraction is key to most metaheuristic classes. Picture for example a metaheuristic working over a permutation of jobs as a representation. This permutation could be the sequence for a single machine or permutation flow shop layout, regardless of the objective and/or constraints present in the problem.

The choice of the representation is crucial to metaheuristic performance. An indirect representation might be more compact, but large parts of the solution space might not be possible to represent. On the contrary, an overdetailed representation might capture the complete solution space but might result in a cumbersome and complex metaheuristic.

An example of a permutation representation of a 1-machine, 20-job problem is given in Fig. 9.11. We can see that the permutation representation contains 20 positions. The solution representation space is formed by all possible permutations in Π of 20 jobs. The solution space has a cardinality of 20!. One possible permutation $\pi \in \Pi$ could be the one shown in the same Fig. 9.11 where the job in the first position of the permutation is $\pi_{(1)} = 7$, the job in the second position is $\pi_{(2)} = 12$, and so on until the job in the 20-th position: $\pi_{(20)} = 8$.

Note that a permutation solution representation could be used for many different scheduling problems, including single machine, flow shop or even hybrid flow shop layouts if some machine assignment rules are employed, as those explained before in Sect. 7.4. However, this representation is not valid for all problems. In parallel machine layouts, it is not easy to separate the machine assignment decision from the sequencing problem. Therefore, other representations are needed. The most common is the 'multi-permutation' where each machine has a permutation of jobs which

1	2	3	4	5	6	7	8	9	10	11	12	13	14	15	16	17	18	19	20
7	12	2	5	13	9	3	1	17	19	10	6	15	20	18	11	16	4	14	8

Fig. 9.11 Example of a permutation encoding with 20 jobs

Fig. 9.12 Example of a multi-permutation encoding with 20 jobs and three machines

	1	2	3	4	5	6	7
Machine 1	7	13	20	14	1	3	6
Machine 2	12	11	5	19	10	8	
Machine 3	2	9	18	17	15	16	4

indicate the order in which they are processed on that machine. Notice that the term is not very accurate as each job has to be assigned to exactly one machine. As a result, each machine does not really contain a permutation of all the jobs, but rather a list of the assigned jobs. An example for a three machine, 20 job representation is shown in Fig. 9.12.

As we can see, every job appears once, but not all machines have the same number of jobs assigned. In the example, we can see that $\pi_{(2,4)} = 19$ i.e. the job in the fourth position of the second machine is the job 19. Notice that for complex problems, like for example, a hybrid job shop lay out with stage skipping and recirculation, much more complex representations are needed. Figure 9.13 shows a possible three stage hybrid flow shop with two, three and two parallel machines at each stage, respectively.

9.4.1.2 Initialisation

In general, a metaheuristic needs at least one starting solution. This initial solution is usually referred to as 'seed'. For some problems, this solution might just be generated at random. However, a random solution is expected to be of a very low quality. In some other cases, a random solution might even be unfeasible. For these reasons, it is usually common in the scheduling methods literature to initialise metaheuristics with good known constructive heuristics. In the very least, there is nothing wrong in using a good dispatching rule to seed a metaheuristic algorithm.

Some metaheuristic methods however, rely on a completely random initialisation as a heuristic initialisation might bias the search process to a specific region of the solution space, possibly hindering to find a global optimum solution in some way or another. Additionally, for some problems, there are no good known heuristics and/or dispatching rules do not offer sufficiently good quality solutions. In all these cases, it is common to randomly initialise metaheuristic algorithms.

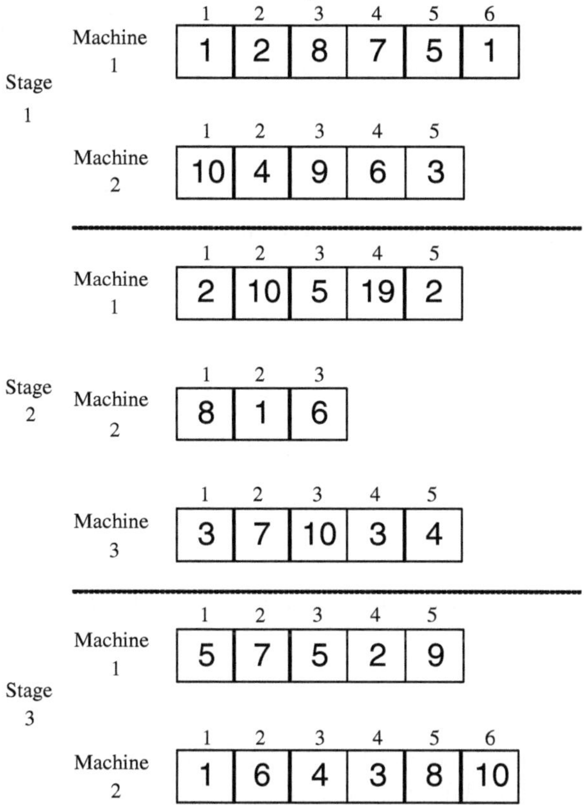

Fig. 9.13 Example of a more complex representation for hybrid layouts

9.4.1.3 Neighbourhood Definition and Search Process

Given the solution representation and also given the impossibility of examining the whole cardinality of the solution representation space, which is often of exponential size, the next concept that is needed in metaheuristics is the neighbourhood of a solution. The neighbourhood of a solution is a set of all possible solutions in the representation space that can be reached by carrying over a small local movement in the solution. For example, given a permutation, a simple neighbourhood is the one in which two adjacent jobs exchange their positions. This neighbourhood is called the adjacent swap. Let us define this adjacent swap neighbourhood in a more formal way. Let us suppose that we have a sequence of jobs expressed as a permutation $\pi = (\pi_{(1)}, \pi_{(2)}, \ldots, \pi_{(n)})$, one adjacent swap neighbourhood is obtained by selecting each possible job in position j and its immediately adjacent job in position k where $j = 1, \ldots, n-1$, $k = j+1$ and swapping the two jobs or also, the jobs in these two positions of the permutation. Given these two positions j and k, and the corresponding jobs in these two positions, namely, $\pi_{(j)}$ and $\pi_{(k)}$ the resulting sequence after the swap will be $\pi' = (\pi_{(1)}, \ldots, \pi_{(j-1)}, \pi_{(k)}, \pi_{(j)}, \pi_{(j+1)}, \ldots, \pi_{(n)})$.

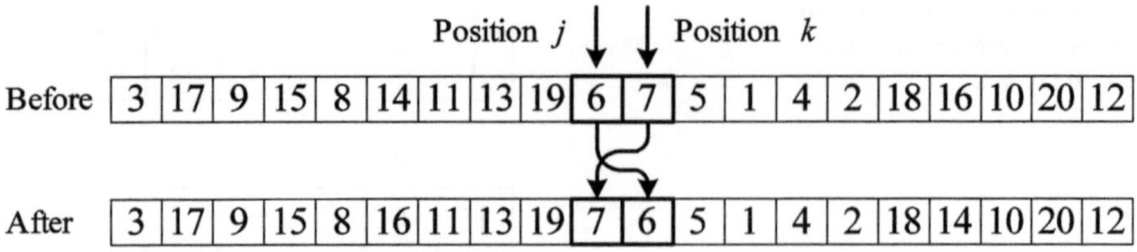

Fig. 9.14 Adjacent swap neighbourhood (A) example

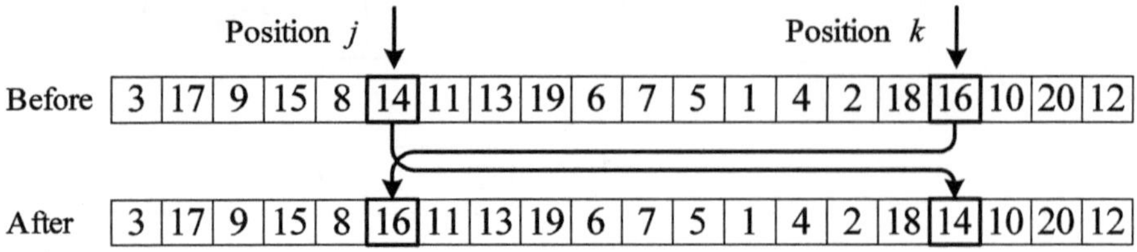

Fig. 9.15 General swap neighbourhood (S) example

More specifically, we can define an adjacent swap 'move' as a duple (j, k) where $j = 1, \ldots, n-1, k = j+1$. The whole set of adjacent swap moves can be defined as $A = ((j, k) : j = 1, \ldots, n-1, k = j+1)$. As we can see, the number of possible moves or cardinality of the adjacent swap neighbourhood is $|A| = n - 1$. This means that any possible permutation or sequence of jobs π has $n - 1$ adjacent swap neighbours. The adjacent swap neighbourhood of a permutation sequence π is finally and formally defined as: $N(A, \pi) = (\pi_v : v \in A)$. A graphical representation of this move is shown in Fig. 9.14.

An immediate generalisation of the adjacent swap neighbourhood is the general swap neighbourhood. In this case, the two positions j and k need not be adjacent. We can then define the general swap neighbourhood as a duple (j, k) where $j = 1, \ldots, n-1, k = j+1, \ldots, n$. The general swap movements are defined as $S = ((j, k) : j = 1, \ldots, n-1, k = j+1, \ldots, n))$. The cardinality of this neighbourhood is greater than that of A since $|S| = n \cdot (n-1)/2$. Formally, the general swap neighbourhood of a sequence π is defined as: $N(S, \pi) = (\pi_v : v \in S)$. Given two jobs in positions j and k where $j = 1, \ldots, n-1, k = j+1, \ldots, n$, the resulting sequence after swapping the jobs in these positions will be: $\pi' = (\pi_{(1)}, \ldots, \pi_{(j-1)}, \pi_{(k)}, \pi_{(j+1)}, \ldots, \pi_{(k-1)}, \pi_{(j)}, \pi_{(k+1)}, \ldots, \pi_{(n)})$. This is graphically depicted in Fig. 9.15.

Another very common neighbourhood employed in permutation sequences is the insertion neighbourhood. This is induced by all local movements where there are two jobs at positions j and k, where $j \neq k$, and the job at position j is extracted and reinserted at position k. More formally, the set of insert movements I is defined by all duples (j, k) of positions where $j \neq k,\ j, k = (1, \ldots, n)$ that, starting from any sequence π result in a sequence

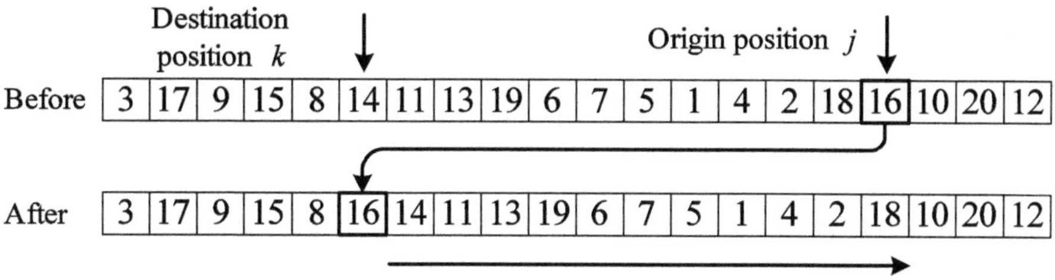

Fig. 9.16 Insert neighbourhood (I) example

$$\pi' = \pi_{(1)}, \ldots, \pi_{(j-1)}, \pi_{(j+1)}, \ldots, \pi_{(k)}, \pi_{(j)}, \pi_{(k+1)}, \ldots, \pi_{(n)}$$

if $j < k$, or in a sequence:

$$\pi' = \pi_{(1)}, \ldots, \pi_{(k-1)}, \pi_{(j)}, \pi_{(k+1)}, \ldots, \pi_{(j-1)}, \pi_{(j+1)}, \ldots, \pi_{(n)}$$

if $j > k$.

The set of insertion moves I is then defined as $I = ((j, k) : j, k = (1, \ldots, n), j <> k)$ and the insertion neighbourhood as $N(I, \pi) = (\pi_v : v \in I)$. Note that the cardinality of I is greater than that of A or S as $|I| = (n-1)^2$. A graphical example of the insertion neighbourhood is given in Fig. 9.16.

Of course, the previous three neighbourhoods are not the only possible ones. There are many other neighbourhoods, like reverse, block swap, block insert, three job insert, etc.

Once the neighbourhood has been defined, a metaheuristic method usually explores neighbours of a given solution in a given way. The general idea is that good solutions are clustered close to each other, i.e. a very good solution will have a hopefully even better one in the neighbourhood. Let us picture an example in the single machine layout with a permutation representation. The objective is the total tardiness minimisation. A very good starting solution has been found by applying the Earliest Due Date (EDD) dispatching rule. It is straightforward to see that an even better solution can be found by selecting a job that is being completed early and swapping it with another job that is being completed late. Normally, the late job will be processed earlier in the sequence and the early job later in the sequence. This is an example of a local movement in the general swap neighbourhood. Hopefully, after this swap, the early job is not tardy and the tardy job is now early. As a result, the total tardiness will have been reduced even further.

With the previous example, neighbours of a solution can be examined sequentially or randomly. Also, given a solution, all neighbours might be examined, and the 'best' selected or the first 'better' neighbour might be selected. This is known as 'best improvement' or 'first improvement', respectively.

There is a known trade-off between the cardinality of the neighbourhood, the quality of the expected solutions and the time it takes to traverse the neighbourhood. Obviously, larger cardinality neighbourhoods are more expensive to examine but higher quality solutions are expected. What to do with new solutions is discussed next in the acceptance criterion.

9.4.1.4 Acceptance Criterion

Usually, neighbouring solutions will result in different objective function values. As such, metaheuristic methods have to implement an acceptance criterion to decide if the incumbent or current solution has to be replaced by any of its neighbours. Again, the underlying idea is to move to the best neighbour and then, from that best neighbour, move again to the best neighbour and continue until no more improvements are found. Accepting only better solutions is not the only possible approach. More elaborated methods allow to move to worse solutions temporarily, in the hope of finding better solutions later on. When accepting a worse solution, there might be probabilistic decisions as how much deterioration in the objective function is allowed. As a result, the acceptance criterion can be highly sophisticated.

9.4.1.5 Termination Criterion

Contrary to constructive heuristics, that commonly finish when a full sequence has been built, metaheuristics iterate until a specific, usually user given, termination criterion is met. The choice of stopping criterion is varied:

- Stop after a predefined number of iterations and/or elapsed time.
- Stop after a predefined number of iterations and/or elapsed time without improvement in the objective function value.
- Stop after a predefined number of iterations and/or elapsed time where no more than an X % improvement in the objective function value has been observed.

Some scheduling problems are known to have tight lower bounds that can be calculated, i.e. best case objective function values that are known a priori. Therefore, it is possible to run a metaheuristic method and to stop it when a given threshold deviation from the best possible optimal solution has been surpassed. However, tight lower bounds are only known for some simple layouts and rather unconstrained scheduling settings.

9.4.2 Simple Descent Search Methods

With all previous elements of metaheuristics, it is possible to define a simple method. Simple descent or iterative descent, also referred to as hill climbing local search, basically starts from an initial feasible solution and searches through a given neighbourhood while improvements are found. The algorithm template is given in Algorithm 3.

It has to be noted that the symbol '$\lhd$' is used as 'better than' in the listing of Algorithm 3. Note that basically, the simple descent local search method starts from an initial job sequence and iterates in a given neighbourhood, looking for better neighbours until all the given neighbours of the best solution found so far are not better, in which case the algorithm stops. The previous listing of Algorithm 3 is actually referred to as a first improvement type of descent local search since given a

Algorithm 3: Simple descent local search

Input: Instance data
Output: Best solution π^* and best objective function value obj^*
begin
  Calculate initial solution π_i;
  Set best known solution $\pi^* := \pi_i$;
  Calculate best objective function value so far $obj^* := Obj(\pi_i)$;
  improving:=**True**;
  while *improving*=***True*** **do**
    improving:=**False**;
    foreach *neighbour* π' *of* π^* *in V* **do**
      Calculate objective function value of neighbour $obj' := Obj(\pi')$;
      if $obj' \lhd obj^*$ **then**
        $\pi^* := \pi'$;
        $obj^* := obj'$;
        improving:=**True**;
  return *Best solution* π^**, best objective function value* obj^*
end

solution, the first improving neighbour replaces the current solution. An alternative would be to examine all neighbours of a given solution sequentially, and replacing the current solution only with the best neighbour.

With the computing power available today, it is fairly straightforward to add a simple descent local search method as a post-processing stage after a solution has been obtained with a dispatching rule or heuristic algorithm. In the worst case scenario, for example in big instances where there might be thousands of improving iterations, the descent local search procedure can be stopped after a given predefined time has elapsed. However, there are many scheduling problems where neighbourhoods are hard to define, and/or each candidate solution is complex to calculate. In such scenarios, iterating through the neighbourhood can be a costly procedure.

9.4.3 Simulated Annealing

One big drawback of the simple descent local search methods is that they might get easily trapped in local optimum solutions. This is expected as they are specifically designed to stop at a solution that has no better neighbours. The problem is that in many cases, after just a few iterations, the algorithm will stop in a fairly mediocre local optimum. The problem of escaping from local optima is approached from many different perspectives. The simplest method could be an iterated simple descent local search where the simple descent local search is iteratively applied to a different random initial solution each time. Although this method will, after some iterations, provide a better solution than the simple descent local search, the probability of finding the optimum solution is slim. Recall that the search spaces in scheduling problems tend to be exponential in size and randomly performing local search is like searching for a needle in a haystack.

One of the earlier methods to cope with the local optimality trapping problems is the simulated annealing metaheuristic. Simulated annealing, as the same implies, emulates the physical annealing process where materials of crystalline structure are slowly cooled off and re-heated in order to alleviate tensions and to foster a perfect crystalline structure. The analogy to local search is the following: During the search, neighbours are always accepted if they are better than the current solution. However, if the first or best neighbour is worse, it will be probabilistically accepted with a diminishing probability that depends on the time that has elapsed (cooling schedule: the more time the algorithm has been running, the lower the probability to accept worse solutions) and also on the difference between the current solution and the best non-improving neighbour (the higher the difference, the lower the probability to accept a worse solution). The template of Simulated Annealing is given in Algorithm 4.

Note that *Random* returns a random uniformly distributed real number between 0 and 1. The probabilistic expression $e^{-\frac{D}{T}}$ returns a diminishing number as D increases and/or as T decreases. Furthermore, $0 < r < 1$ is referred as the cooling rate and it determines the cooling schedule.

As one can see, a high initial temperature and an r value close to 1 will ensure that a large number of iterations will be carried out. Some more advanced designs add a number of iterations L at each temperature level and other complex settings.

Simulated annealing was proposed for optimisation problems by Kirkpatrick et al. (1983) and by Černý (1985). Scheduling applications appeared in the late eighties and early nineties with the works of Osman and Potts (1989) and Ogbu and Smith (1990) to name just a few.

Simulated Annealing, as many other metaheuristics, employ different parameters that might affect the performance of the algorithm. The listing of Algorithm 4 shows some parameters as the initial temperature T_{ini} and the cooling rate r, apart from the initial solution. These parameters have to be tuned for performance and efficiency and there is a whole area in experimental algorithms devoted to the tuning of metaheuristics. Some relevant texts are those of Birattari (2005) and Hoos et al. (2005).

9.4.4 Tabu Search

Tabu Search or Taboo Search is actually a family of methods that, similarly to simulated annealing, include some mechanisms to escape from local optima. A key difference with simulated annealing is that, in tabu search, the incumbent solution is always replaced with the best neighbourhood, regardless of whether this neighbour is actually better or worse than the current solution. Obviously, this approach can lead to infinite loops in which the search moves from one solution π_1 to another neighbouring solution π_2 and from that one again back to π_1. In order to avoid this, tabu search methods keep a list of recently visited solutions. This list is what gives

Algorithm 4: Simulated Annealing

```
Input: Instance data
Output: Best solution π* and best objective function value obj*
begin
    Calculate initial solution π_i;
    Set best known solution π* := π_i;
    Calculate best objective function value so far obj* := Obj(π_i);
    Set current solution π_c := π*;
    Set initial temperature T := T_ini;
    while Stopping criterion is not satisfied or not frozen do
        foreach neighbour π' of π_c in V do
            Calculate objective function value of neighbour obj' := Obj(π');
            if obj' ◁ obj* then
                π* := π';
                obj* := obj';
            else
                Calculate difference in objective function D := obj' − obj*;
                Accept worse solution probabilistically;
                if Random ≤ e^(−D/T) then
                    π_c := π';
        T := r · T;
    return Best solution π*, best objective function value obj*
end
```

the name to this family of metaheuristics as it is called the Tabu list. The tabu list usually does not contain full visited solutions, but rather elements of the visited solutions or the movements that were carried out in the solution elements at the most recent iterations. Therefore, when moving from one solution to the best neighbour, the movement is not accepted if it leads to a tabu solution. The result is a methodology that has a short term memory that avoids the formation of cycles in the search.

A comprehensive reference on the mechanisms of Tabu Search is Glover and Laguna (1997), while some examples of its application to manufacturing scheduling are Widmer and Hertz (1989); Nowicki and Smutnicki (1996) and Nowicki and Smutnicki (1998) to name just a small set.

9.4.5 Genetic Algorithms

Genetic algorithms were initially proposed by Holland (1975) and later by Goldberg (1989). More recent books are those of Michalewicz (1996) and Davis (1996). The underlying idea greatly departs from the previous simulated annealing or tabu search methods. Genetic algorithms mimic evolutionary biological processes such as inheritance, mutation, selection, crossover, natural selection and survival of the fittest.

Genetic algorithms are also different from earlier metaheuristics in that they are population based. The search is not carried out over one single solution but rather over a set of solutions. This set of solutions is referred to as 'population' and each member of the 'population' is called an individual. Individuals are actual solutions to the problem, or rather, solution representations. In the genetic algorithm jargon, each individual is formed by a 'chromosome' in which the solution elements are encoded. For scheduling problems, chromosomes can just be the solution representations, for example, job permutations that form sequences.

The population is evaluated and all individuals are assigned a 'fitness value'. This fitness value is directly related with the objective function value, in such a way that better objective function values are assigned higher fitness values. A selection operator is introduced. This operator randomly chooses individuals from the population with a strong bias towards fitter solutions. As a result, the fittest individuals have greater probabilities of being selected. Selected individuals undergo a crossover procedure in which new individuals are generated. The crossover operator tries to generate new solutions by combining the best characteristics of the fitter selected individuals. A further mutation operator is introduced to randomly alter some characteristics of the newly created individuals. Lastly, the new solutions are evaluated and introduced in the population. The previous process is repeated generation after generation.

Algorithm 5: Genetic Algorithm

Input: Instance data
Output: Population *Pop* of solutions, including Best solution π^* and best objective function value obj^*
begin
  Initialise a population *Pop* of P_{size} individuals;
  Evaluate individuals;
  while *Stopping criterion is* **not** *satisfied* **do**
    Select individuals from *Pop*;
    Cross individuals;
    Mutate individuals;
    Evaluate individuals;
    Include individuals in the population;
  return *Pop, Best solution* π^**, best objective function value* obj^*
end

Genetic algorithms are exotic as far as metaheuristic methods go. When first exposed to them, the natural reaction is to wonder how this seemingly complex mechanism could work. Actually, once individuals are seen as simple representations of actual job schedules, the process is much better understood.

For example, a simple genetic algorithm using job permutations, for example, for a single machine layout scheduling problem could be based on the template shown in Algorithm 5.

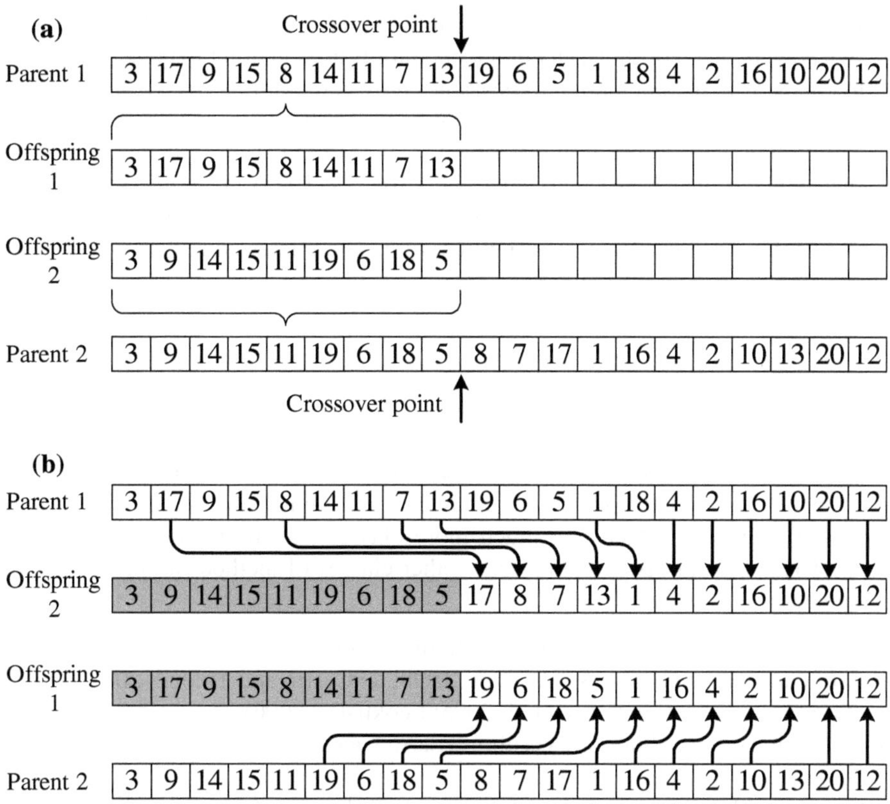

Fig. 9.17 One point order crossover in genetic algorithms. **a** Offspring inherit the jobs prior to the random cut point from the direct parent. **b** Remaining jobs are inherited from the other parent in their relative order

A population of, lets say, 50 individuals could be randomly generated as simple random job permutations. Fitness evaluation could simply be the objective function value. The most simple selection scheme is called the binary tournament selection, where two random individuals are selected from the population and the one with the best objective function value is selected as the first parent. The process is repeated to select a second parent. These two parents are crossed over to generate two new solutions as for example shown in Fig. 9.17 for what is called the one point order crossover.

As we can see, a random crossover point is generated and the jobs prior to that point are copied to one of the new solutions from a direct parent. Then, the jobs not already present in the permutation, are copied from the other parent in the order in which they appear. It is interesting to see how the new solutions contain characteristics of both previous solutions. This can be seen as a complex form of local search.

The mutation operator can be, for example as simple as a single adjacent job swap, or a single job insertion. After crossover and mutation, the new solutions are evaluated and usually, replace the old solutions that were selected. In some more advanced genetic algorithms, an elitism operator is used in order to preserve the best individuals from the population to be replaced by new solutions of worse quality.

Applications of genetic algorithms to scheduling problems appeared as early as in the middle 1980s in Davis (1985) and also at Mattfeld (1996); Syswerda (1996); Biegel and Davern (1990); Chen et al. (1995); Reeves (1995) and more recently, Ruiz et al. (2006). Note that this is just a small list as genetic algorithms are among the most prolific metaheuristics from the literature. The interested reader is redirected to such texts for more information.

9.4.6 Other Metaheuristics

There are countless metaheuristic methods or classes of metaheuristics. We have previously commented briefly over simulated annealing, tabu search and genetic algorithms. However, there are much more classes and the number continues to grow. The following list contains some other classes of methods along with some references.

- Ant Colony Optimisation (Dorigo and Stützle 2004).
- Artificial Immune Systems (Dasgupta 1998; de Castro and Timmis 2002).
- Differential Evolution (Price et al. 2005; Feoktistov 2006).
- Estimation of Distribution Algorithms (Larrañaga and Lozano 2002; Lozano et al. 2006).
- Greedy randomised adaptive search procedure (Feo and Resende 1989, 1995).
- Harmony Search (Geem et al. 2001).
- Particle Swarm Optimisation (Kennedy et al. 2001; Clerc 2006; Engelbrecht 2006).
- Scatter Search (Laguna and Martí 2003).
- Variable Neighbourhood Search (Mladenovic and Hansen 1997; Hansen and Mladenovic 2001).

The previous list is not comprehensive, as there are other not as well known methods like Iterated Local Search, Threshold Accepting, Constructive Genetic Algorithms, Iterated Greedy, Dynamic Local Search, Bees Algorithm, as well as a whole myriad of hybrid methods. Additionally, there is a host of general texts on metaheuristics like Glover and Kochenberger (2003); Resende et al. (2004); Hoos et al. (2005); Ibaraki et al. (2005) and Doerner et al. (2007). As of late, the field of metaheuristics is evolving to a more holistic paradigm which is being known as 'computational intelligence' with some books already published, like those of Konar (2005); Eberhart and Shi (2007) or Engelbrecht (2007).

9.5 Other Techniques

This chapter presented some main ideas to design and to evaluate approximate algorithms for manufacturing scheduling problems. However, the large variety of such algorithms prohibits their comprehensive overview or even a comprehensive overview of the respective ideas. For instance, it should be mentioned here again that simply stopping a (possibly, but not necessarily exact) optimisation approach before its *pre-defined* termination, e.g. for time or other resource reasons, obviously defines a class of heuristics as well.

In addition, apart from the different methods reviewed in this chapter, it should be mentioned that additional heuristic solution procedures might result from automated integrated or non-integrated decision support systems, such as solution modules provided by approaches from artificial intelligence or agent-based systems. These methods will not be addressed here, but instead some further remarks on a general technique for generating approximate solutions (Lagrangean relaxation), and the combination of exact approaches with metaheuristics) are presented.

9.5.1 Lagrangean Relaxation

As already mentioned, a well-known (sometimes exact, sometimes only approximate) approach to the solution of optimisation problems is Lagrangean relaxation, i.e. the dualisation of constraints into the objective function by penalizing violations of these constraints in the objective function value using Lagrangean multipliers as unit costs of violation. Since manufacturing scheduling models usually are of combinatorial, i.e. binary or integer type, a non-zero duality gap will make Lagrangean relaxation almost ever an approximate and not an exact method in scheduling.

Apart from the fact that this approach basically can be applied in a formal way to more or less every constraint of every optimisation problem (at least as an approximate approach), in manufacturing scheduling there are settings where this approach is/can be used implicitly. Next we sketch this approach for the specific case of flowtime minimisation with due dates/deadlines of jobs. This problem could be solved in (at least) three manners:

1. The problem can be solved as a pure flowtime problem (without considering the due dates). The output would maybe yield that there is no feasible or acceptable solution because of the tight and strict deadlines.
2. The problem can be solved as a 2-objective problem including flowtime and tardiness as objectives. This approach would disclose the trade-off between flowtime and tardiness, probably by some kind of Pareto analysis (see Chap. 10).
3. The violation of due dates can also be added to flowtime in the objective function including adequate Lagrangrean multipliers which penalise the violation of due dates.

From a formal point of view, Lagrangean relaxations often are able to provide bounds on the optimal objective function value which enables the decision maker to estimate the maximum deviation of a current solution to the optimum. (However, the integrality requirements with respect to the bound interpretation have to be considered carefully, at least if mathematically valid bounds should be derived.)

9.5.2 Combination of Exact Approaches with (Meta-) Heuristics

At a first glance, it seems to be absurd to consider the combination of exact approaches with heuristics since exact approaches solve the problem under consideration exactly and no further processing is needed. However, the second glance offers at least two general perspectives of how exact approaches and heuristics can be combined in manufacturing scheduling.

On one hand, as already mentioned in Chap. 7, exactness of exact approaches have to be interpreted in a rather strict and narrow context relative to a formal problem under consideration. Almost every deviation and/or generalisation from this formal problem, usually pointing into the direction of the real-world problem to be solved and including more complexity will give reason for (heuristic) modifications and extensions of the 'exact' solution deduced from the formal problem which is to be solved by the exact procedure. We point to this perspective but will not discuss it further here since the respective heuristic approaches are usually rather specific with respect to the real-world problem under consideration.

On the other hand, while staying completely in the 'formal' sphere of manufacturing scheduling problems, the limited 'exact' solvability of many formal scheduling problems for complexity reasons implies the use of heuristics. Apart from the two 'pure' heuristic approaches, i.e. premature truncation of an exact approach, e.g. after reaching an accepted error level and/or a time limit, or exclusively referring to (constructive and/or meta-) heuristics, both pure approaches can be combined in several ways. We describe some basic types of these combined approaches.

1. The performance of a non-polynomial exact procedure will usually be increased if a good (and feasible) initial solution is available which might be obtained by a heuristic approach. On one hand, this solution gives a valid and (hopefully) good upper bound for the optimal objective function value. Obviously, the better this bound is the better the remaining potential for further optimisation can be estimated, e.g. in branch and bound approaches. On the other hand, the solution itself might be a good initial solution for starting a branch and bound approach where a feasible initial solution is required, especially in its first-depth version.
2. If an exact approach has to be stopped prematurely, some 'post optimisation' might take place. E.g., having determined one or several good solutions by a prematurely stopped exact procedure, these solutions can be used as starting point for improvement heuristics. This approach will often include deterministic or non-deterministic neighbourhood search and/or will use the best solutions determined

so far as initial population for a genetic algorithm or some other meta-heuristic approach.

3. Decomposing the overall problem into several exactly solvable subproblems, solving these subproblems exactly and finally heuristically composing the subproblems' solution into an overall solution characterises a third way of combining exact and heuristic approaches. Usually, the determination of exactly solvable subproblems is based on the determination of polynomially solvable subproblems, e.g. specific 1-machine/1-stage problems. However, also the number of jobs can be reduced to a number which allows also more complicate problems to be solved exactly. E.g., a standard 50 jobs flow shop problem (with an arbitrary objective function) might be separated into five 10-job problems. These smaller problems can be solved to optimum. Afterwards the five optimal solutions of the subproblems are combined by looking for the best among the $5! = 120$ (macro-) sequences of these 10-job subsequences. Obviously, the crucial component of this approach is the assignment of jobs to 10-job subproblems (as well as the determination of the number 10 as size of the subproblems). This job assignment might be induced by 'similar' jobs, e.g. by similarity in the processing time profile, by similar setups required etc.
4. A subset of decomposition/composition approaches represent those approaches who include aggregation/disaggregation components. In aggregation and disaggregation approaches the number of objects to be considered is reduced by aggregating objects into aggregated objects solving a problem on the aggregate level and then disaggregating the solution to a solution of the detailed formal problem under consideration. Objects to be aggregated could be jobs (into macro-jobs), machines/stages or a combination of both. The resulting aggregate problem (with less machines/stages and/or jobs) should be solvable exactly and the disaggregation step then consists of refining the aggregate result into disaggregate information for the detailed problem, maybe combined with some post-optimisation heuristic on the disaggregate level.
   A famous representative of this approach, considering machine aggregation, is the shifting bottleneck job shop heuristic, Sect. 9.2.2. Here scheduling on one machine is considered in detail while scheduling on the other machines is included only by release times (coarsely representing the lead time of a job's operations to be processed before the operation of this job on the machine under consideration may take place) and follow-up times (coarsely representing the time needed for the remaining operations of a job). The shifting bottleneck heuristic considers iteratively all machines/stages in the job shop. However, if a clear bottleneck can be identified in the system, the above consideration also can be limited only to this bottleneck machine/stage. An aggregation/disaggregation approach for flow shop scheduling is the well-known heuristic by Campbell et al. (1970). Here, the machines are aggregated into two machines, the two machine problem is solved to optimality by Johnson's algorithm and the resulting solution is adapted as an approximation for the original flow shop problem.
   Similar approaches can be derived for job aggregation/disaggregation where similar jobs are aggregated into one job. The similarity of jobs can be defined in many

ways which are usually problem-dependent. E.g., similar setup requirements or similar due dates might be approaches to substantiate the similarity of jobs.
More generally, in aggregation/disaggregation approaches, a common way of determining aggregation objects is to aggregate non-bottleneck objects (jobs and/or machines/stages) and to consider the bottleneck(s) (jobs and/or machines/ stages) in detail.
5. Most of the approaches mentioned before might be applied as well iteratively, e.g. by using the solution/the schedule of the first phase as input for the second phase and vice versa.

9.6 Conclusions and Further Readings

In this chapter we have dealt with approximate algorithms. Approximate algorithms intend to reduce the complexity of a scheduling model and/or solution method. Regarding taylored approximate algorithms (i.e. heuristics), there is no template that can be used to design heuristics, so the main ideas and concepts behind heuristics have been discussed. Nevertheless, given the large variety of these methods and underlying ideas, such an exposition is far from being comprehensive.

In the last decades there has been a tremendous interest in metaheuristic techniques. While the templates in which metaheuristic methods are based are general and can be applied to most optimisation problems, the instantiation process to any scheduling problem is often not straightforward. A big advantage is that effective metaheuristics often obtain solutions that are very close to optimal sequences in most settings, which comes at a high computational cost when compared to other methodologies albeit they are often faster than exact methods.

One basic reference on approximate methods is the book of Morton and Pentico (1993). Although this book is fairly old meanwhile, it gives a structured overview of many approximate methods according to the state-of-the-art by the beginning of the 1990s. For more recent methods the interested reader is referred to the numerous references in scientific journals, may they present detailed approaches themselves or surveys on specific classes of scheduling problems. The performance of the NEH has been acknowledged by many authors, including Turner and Booth (1987); Taillard (1990); Ruiz and Maroto (2005). Furthermore, NEH is actively used as a seed sequence for metaheuristic, such as in Reeves (1995); Chen et al. (1995); Nowicki and Smutnicki (1996); Reeves and Yamada (1998); Wang and Zheng (2003); Rajendran and Ziegler (2004); Grabowski and Wodecki (2004); Ruiz and Stützle (2007), among others.

The literature on metaheuristics is overwhelming. Only listing some of the most interesting books is already a hard task. Our recommendation goes to the general texts of Glover and Kochenberger (2003); Resende et al. (2004); Hoos et al. (2005); Ibaraki et al. (2005) and Doerner et al. (2007). Basically, each type of metaheuristic has its own classical books and recent references and previous sections contain some interesting books and further reading pointers.

References

Adams, J., Balas, E., and Zawack, D. (1988). The shifting bottleneck procedure for job shop scheduling. *Management Science*, 34(3):391–401.

Biegel, J. E. and Davern, J. J. (1990). Genetic Algorithms and Job Shop Scheduling. *Computers and Industrial Engineering*, 19(1):81–91.

Birattari, M. (2005). *The Problem of Tuning Metaheuristics as seen from a machine learning perspective*. Intelligence-Infix, Berlin.

Campbell, H. G., Dudek, R. A., and Smith, M. L. (1970). A Heuristic Algorithm for the n Job, m Machine Sequencing Problem. *Management Science*, 16(10):B-630–B-637.

Chen, C.-L., Vempati, V. S., and Aljaber, N. (1995). An application of genetic algorithms for flow shop problems. *European Journal of Operational Research*, 80(2):389–396.

Clerc, M. (2006). *Particle Swarm Optimization*. Wiley-ISTE, New York.

Dannenbring, D. G. (1977). An evaluation of flowshop sequencing heuristics. *Management Science*, 23:1174–1182.

Dasgupta, D., editor (1998). *Artificial Immune Systems and Their Applications*. Springer, New York.

Davis, L. (1985). Job Shop Scheduling with Genetic Algorithms. In Grefenstette, J. J., editor, *Proceedings of the First International Conference on Genetic Algorithms and their Applications*, pages 136–140, Hillsdale. Lawrence Erlbaum Associates.

Davis, L., editor (1996). *Handbook of Genetic Algorithms*. International Thomson Computer Press, London.

de Castro, L. N. and Timmis, J. (2002). *Artificial Immune Systems: A New Computational Intelligence Approach*. Springer, London.

Doerner, K. F., Gendreau, M., Greistorfer, P., Gurjahr, W. J., Hartl, R. F., and Reinmann, M., editors (2007). *Metaheuristics: progress in complex systems optimization*. Operations Research/Computer Science Interfaces. Springer, New York.

Dorigo, M. and Stützle, T. (2004). *Ant Colony Optimization*. Bradford Books, USA.

Eberhart, R. C. and Shi, Y. (2007). *Computational Intelligence: Concepts to Implementations*. Morgan Kaufmann, San Francisco.

Engelbrecht, A. P. (2006). *Fundamentals of Computational Swarm Intelligence*. John Wiley & Sons, New York.

Engelbrecht, A. P. (2007). *Computational Intelligence: An Introduction*. John Wiley & Sons, New York, second edition.

Feo, T. A. and Resende, M. G. C. (1989). A probabilistic heuristic for a computationally difficult set covering problem. *Operations Research Letters,* 8(2):67–71.

Feo, T. A. and Resende, M. G. C. (1995). Greedy randomized adaptive search procedures. *Journal of Global Optimization,* 6:109–133.

Feoktistov, V. (2006). *Differential Evolution. In Search of Solutions*. Springer, New York.

Framinan, J. M., Leisten, R., and Rajendran, C. (2003). Different initial sequences for the heuristic of Nawaz, Enscore and Ham to minimize makespan, idletime or flowtime in the static permutation flowshop sequencing problem. *International Journal of Production Research,* 41(1):121–148.

Geem, Z. W., Kim, J. H., and Loganathan, G. V. (2001). A new heuristic optimization algorithm: Harmony search. *Simulation,* 76(2):60–68.

Glover, F. and Kochenberger, G. A., editors (2003). *Handbook of Metaheuristics*. Kluwer Academic Publishers, Dordrecht.

Glover, F. and Laguna, M. (1997). *Tabu Search*. Kluwer Academic Publishers, Dordrecht.

Goldberg, D. E. (1989). *Genetic Algorithms in Search, Optimization and Machine Learning*. Addison-Wesley, Reading.

Grabowski, J. and Wodecki, M. (2004). A very fast tabu search algorithm for the permutation flow shop problem with makespan criterion. *Computers & Operations Research,* 31(11):1891–1909.

Hansen, P. and Mladenovic, N. (2001). Variable neighborhood search: Principles and applications. *European Journal of Operational Research,* 130(3):449–467.

Holland, J. H. (1975). *Adaptation in Natural and Artificial Systems.* The University of Michigan Press, Ann Arbor.

Hoos, Holger, H. and Stützle, T. (2005). *Stochastic Local Search: Foundations and Applications.* Morgan Kaufmann, San Francisco.

Ibaraki, T., Nonobe, K., and Yagiura, M., editors (2005). *Metaheuristics: progress as real problem solvers.* Operations Research/Computer Science Interfaces. Springer, New York.

Kennedy, J., Eberhart, R. C., and Shi, Y. (2001). *Swarm Intelligence.* Academic Press, San Diego.

Kirkpatrick, S., Gelatt, C. D., and Vecchi, M. P. (1983). Optimization by simulated annealing. *Science,* 220(4598):671–680.

Konar, A. (2005). *Computational Intelligence: Principles, Techniques and Applications.* Springer, New York.

Laguna, M. and Martí, R. (2003). *Scatter search: methodology and implementations in C.* Operations Research/Computer Science Interfaces. Kluwer Academic Publishers, New York.

Larrañaga, P. and Lozano, J. A., editors (2002). *Estimation of distribution algorithms: A new tool for evolutionary computation.* Kluwer Academic Publishers, Dordrecht.

Lozano, J. A., Larrañaga, P., Inza, I. n., and Bengoetxea, E., editors (2006). *Towards a New Evolutionary Computation: Advances on Estimation of Distribution Algorithms.* Springer, New York.

Mattfeld, D. C. (1996). *Evolutionary Search and the Job Shop; Investigations on Genetic Algorithms for Production Scheduling.* Production and Logistics. Springer/Physica Verlag, Berlin.

Michalewicz, Z. (1996). *Genetic Algorithms + Data Structures = Evolution Programs.* Springer-Verlag, Berlin, tercera edition.

Mladenovic, N. and Hansen, P. (1997). Variable neighborhood search. *Computers & Operations Research,* 24(11):1097–1100.

Morton, T. E. and Pentico, D. W. (1993). *Heuristic Scheduling Systems With Applications to Production Systems and Project Management.* Wiley Series in Engineering & Technology Management. John Wiley & Sons, Hoboken.

Nawaz, M., Enscore, Jr, E. E., and Ham, I. (1983). A Heuristic Algorithm for the m-Machine, n-Job Flow-shop Sequencing Problem. *OMEGA, The International Journal of Management Science,* 11(1):91–95.

Nowicki, E. and Smutnicki, C. (1996). A fast tabu search algorithm for the permutation flow-shop problem. *European Journal of Operational Research,* 91(1):160–175.

Nowicki, E. and Smutnicki, C. (1998). The flow shop with parallel machines: A tabu search approach. *European Journal of Operational Research,* 106(2–3):226–253.

Ogbu, F. A. and Smith, D. K. (1990). The Application of the Simulated Annealing Algorithm to the Solution of the $n/m/C_{\max}$ Flowshop Problem. *Computers and Operations Research,* 17(3):243–253.

Osman, I. H. and Potts, C. N. (1989). Simulated Annealing for Permutation Flow-shop Scheduling. *OMEGA, The International Journal of Management Science,* 17(6):551–557.

Park, Y. B., Pegden, C., and Enscore, E. (1984). A survey and evaluation of static flowshop scheduling heuristics. *International Journal of Production Research,* 22(1):127–141.

Pinedo, M. L. (2012). *Scheduling: Theory, Algorithms, and Systems.* Springer, New York, fourth edition.

Price, K., Storn, R. M., and Lampinen, J. A. (2005). *Differential Evolution: A Practical Approach to Global Optimization.* Springer, New York.

Rajendran, C. and Ziegler, H. (2004). Ant-colony algorithms for permutation flowshopn scheduling to minimize makespan/total flowtime of jobs. *European Journal of Operational Research,* 155(2):426–438.

Reeves, C. and Yamada, T. (1998). Genetic algorithms, path relinking, and the flowshop sequencing problem. *Evolutionary Computation,* 6(1):45–60.

Reeves, C. R. (1995). A genetic algorithm for flowshop sequencing. *Computers & Operations Research,* 22(1):5–13.

Resende, M. G. C., Pinho de Sousa, J., and Viana, A., editors (2004). *Metaheuristics: computer decision-making.* Kluwer Academic Publishers, Dordrecht.

Ribas, I., Companys, R., and Tort-Martorell, X. (2010). Comparing three-step heuristics for the permutation flow shop problem. *Computers & Operations Research,* 37(12):2062–2070.

Ruiz, R. and Maroto, C. (2005). A comprehensive review and evaluation of permutation flowshop heuristics. *European Journal of Operational Research,* 165(2):479–494.

Ruiz, R., Maroto, C., and Alcaraz, J. (2006). Two new robust genetic algorithms for the flowshop scheduling problem. *OMEGA, The International Journal of Management Science,* 34(5):461–476.

Ruiz, R. and Stützle, T. (2007). A simple and effective iterated greedy algorithm for the permutation flowshop scheduling problem. *European Journal of Operational Research,* 177(3):2033–2049.

Syswerda, G. (1996). Scheduling Optimization Using Genetic Algorithms. In Davis, L., editor, *Handbook of Genetic Algorithms,* pages 332–349, London. International Thomson Computer Press.

Taillard, E. (1990). Some efficient heuristic methods for the flow shop sequencing problem. *European Journal of Operational Research,* 47(1):67–74.

Turner, S. and Booth, D. (1987). Comparison of Heuristics for Flow Shop Sequencing. *OMEGA, The International Journal of Management Science,* 15(1):75–78.

Černý, V. (1985). A thermodynamical approach to the travelling salesman problem: an efficient simulation algorithm. *Journal of Optimization Theory and Applications,* 45(1):41–51.

Wang, L. and Zheng, D. Z. (2003). An effective hybrid heuristic for flow shop scheduling. *The International Journal of Advanced Manufacturing Technology,* 21(1):38–44.

Widmer, M. and Hertz, A. (1989). A new heuristic method for the flow shop sequencing problem. *European Journal of Operational Research,* 41(2):186–193.

Woo, H.-S. and Yim, D.-S. (1998). A heuristic algorithm for mean flowtime objective in flowshop scheduling. *Computers & Operations Research,* 25(3):175–182.

Chapter 10
Multi-Objective Scheduling

10.1 Introduction

Starting with Chap. 3, where scheduling models were presented, followed by Chaps. 4 and 5 where constraints and scheduling objectives were outlined, a common assumption so far in this book has been that every scheduling problem and its corresponding model has one single objective or criterion to optimise. However, internal goals, e.g. minimising tied-up capital and the related consequences with respect to utilisation and work in process, are in conflict with other internal goals or with externally oriented goals, e.g. maximising service level with its relation to due date-oriented objectives. Therefore, it is not surprising that manufacturing scheduling models have been the subject of many publications and considerations dealing with multi-criteria aspects.

The consideration of multiple criteria greatly increases the types of models to be considered, as well as the approaches and methods to solve these models. Additionally, the assessment of the quality of the solutions obtained by these methods is far from straightforward in many cases.

In this chapter we

- discuss the multi-objective nature of many manufacturing scheduling decision problems (Sect. 10.2),
- give the main definitions and notation required for dealing with multiple objectives (Sect. 10.3),
- present the most employed multi-criteria models (Sect. 10.4),
- introduce the different approaches for addressing multi-criteria scheduling (Sect. 10.5),
- address the issue of assessing the quality of algorithms in Pareto optimisation (Sect. 10.6), and
- present interfering jobs scheduling models as a general case of multi-criteria scheduling models (Sect. 10.7).

J. M. Framinan et al., *Manufacturing Scheduling Systems*,
DOI: 10.1007/978-1-4471-6272-8_10,

10.2 Multi-Objective Nature of Many Scheduling Problems

Strictly speaking, one could say that a multi-objective problem is no more than a poorly defined single objective problem. In other words, if one needs to optimise a given scheduling problem with several simultaneous objectives is because it was not possible to verbalise and to rationalise a single-objective function—no matter how complex this function might be. However, agreeing upon a single objective is by no means an easy task in real production scheduling settings. As it has been mentioned, real-life scheduling problems are never single objective. This might seem as a gratuitous assessment. However, a close examination of how scheduling works in practice should convince the reader that this is, in fact, the truth. Below we state several issues concerning single- versus multi-objective optimisation in real production scheduling settings:

- Well-defined objectives are hard to find in practice. Shop managers have a fair amount of intuition of what makes a good or a bad schedule, but not a precise and specific objective function. Moreover, even if a well-defined objective exists, its actual value might be of little relevance in practice. Even simple objectives like minimising the number of tardy jobs can quickly turn into a debatable measure. Some jobs are not as important as others. Therefore, being tardy for some jobs is not as critical. Trying to add weights in order to cope with this issue is only a temporary fix since weights are often not enough. Things like 'this job cannot be tardy. I have added a weight of nine and the scheduling method still results in this job being tardy' easily pop out. In other occasions, a simple phone call to a client in order to advance that the delivery date for a job or order is going to change suffices to make up for a tardy job.
- Preferences or objectives vary over time. They do depend on past and current shop performance, on the mix of pending production and on market situation. Therefore, some days, the focus might be at reducing lead times, while other days machine utilisation might be of paramount importance.
- Decisions on schedules are often taken in a committee rather than by a single individual. Agreeing on a single objective in these cases is seldom feasible. Moreover, different managers at the company will have various and often conflicting goals. Top management often desires to increase revenue. Sales force desires to meet client's due dates while plant managers care about machine utilisation, to name just a few examples. As a result, even deciding on an objective to optimise can be challenging.
- The concept of 'optimal solution' is often not well understood. Plant managers have a fair understanding of what constitutes a bad or a good schedule. However, discerning between a good and an optimal schedule is not within plant personnel capabilities and is often a moot issue. Furthermore, an optimal solution for one objective might degrade another objective more than necessary.
- Problem constraints might be relaxed in view of the performance of the schedule. For example, working shifts might be reduced by sending workers home or augmented via overtime hours. Similarly, production might be subcontracted in

order to cope with additional orders. Given this potential relaxation of problem constraints, the mere concept of optimal solution is debatable at best.

- Human schedulers are more oriented towards goals or minimum/maximum achievement levels for a set of preferences rather than towards complex mathematical formulations and precise objectives. Verbalising objectives as 'reaching an 80 % service level' is much more meaningful than saying that the total tardiness is 1,700,000 time units.
- Real settings are marred with portions of data that are unknown, imprecise, constantly changing and even wrong. Under these circumstances, focusing on improving a single criterion by a few percentage points makes little, if any, sense.

Although it turns out that, in some cases, it is desirable to consider more than one criteria when scheduling, multi-criteria scheduling (and multi-criteria decision-making, in general) is substantially more difficult than dealing with a single criterion. In the latter case, for a formal deterministic model involving the minimisation of objective f, establishing if schedule S is better than S' reduces to checking if $f(S) < f(S')$. However, this is not possible, in general, for more than one objective. If we consider the minimisation objectives f_1 and f_2, and it turns out that $f_1(S) < f_1(S')$ but $f_2(S) > f_2(S')$, then it is clear that we cannot say that S is better than S' or vice versa. Therefore, in order to advance into multi-criteria scheduling, we first need some specific definitions. These are provided in the next section.

10.3 Definitions and Notation

Let us assume that we are dealing with a scheduling model with a number H of minimisation[1] objectives, for which two solutions x_1 and x_2 exist. Let us derive the following definitions:

Strong domination: A solution x_1 strongly dominates solution x_2 ($x_1 \prec\prec x_2$) if:

$$f_h(x_1) < f_h(x_2) \quad \forall h$$

Therefore, x_1 is said to strongly dominate x_2 if x_1 is strictly better than x_2 for all objectives. This is often also defined as strict domination.

(Regular) domination: Solution x_1 (regularly) dominates solution x_2 ($x_1 \prec x_2$) if the following two conditions are simultaneously satisfied:

$$(1)\, f_h(x_1) \leq f_h(x_2) \quad \forall h$$
$$(2)\, f_h(x_1) < f_h(x_2) \text{ for at least one } h = 1, 2, \ldots, H$$

[1] This does not pose a loss of generality, since all definitions can be easily adapted. Furthermore, most scheduling objectives, as shown in Chap. 5, are of the minimisation type.

This implies that x_1 is not worse than x_2 for all objectives except for one in which it is strictly better.

Weak domination: Solution x_1 weakly dominates solution x_2 ($x_1 \preceq x_2$) if:

$$f_h(x_1) \leq f_h(x_2) \;\; \forall h$$

This last definition implies that x_1 is not worse than x_2 for all objectives.

Incomparability: Solutions x_1 and x_2 are said to be incomparable, and is denoted by $x_1 \| x_2$ or by $x_2 \| x_1$ if the following two conditions are simultaneously satisfied:

$$(1)\; x_1 \text{ does not weakly dominates } x_2 (x_1 \npreceq x_2)$$
$$(2)\; x_2 \text{ does not weakly dominates } x_1 (x_2 \npreceq x_1)$$

In other words, if two solutions do not weakly dominate each other, they are said to be incomparable.

Given a set of solutions A, there might be relations between the solutions contained. More specifically, one is usually interested in the dominant solutions from the set:

Weak Pareto optimum: a solution $x \in A$ is a weak Pareto optimum if and only if there is no $x' \in A$ for which $x' \prec\prec x$. In other words, $x \in A$ is a weak Pareto optimum if no other solution in set A strongly dominates x. 'Weakly efficient solution' is an alternative name for a weak Pareto optimum.

Strict Pareto optimum: a solution $x \in A$ is a strict Pareto optimum if and only if there is no $x' \in A$ for which $x' \prec x$. In other words, $x \in A$ is a strict Pareto optimum if no other solution in set A dominates x. 'Efficient solution' is an alternative name for a strict Pareto optimum.

Pareto set: a set $A' \subseteq A$ is denoted a Pareto set if and only if it contains only and all strict Pareto optima. Since it contains only efficient solutions, this set is denoted as *efficient set*. Alternative names are Pareto front or Pareto frontier.

All previous definitions can be more clearly understood by a graphical representation. Let us consider a scheduling problem with two minimisation objectives. We depict the values of the objective functions along two axis: X-axis represents the makespan ($C_{\max}$) values and the Y-axis the total tardiness ($\sum T_j$) values. Note that every possible solution x_i to this biobjective scheduling problem can be represented by a point in this graphic. We have represented some possible solutions in Fig. 10.1.

By looking at Fig. 10.1, we can also give some further definitions:

Utopian (or ideal) solution: Is a given solution with the best possible objective function value for all objectives. Often, this ideal solution does not exist as the multiple objectives are often conflictive. It has been represented in Fig. 10.1 as solution 'U'.

Nadir solution: It is the contrary to the utopian solution. It is a solution with the worst possible objective function value for all objectives. Luckily, such solution is

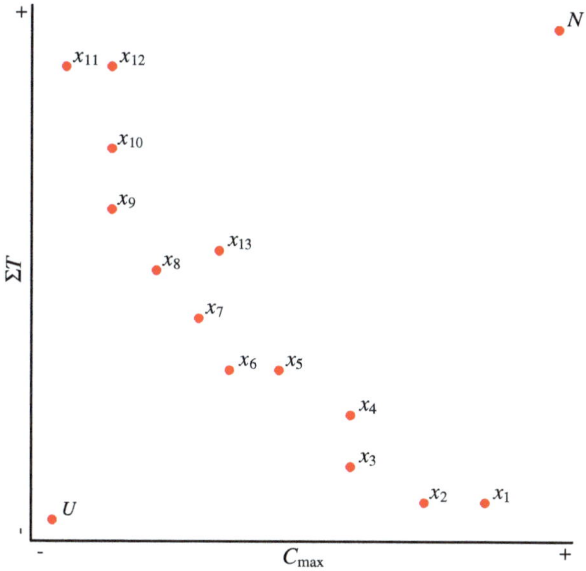

Fig. 10.1 Representation of the function objectives' values of several possible solutions

not likely to exist either. In Fig. 10.1 the nadir solution has been represented by 'N'. Solutions U and N are many times used for bounding (topologically speaking) the sets of solutions for multi-objective problems.

Following with Fig. 10.1 we have the following relations:

- Picture, for example, solutions x_1 and x_2. Both solutions have the same total tardiness value. However, solution x_2 has a lower makespan value than x_1. Therefore, we say that solution x_2 regularly dominates solution x_1 or $x_2 \prec x_1$.
- Following the previous example we also have that $x_3 \prec x_4, x_6 \prec x_5, x_9 \prec x_{10} \prec x_{12}$ and $x_{11} \prec x_{12}$.
- It can also be seen that both x_7 and x_8 have both a lower makespan and a lower total tardiness than x_{13}. Therefore, it can be stated that $x_7 \prec\prec x_{13}$ and $x_8 \prec\prec x_{13}$.
- There are some incomparable solutions in the figure. Let us pick, for example, x_1 and x_3. x_1 has a lower total tardiness value than x_3 but also it has a worse makespan value. Therefore, we cannot conclude, in a multi-objective sense, whether x_1 is better than x_3 or viceversa. We then state that x_1 and x_3 are incomparable or $x_1 \| x_3$.
- The Pareto front for the example is formed by solutions x_2, x_3, x_6, x_7, x_8, x_9, and x_{11}. This is further depicted in Fig. 10.2. Note the stepwise 'frontier'. This is indeed the case as straight lines in between Pareto solutions would imply that other solutions in between have been found.

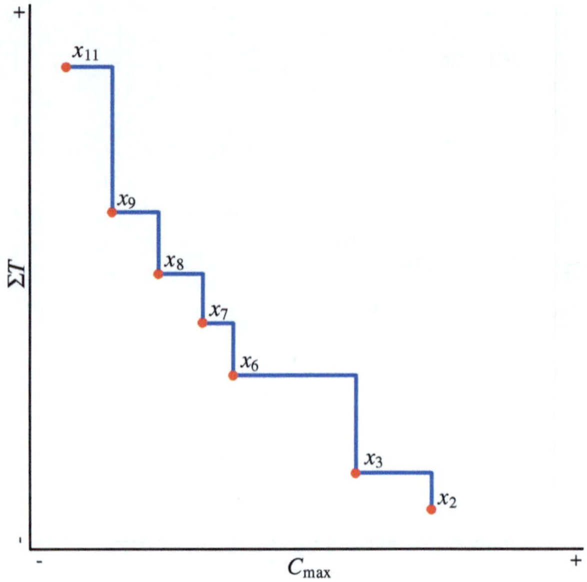

Fig. 10.2 Representation of the Pareto frontier

10.4 Multi-Objective Models

When dealing with multi-objective scheduling models, different classes of multi-objective models can be formulated depending on how the different objectives are considered in the decision problem. For instance, the decision-maker may accept some trade-off between the different objectives considered, or may be interested in that the schedule has at least a (minimum) performance with respect to one or several objectives considered.

There are many classes of models, being the most employed in manufacturing scheduling the following classes:

- Weighted combination of objectives. In this class of models, an explicit importance (weight) of each objective can be determined. As we will see, this transforms the multi-objective model into a single-objective model.
- Lexicographical optimisation. In this class of models, a ranking or importance order of the objectives exists, so they should be optimised one after another.
- ε-constraint optimisation. Only one objective is optimised at a time, while all others are kept as constraints with a known bound.
- Goal programming. In this class of models, a given goal or target value is set for each objective considered. This class of models can be seen as a multi-objective version of a feasibility model.
- Pareto optimisation. In this class of models, the goal is finding the Pareto set with respect to the objectives considered.

In Chaps. 3–5, the well-known $\alpha|\beta|\gamma$ notation was introduced. As already indicated, the third field γ covers the criterion for the manufacturing scheduling model. In order to be able to represent the different classes of multi-objective scheduling models, this notation is extended and presented in the corresponding class.

10.4.1 Weighted Combination of Objectives

The simplest approach to multi-objective scheduling problems is to combine all objective functions into a single-objective, usually by adding weights to each objective. One example is the makespan and total tardiness minimisation as follows: $\alpha C_{\max} + (1-\alpha)\sum T_j$, where $0 \leq \alpha \leq 1$. Note that this is a convex linear combination, a reason for which this class of models is also known as Linear Convex Combination (LCC).

For three objectives F_1, F_2 and F_3, one can write something like $\mathrm{a}F_1 + \mathrm{b}F_2 + \mathrm{c}F_3$ where $0 \leq \mathrm{a, b, c} \leq 1$ and $\mathrm{a} + \mathrm{b} + \mathrm{c} = 1$. The advantage of such an approach is that at the end one has effectively one single-objective (a weighted sum) and therefore, existing single-objective approaches can be used with little adaptation. Indeed, some authors do not even consider this method a 'true' multi-objective approach.

Multi-criteria scheduling models of the weighted combination of objectives class are denoted as follows:

$\gamma = F_l(F_1, F_2, \ldots, F_H)$: weighted combination of objectives $F_1, F_2, \ldots, F_H$.

Therefore, $1|d_j|F_l(\sum T, \sum E)$ indicates a one machine model where each job has a due date and the objective is to minimise a LCC of the total tardiness and the total earliness.

10.4.2 Lexicographical Optimisation

Lexicographical optimisation basically entails a definition of an order among the considered objectives. Once this order is determined, the scheduling problem is optimised as regard the first objective, with no trade-offs between all other objectives. Once a solution is obtained, the second objective is optimised subject to no deterioration in the first, i.e. the second objective is optimised and a constraint is imposed so that the objective function value for the first objective is not worsened. Lexicographical optimisation is 'easy' in the sense that one has as many optimisation runs as objectives to consider. In the context of mathematical programming the first mathematical model to optimise is the following:

$$\min F_1$$

If we denote by f_1 the value obtained by the first minimisation objective, the second mathematical model goes as follows:

$$\min F_2$$
$$s.t.$$
$$F_1 \le f_1$$

The third run would be:

$$\min F_3$$
$$s.t.$$
$$F_1 \le f_1$$
$$F_2 \le f_2$$

The procedure continues until the last objective is optimised. The final solution is the lexicographic optimal solution for all considered objectives in order.

Multi-criteria scheduling models of the lexicographical class are denoted as follows:

$\gamma = Lex(F_1, F_2, \ldots, F_H)$: lexicographical optimisation of objectives $F_1, F_2, \ldots,$ F_H. The order indicates that F_1 is optimised first, then F_2 is optimised subject to no deterioration in F_1 and so on until F_H.

Note that the order of the objectives in the γ field in these models is very important, as different orders define different models.

According to this notation, $Fm|prmu, d_j|Lex(\max T_j, C_{max})$ indicates a model of a permutation flowshop with due dates where the maximum tardiness is minimised first and then, the makespan is minimised as long as there is no deterioration in the maximum tardiness value found first. Obviously, $Fm|prmu, d_j|Lex(C_{max}, \max T_j)$ constitutes a different model.

10.4.3 ε-Constraint Optimisation

The ε-constraint is another interesting approach in which only one objective is optimised at a time, while all others are kept as constraints with a known bound on their minimum or maximum possible values. In other words, it optimises one objective while maintaining maximum or minimum acceptable levels for all others. These levels are known as the ε-values or ε-bounds for each objective. These values are hard to set appropriately and surely depend on the combination of objectives and on the scheduling problem at hand. In general, large (or small) values for the ε-bounds allow for larger feasible regions when optimising the chosen objective and therefore, better objective values. However, this usually sacrifices all other ε-bounded criteria. Equally important is the choice of which objective to optimise first. Some of these disadvantages are clearly onset by the fact that in this method a single objective is optimised at each run, needing only an existing effective single-objective algorithm.

An example with three minimisation objectives is given next:

$$
\begin{aligned}
&\min F_1\\
&s.t.\\
&F_2 \le \varepsilon_2\\
&F_3 \le \varepsilon_3
\end{aligned}
$$

Multi-criteria scheduling models of the ε-constraint class are denoted as follows:
$\gamma = \varepsilon(F_1/F_2, F_3, \ldots, F_H)$: F_1 is optimised subject to some lower or upper bounds in the remaining objectives.

Again, the order of the objectives is important: While $1|d_j|\varepsilon(\sum C_j/\max T_j)$ defines a one machine model where the objective is to minimise the total completion time subject to an upper bound on the maximum tardiness, $1|d_j|\varepsilon(\max T_j/\sum C_j)$ seeks to minimise the maximum tardiness subject to an upper bound on the total completion time. Note, however, that the importance of the order only affects to the objective that is minimised, and not to the rest, i.e. $1|d_j|\varepsilon(\sum C_j/\max T_j, \max E_j)$ is the same model as $1|d_j|\varepsilon(\sum C_j/\max E_j, \max T_j)$.

10.4.4 Goal Programming

Goal programming (sometimes referred to as goal-attainment approach) is an interesting methodology, similar, but intrinsically different to the previous lexicographic or ε-constraint approaches. It can be viewed as a generalisation where each objective is associated with a given goal or target value to be achieved. A new objective function—referred to as achievement function—is constructed where the deviations from these set of target values are minimised. Note that it is accepted that the decision-maker is going to assume that some sacrifices will be necessary as achieving all goals will not be possible. From the perspective of mathematical programming, an example with one minimisation objective (F_1), another maximisation objective (F_2) and an equality objective (F_3) is given:

$$
\begin{aligned}
&\min E_1 + D_2 + E_3 + D_3\\
&s.t.\\
&F_1 = G_1 + E_1\\
&F_2 = G_2 - D_2\\
&F_3 = G_3 + E_3 - D_3
\end{aligned}
$$

Note that G_1, G_2 and G_3 are the target values of objectives F_1, F_2 and F_3, respectively. Similarly, the variables 'D' and 'E' modelise the defect and excess values below or above the desired target values, respectively. Therefore, by minimising the sum of $E_1 + D_2 + E_3 + D_3$ we ensure that all three objectives are as close as possible to their desired goals.

Goal programming shares the same dimensionality problems with the weighted combination of objectives. In the previous example, the objective min $E_1+D_2+E_3+D_3$ assumes that all deviations are measured in the same units, or alternatively, the orders of magnitude in which each objective is measured are comparable. If this is not the case, the same normalisation problems appear. There are many extensions of goal programming, like when weights are added in order to penalise some objectives in favour of others.

Multi-criteria scheduling models of the goal programming class are denoted as follows:

$\gamma = GP(F_1, F_2, \ldots, F_H)$: Goal Programming optimisation of objectives F_1 to F_H.

Here, the order of the objectives is not relevant for defining the model. According to this notation, $F2|prmu, d_j|GP(C_{\max}, \sum U_j)$ indicates a two-machine permutation flow shop model with due dates and the objectives of finding schedules for which their makespan and number of tardy jobs are below given bounds.

10.4.4.1 Pareto Optimisation

Pareto optimisation models provide the decision-maker with the Pareto front. Therefore, not only one solution, but many, are given to the decision-maker.

We have already stated the many drawbacks of the Pareto optimisation approach. First and foremost, obtaining a good Pareto set approximation is extremely complicated. Note that many scheduling models with a single-objective are already of the NP-hard class. Therefore, obtaining many Pareto solutions is usually hard for most scheduling settings. Second, the cardinality of the Pareto front might be extremely large, making the choice of the solution to be implemented in practice hard.

Multi-criteria scheduling models of the Pareto optimisation class are denoted as follows:

$\gamma = \#(F_1, F_2, F_3, \ldots, F_H)$: Pareto optimisation of objectives F_1 to F_H.

Note that in these models, the order of the objectives is not relevant. According to the notation, $Fm||\#(C_{max}, \sum C_j)$ denotes a model of a m-machine flowshop where the goal is finding the Pareto set for the makespan and the total completion time.

10.5 Multi-Objective Methods

In this section, we discuss how multi-criteria scheduling decision problems can be addressed. We confine our discussion to the formal sphere (see Sect. 1.5), as issues dealing with transferring the real-world problem to a formal model are roughly the same as in problems with a single criterion. We first discuss several approaches to address the problems, and then give some insights on the algorithms that can be employed.

10.5.1 Approaches to Multi-Objective Models

Approaches to multi-criteria problems can be broadly classified into three categories, which mainly depend on when and how the user intervenes:

- 'A priori' approaches. There is user intervention before the optimisation run. This information is in the form of user preferences.
- 'Interactive' optimisation methods. The user intervention occurs during the optimisation run.
- 'A posteriori' methods. The user is presented with a complete as possible Pareto front. The user selects the desired trade-off solution to be implemented in practice from the front.

In the 'a priori' category, the user must provide some information in order to guide the search. This information often entails objective preferences or weights. The 'a priori' optimisation methods often need a single optimisation run and usually one single solution is obtained as a result. This single solution already considers the information and preferences given by the user. A priori multi-objective methods are very interesting as much of the work and results obtained in the single-objective scheduling literature can be reused. The biggest drawback is that the 'a priori' information must be precise and must reflect the preferences of the user accurately. Many times it is not possible to provide this information without any knowledge of the resulting sequences and solutions. A possible workaround to this problem is to enter in a loop in which the user provides information, a solution is obtained and later analysed. Should this solution fail to address the user's preferences, the 'a priori' information can be updated accordingly and a new solution can be obtained.

Conversely, in the 'a posteriori' approach, the user is presented with the Pareto front (or at least an approximation of such Pareto front). The user can then examine as many solutions as desired with the aid of graphical tools so as to select the solution to be implemented in practice. This is an ideal approach as no preferences or any a priori information must be provided. However, this approach is not exempt of problems. First, obtaining an approximation of the Pareto front might be computationally costly and/or challenging. Let us not forget that any a posteriori multi-objective optimiser has to obtain not only one but also many dominant solutions that sufficiently cover the objective space. Due to these problems, most a posteriori optimisation methods are often limited to two simultaneous objectives. Second, there is a lingering problem of over information. Given a load of dominant solutions, the user can easily become overwhelmed by the quantity of information. However, there are some techniques that can be used to cluster solutions in the objective space so to present the user a given number of selected solutions.

Lastly, the interactive approaches seek to amalgamate the advantages of the two previous categories. With some limited information, an optimisation run is carried out and a solution is presented to the user. After examining the solution, the user provides further information, like where to go, what to avoid or what to enforce. A second optimisation is run and another solution is presented to the user. The process goes on as much as the user desires or until he or she finds an acceptable solution. With proper training, very good solutions (albeit usually not probably Pareto optimal) can be obtained. This methodology allows to include advanced human knowledge in the process. Surprisingly, not many interactive approaches have been proposed in the multi-objective scheduling literature.

The different classes of models discussed in Sect. 10.4 can be analysed under the light of the approaches in which these models may be embedded:

- In principle, LCC models fall within the *a priori* approach. Namely, the weights of the different objectives must be given to formulate the model. However, these models can be embedded into an interactive approach where, at each iteration, the weights are changed according to the opinion of the decision-maker in order to direct the search towards new solutions.
- Lexicographical optimisation requires very little data, but it should be still considered as an 'a priori' technique since the user must provide, in the very least, the order of the objectives to optimise.
- ε-constraint models are often regarded as adequate for *a priori* approaches. However, many authors also pinpoint the need to run the method several times when adjusting the ε-bounds, which effectively turns the technique into an interactive approach. Under specific scenarios and following certain—actually, strict—conditions (T'Kindt and Billaut 2006) the ε-constraint method can generate the Pareto set.
- Goal Programming is clearly an 'a priori' method since aspiration levels or goals for each objective have to be set.
- Pareto optimisation is the typical example of an 'a posteriori' approach. Once the decision-maker is confronted with a set of dominant solutions, he or she knows that each given solution is a non-dominated solution and that improving one objective always entails a deterioration of the other objectives and in such a case, the closest non-dominated solution is available in the given Pareto front approximation. The 'a posteriori' decision is then reduced to choosing one of the non-dominated solutions to be implemented in practice.

Recall from Sect. 7.3.3 that it is possible to establish a hierarchy of the computational complexity of different scheduling problems and thus construct reductions trees such as the ones in Figs. 7.1, 7.2 and 7.3. Clearly, the two first reduction trees, which corresponding to layout and constraints, are valid in the multi-criteria context. With respect to the third reduction tree, it is valid for the different objectives within the same model. In addition, another reduction tree with respect to the models can be devised.

On one hand, it is clear that multi-criteria scheduling models are computationally more complex than their single-criterion counterpart, therefore any model $\alpha|\beta|C$ in a single-criterion model can be reduced to model $\alpha|\beta|\varepsilon(C/\ldots)$ by considering as upper bounds of the rest of the objectives a very large number. Also quite obviously, the solution obtained by any $\alpha|\beta|Lex(C,\ldots)$ model serves to obtain the solution for the single-criterion model.

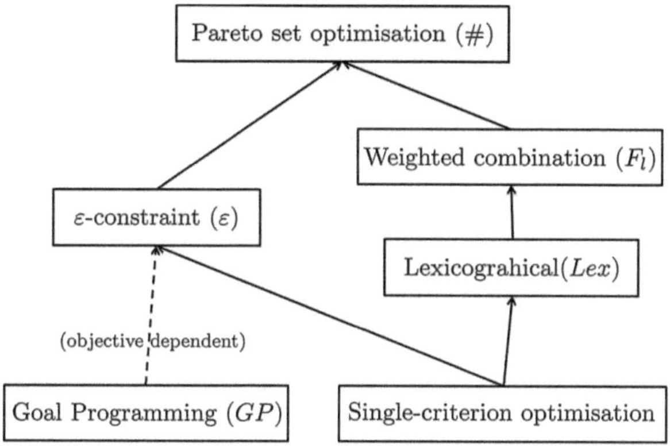

Fig. 10.3 Complexity hierarchy of multicriteria models

On the other hand, it is possible to establish a hierarchy among different classes of multi-criteria problems. More specifically, the model $\alpha|\beta|Fl(F_1, \ldots, F_H)$ reduces to $\alpha|\beta|\#(F_1, \ldots, F_H)$, as the Pareto set of the latter contains the solution given by the former, which is indeed one particular trade-off among the objectives. A similar reasoning can be done to see that $\alpha|\beta|\varepsilon(F_1, F_2, \ldots, F_H)$ reduces to $\alpha|\beta|\#(F_1, \ldots, F_H)$.

Other reductions are far from being straightforward and, in some cases, depend on the objectives considered. In Fig. 10.3 we give the reduction tree for multi-criteria models.

10.5.2 *Algorithms for Multi-Criteria Scheduling Models*

Given the host of different multi-objective techniques, the existing methods and algorithms for solving multi-objective scheduling problems are incredibly numerous and, in most cases, extremely model- and method-specific. Discussing them or even classifying them is beyond the scope of the book, and we refer the reader to some specific sources in Sect. 10.8. In what follows, we simply refer to one specific type of models (Pareto optimisation models) and algorithms (metaheuristics). The reason for picking Pareto optimisation models is, as we have seen, the most general approach, in the sense that some other approaches reduce to these models. Analogously, metaheuristic are chosen as they are the less model-specific type of algorithms and therefore, their study will be more valuable for a reader approaching this field for the first time.

10.5.2.1 Approximate Multi-Criteria Metaheuristics

Among the different types and variants of metaheuristics discussed in Sect. 9.4, Genetic Algorithms seems to be particularly well-suited for Pareto optimisation

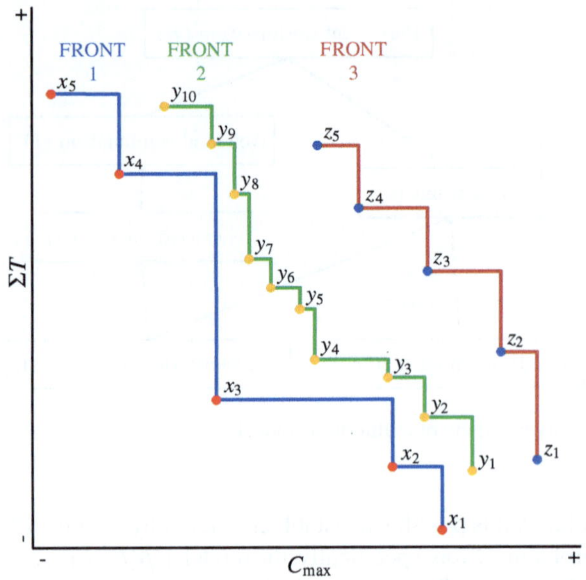

Fig. 10.4 An example of a set of solutions and the non-dominated sorting procedure

models since they are population-based approaches and keep a set of solutions. The efficient solutions within this set can be selected so this subset yields an *approximation* of the Pareto front. As a result, GAs are a natural choice for multi-objective problems.

Let us illustrate the main features of these algorithms by describing the well-known Non-dominated Sorting Genetic Algorithm (NSGA) by Srinivas and Deb (1994) and its subsequent version NSGA-II by Deb (2002). These two versions of GA have attracted a lot of interest from researchers and it is considered a benchmark algorithm for Pareto optimisation models.

NSGA basically differs from a simple GA only in the selection operator. As the name of the algorithm implies, there is a non-dominated sorting procedure (*NDS*) that is applied to the current population at each generation. Randomly selecting any individual from the population to do crossover and mutation is generally a bad idea in genetic algorithms. In multi-objective optimisation, a selection bias should be given to non-dominated solutions. *NDS* iteratively divides the entire population into different sets or fronts. The first front is formed by the non-dominated individuals in the population. Then, the individuals in the first front are removed from the population and, from the remaining individuals, the non-dominated ones are extracted again to constitute the second front. The procedure continues until all individuals have been assigned a front. An example of this procedure is graphically shown in Fig. 10.4.

As we can see from Fig. 10.4, the population is formed by 20 individuals. Individuals $x_1, x_2, \ldots, x_5$ form the first front (denoted in the figure as FRONT 1) as they are non-dominated by all others. If we remove these 5 individuals, the remaining population of 15 individuals is dominated by solutions $y_1, y_2, \ldots, y_{10}$, which would constitute the second front (FRONT 2). Repeating the procedure would yield the third front (FRONT 3) with solutions $z_1, z_2, \ldots, z_5$.

Although FRONT 2 and FRONT 3 are not useful regarding to the final result that should be given to the decision-maker, they serve to maintain the diversity of solutions in the different iterations of the algorithm, which helps to provide the decision-maker with a diverse approximation of the Pareto front.

In Fig. 10.4, we can see how there are large gaps in the objective space between solutions x_2 and x_3 and between solutions x_3 and x_4 from FRONT 1. Actually, FRONT 1 is rather small with only five non-dominated solutions. Comparatively, FRONT 2 has twice as many solutions and, although their solutions are dominated by those in FRONT 2, they are more evenly spread out.

As mentioned before, NSGA keeps all fronts in the population since, in future iterations, it might be possible to obtain non-dominated solutions from, let us say, solutions y_5, y_6, y_7 or y_8 in FRONT 2 so to fill in the gap between solutions x_3 and x_4 in FRONT 1, to name a simple example.

The *NDS* first assigns the same large dummy fitness value to all solutions in FRONT 1. Then, a lower dummy fitness value is assigned to all solutions in FRONT 2, and so on. In order for the GA to distinguish between solutions in the same front as far as the fitness value is concerned, a 'sharing' operator is also performed. Basically, the fitness value of each individual is modified by a factor that is calculated according to the number of individuals crowding a portion of the objective space. A sharing parameter σ_{share} is used in this case. Therefore, solutions that have many other 'surrounding' individuals are penalised whereas isolated solutions in the objective space are left with their fitness values intact. Following the example given in Fig. 10.4, solution x_3 is far away from all other solutions in the first Pareto front and will be left untouched. Conversely, solution y_6 is occupying a crowded region in the second Pareto front and its fitness value would be reduced. Clearly, this mechanisms seeks to reinforce the diversity of solutions.

All other components on the NSGA method of Srinivas and Deb (1994) are basically identical to those found on GA for regular single objective problems. NSGA uses a stochastic remainder proportionate selection using the dummy fitness value returned by the *NDS* procedure and the sharing mechanism. Since NSGA was originally proposed for continuous optimisation, a binary representation and thus, binary mutation and crossover operators are used. A pseudo-algorithm listing of the Non-dominated Sorting procedure is given in Algorithm 1.

NSGA, although profusely used in both theory and in practice, suffered from some shortcomings:

- As it happens in genetic algorithms for single-objective optimisation, degradation of solutions often occurs in NSGA, especially in the later stages of the algorithm. NSGA enforces selection of non-dominant solutions in less crowded areas of the first Pareto front. These selected solutions cross and mutate and result in new solutions (offspring) that are not guaranteed to be better. Therefore, often, important genetic information is lost. In single- objective genetic algorithms, this is solved by

---

Algorithm 1: Non-dominated Sorting Genetic Algorithm of Srinivas and Deb (1994).

Input: instance data
Output: Population *Pop* of solutions
begin
  Initialise a population *Pop* of P_{size} individuals;
  Apply Non-Dominated Sorting;
  Apply Sharing at each obtained front;
  while *Stopping criterion is* ***not*** *satisfied* **do**
    Select individuals from *Pop* according to dummy fitness values;
    Cross individuals;
    Mutate individuals;
    Evaluate individuals;
    Include individuals in the new population;
    if *new population contains Pop individuals* **then**
      Copy new population into old population;
      Apply Non-Dominated Sorting;
      Apply Sharing at each obtained front;
  return *Pop, first Pareto front*
end

---

using elitism or other ways of preserving the good individuals from one generation to another.
- *NDS* is actually a computationally demanding procedure. Its computational complexity is $O(MPop^3)$, where M is the number of objectives and *Pop* is the size of the population. Recall that this procedure has to be applied at each generation of the NSGA method.
- Setting the sharing parameter σ_{share} has proven to be far from trivial and problem dependent.

All these three problems motivated the design of NSGA-II, an improved version of NSGA. NSGA-II deals with the three aforementioned caveats of the NSGA algorithm. NSGA-II includes elitism since the old and new populations are combined at the end of each generation to form a new augmented population of size $2 \cdot Pop$. A more effective Fast Non-Dominated Sorting procedure (*FNDS*) is applied. This operator is used in order to select the best fronts for the final population that is again of size *Pop*. Since all previous individuals from the old generation are considered, elitism is preserved. *FNDS* has a lower computational complexity of $O(MPop^2)$. A rank value is assigned to each individual of the population according to the front they are in. Two additional fast operators, the Crowding Distance calculation and assignment and the Crowded-Distance Sorting Operator are employed to do without the σ_{share} and to achieve fitness sharing. NSGA-II performs largely better than NSGA and than most well known multi-objective approaches. For many continuous optimisation problems, it actually achieves state-of-the-art results. Given *FNDS*, the implementation of NSGA is not difficult. All these features are key for the success of the NSGA-II in the literature.

Next, we show a metaheuristic that is an adaptation of a single-objective metaheuristic that, in contrast to GA, it employs a single solution. As we will see, the adaptation is less straightforward and special care has to be taken in the design of the metaheuristic in order to achieve a correct diversity of solutions so the algorithm does not get stuck.

The presented method is referred to as Restarted Iterated Pareto Greedy or RIPG Minella et al. (2011). In order to understand RIPG, some basic notions about the Iterated Greedy (IG) method of Ruiz and Stützle (2007) should be given. IG was originally proposed for the m-machine permutation flow shop problem with makespan criterion or $F|prmu|C_{\max}$. It basically consists of:

- destruction: d elements of a solution are iteratively removed.
- reconstruction: the d removed elements are reincorporated to the solution by means of a greedy phase.
- improvement: the solution obtained after the reconstruction is improved by means of a local search procedure.

A simulated annealing-like acceptance criterion is applied so as to temporarily accept worse solutions as incumbent ones. IG is a very simple local search-based method and it has been shown to produce state-of-the-art results for the permutation flow shop problem and makespan criterion.

The RIPG algorithm is a Pareto optimisation extension of the IG, including the following differences:

- Use of a population of solutions. While the standard IG works with only one incumbent solution and a best solution found so far, RIPG handles a population (of variable size) of non-dominated solutions as a working set. In order to provide an initial population, constructive heuristics for each one of the objectives considered can be used. Note that there is no guarantee to be non-dominated, and a greedy phase of IG is applied to these solutions in order to obtain an initial set of non-dominated solutions.
- Selection operator. Since the main logic of the IG, namely, the destruction/reconstruction, is still carried out over a single-selected solution from the working set, a selection operator is needed. RIPG employs a modified version of the Crowding Distance operator of NSGA-II where there is a selection counter keeping track of the number of times that a solution has been already selected in order to add diversity.
- Reconstruction. In RIPG, instead of fully reconstructing the selected solution after the destruction, a whole set of non-dominated solutions is built by inserting each removed job into a population of partial solutions. After the greedy phase, the working set is updated with the recently created set of non-dominated solutions from the reconstruction procedure. After each upgrade of the working set, dominated solutions are removed.
- Local search. The local search phase consists in randomly choosing one element from the selected solution, removing and reinserting it into a number of adjacent

positions (which can be seen as a curtailed local search). Furthermore, this procedure is repeated as many times as the number of times the solution has been previously selected (selection counter). This way, a deeper local search is carried out to stronger local optimal solutions.
- Restarting. In order to avoid that the method tends to get stuck, a restart operator is proposed. This operator consists in archiving the current working set, and then creating a new one with randomly generated solutions. Note that random generation of the new working set is both fast and simple. The restart mechanism is triggered after a number of iterations without changes in the cardinality of the working set.

With all previous explanations, we can see a sketch outlining of the RIPG method in Algorithm 2.

Algorithm 2: Restarted Iterated Pareto Greedy (RIPG) Algorithm of Minella et al. (2011).

Input: instance data
Output: Archive of non-dominated solutions
begin
  Obtain $2 \cdot M$ good initial solutions applying two effective heuristics to each objective;
  Apply the Greedy Phase to each $2 \cdot M$ initial solution. Generate $2 \cdot M$ non-dominated sets of solutions;
  Include all $2 \cdot M$ non-dominated sets of solutions in the working set;
  Remove dominated solutions from the working set;
  Update Archive of non-dominated solutions with the working set;
  while *Stopping criterion is **not** satisfied* **do**
    Select solution from the working set (Modified Crowding Selection Operator);
    Greedy Phase (destruction/reconstruction). Generate set of non-dominated solutions;
    Include set of non-dominated solutions from the Greedy Phase in the working set;
    Remove dominated solutions from the working set;
    Select solution from the working set (Modified Crowding Selection Operator);
    Local Search Phase. Generate set of non-dominated solutions;
    Include set of non-dominated solutions from the Local Search Phase in the working set;
    Remove dominated solutions from the working set;
    if *Cardinality of the working set unchanged for* $2 \cdot n$ *iterations* **then**
      Update Archive of non-dominated solutions with the working set;
      Randomly generate a new working set;
  return *Archive of non-dominated solutions*
end

10.6 Comparison of Pareto Optimisation Algorithms

We already outlined in Sect. 10.2 that comparing two sets of solutions, coming from different Pareto optimisation algorithms is not an easy task, at least as compared to the evaluation of the performance of algorithms for the single-criterion case. As we have seen, the result of a Pareto optimisation (approximate) algorithm is (an approximation of) the Pareto front. This approximation might potentially have a large number of solutions. A significant added complexity is that, as we have seen, many existing Pareto optimisation algorithms include stochastic elements, meaning that each run of these algorithms will produce, in general, different Pareto front approximations. To these problems, those also present in the single criterion case persists here.

Although there are more aspects of interest when comparing two Pareto optimisation algorithms (computation effort, among them), here we restrict to two aspects:

- Quality of the solutions offered to the decision-maker. Clearly, the decision-maker is interested in solutions yielding, in general, good values with respect to the objective functions considered.
- Quantity and diversity of these solutions. Since Pareto optimisation falls within the *a posteriori* approach, the decision-maker may consider a trade-off among the different objectives, although he/she does not know in advance such specific trade-off. Therefore, offering him/her a higher number of options (i.e. trade-offs) that are different (i.e. represent different specific balances among objectives) may be of great interest.

Regarding the quality of the solutions offered to the decision-maker, the only case in which the comparison of the outcomes of two Pareto optimisation algorithms is more or less straightforward is when the approximation of the Pareto front of the first algorithm (set A) completely dominates that of the second (set B), i.e. all and each solution in B is dominated by at least one solution in A. We can see an example in the previous Fig. 10.4. In that figure, we can assume that the FRONT 1 comes from a given algorithm A and that FRONT 2 comes from a second algorithm B. It is clear—again from the viewpoint of the solutions to be offered to the decision-maker—that algorithm A is better than algorithm B, as all solutions in FRONT 2 are dominated by at least one solution in FRONT 1.

However, as we have already noted, the quantity and diversity of the solutions offered to the decision-maker is also important. From this perspective, FRONT 2 doubles the number of solutions than FRONT 1. Furthermore, there are large gaps in the objective space between solutions x_2 and x_3 and between solutions x_3 and x_4 in FRONT 1.

Both aspects (quality and quantity/diversity) have to be balanced. For instance, while it is clear that solution y_4 from FRONT 2 is dominated by solution x_4 from FRONT 1, this dominance might be just by a few percentage points of improvement at each objective and therefore of little practical relevance for the decision-maker. He/she might be very well interested in a set with solutions of lower quality as long as it has more solutions and more evenly spread out so to have a richer basis for decision-making.

As we can see, assessing performance in multi-objective optimisation is a very complex issue in its own right. There are actually two interrelated problems involved when assessing the performance of multi-objective scheduling methods:

1. Measuring the quality of the Pareto front approximations given by a certain algorithm 'A'.
2. Comparing the out-performance of one algorithm 'A' over another algorithm 'B'.

Both issues can be dealt with the same techniques, which fall within one of the following three approaches:

- Performance measures. The set produced by an algorithm is summarised by a number indicating its quality/quantity/diversity.
- Dominance ranking. It is possible to rank a given algorithm over another based on the number of times that the resulting sets dominate each other.
- Empirical Attainment Functions (EAF). Attainment functions (AF) provide, in the objective space, the relative frequency that each region or part of the objectives space is attained by the set of solutions given by an algorithm.

Each of these three approaches present a number of advantages and disadvantages. These are discussed in the following sections.

10.6.1 Performance Measures

In order to compare two approximations of Pareto fronts, one option is to first represent each set by a single figure (*performance measure*) representing its performance, and then compare these figures. In order to provide some examples of these performance measures, we present some (rather simple and, in some cases, potentially misleading) performance measures employed in scheduling. We can make a coarse distinction between *generic* performance measures and *pairwise* performance measures. The first group of measures use a reference set S^* of solutions constituted either by the Pareto set (if these can be obtained) or a set of non-dominated solutions. The performance of a given algorithm is then measured in terms of the quality of the solutions obtained by this algorithm with respect to the solutions in S^*. Several indicators for the quality of these solutions have been proposed, being perhaps the simplest the following:

- a the number of solutions found by the algorithm,
- b^* the number of solutions in S^* found by each algorithm,
- b^*/a the percentage of non-dominated solutions (the quotient of the two first indicators).

Note that the last two measures can be also employed for comparing the performance of one algorithm relative to another (pairwise measures). In addition, other measures can be recorded, such as b the number of non-dominated solutions of each algorithm after the two algorithms are compared with each other, and b/a the percentage of non-dominated solutions (the quotient between b and a).

Note that there is an intrinsic loss of information when using performance measures as it is obviously not the same comparing two Pareto fronts with (possibly) a large number of solutions each one than just comparing two performance measures. In addition, not all metrics—including some of those usually employed in multi-objective scheduling—are so-called *Pareto-compliant*, i.e. in some cases, a non-Pareto-compliant performance measure can assign a better value to a set B than to set A even if solutions in A dominate at least one solution in B. Again, this result highlights the fact that comparing Pareto fronts it not an easy task at all.

Some Pareto-compliant performance measures are the hypervolume (I_H) and the binary/unary Epsilon (I_ε^1) indicators. For the case of two objectives, the hypervolume indicator I_H measures the area covered by the approximated Pareto front given by an algorithm. A reference point (usually a solution with bad objective functions' values) is used for the two objectives in order to bound this area. A greater value of I_H indicates both a better convergence to as well as a good coverage of the Pareto front. Unfortunately, calculating the hypervolume can be costly. In order to have similar measures over a large benchmark, normalised and scaled variants of I_H are often employed.

The binary epsilon indicator I_ε is calculated as follows: Given two Pareto optimisation algorithms A and B, each one producing a set S_A and S_B, respectively, the indicator is defined as:

$$I_\varepsilon(A, B) = \max_{x_B} \min_{x_A} \max_{1 \le h \le H} \frac{f_j(x_A)}{f_j(x_B)}$$

where x_A and x_B are each of the solutions in S_A and S_B, respectively.

As we can see, I_ε needs to be calculated for all possible pairs of compared algorithms, something that might be impractical. An alternative unary I_ε^1 version can be proposed, where the Pareto approximation set S_B is substituted by the best-known non-dominated set or by the Pareto front if this is known. Furthermore, the interpretation of this unary version is easier since it tells us how much worse (ε) an approximation set is with respect to the best case. Therefore 'ε' gives us a direct performance measure.

An additional problem of these indicators is that in manufacturing scheduling, and particularly for due date-based criteria, the objective function might be zero. Therefore, operations of translation and/or normalisation have to be carried out.

The comparison of algorithms by means of performance measures is severely limited since a large amount of information is lost when calculating the performance measure. Picture for example the hypervolume indicator I_H. Clearly, a larger hypervolume indicates that an algorithm 'covers' a larger region of the objective space. Another algorithm might have a lower hypervolume but possibly a much more concentrated and better performance for one of the two objectives. Hence, although Pareto-compliant performance measures indicate whether an approximated Pareto front dominates or is dominated by another one, there is no way to infer in which part of the objective space each algorithm performs better. The EAF is a possible answer for this issue.

10.6.2 Dominance Ranking

Dominance ranking is not really a performance measure as there is no loss of information. It can be best observed when comparing one algorithm against another. By doing so, the number of times the solutions given by the first algorithm strongly, regularly or weakly dominate those given by the second gives a direct picture of the performance assessment among the two. However, this approach is limited because it has to be applied for each run on each instance and involves only two algorithms. Comparing many stochastic algorithms on large-sized test instances becomes quickly impractical.

10.6.3 Empirical Attainment Functions

EAF basically depict the probability for an algorithm to dominate a given point of the objective space in a single run. As many multi-objective algorithms are stochastic, different approximations to the Pareto front might be obtained in different runs even for the same instance. EAFs use colour gradients to show the relative number of times that each region of the objective space is dominated. Different from the previous performance measures, EAFs do not condense information and the behaviour over the whole Pareto front can be observed.

More formally, AFs are defined as follows: Let SS be a set of non-dominated solutions for a scheduling problem. $AF(x)$ the Attaintment Function of a solution x is defined as $AF(x) = P(\exists s \in SS\colon s \preceq x)$ which describes the probability for a given Pareto optimisation algorithm of producing, in a single run, at least one solution that weakly dominates x.

In the case of stochastic algorithms, it is not possible to express this function in a closed form but one can approximate it empirically using the outcomes of several runs. This approximation is referred to as EAF. EAFs graphically show, for a certain instance, which area is more likely to be dominated by a given algorithm. A plot is given in Fig. 10.5 taken from Minella et al. (2011) for the application of the algorithm MOSAII$_M$ of Varadharajan and Rajendran (2005) for the $Fm|prmu|\#(C_{\max}, T_{\max})$ model.

As we can see, the MOSAII$_M$ algorithm manages at obtaining a good set of solutions which is evenly spread out. The solid blue areas indicate that those parts of the objective space are dominated by solutions obtained by MOSAII$_M$ 100 % of the times. White solid areas indicate the opposite. We also see how the transition from solid blue to solid white is quick, which means that MOSAII$_M$ is robust and that obtains fairly similar Pareto fronts at each run.

EAFs show the result, over many different runs, of an algorithm over one instance. In order to compare two algorithms, Differential EAF or Diff-EAF can be employed. In short, Diff-EAF basically subtract two EAFs generated by two different algorithms for the same instance and puts the results in the same chart. By analysing Diff-EAF images, one can see in which part of the solution space an algorithm is better than the other.

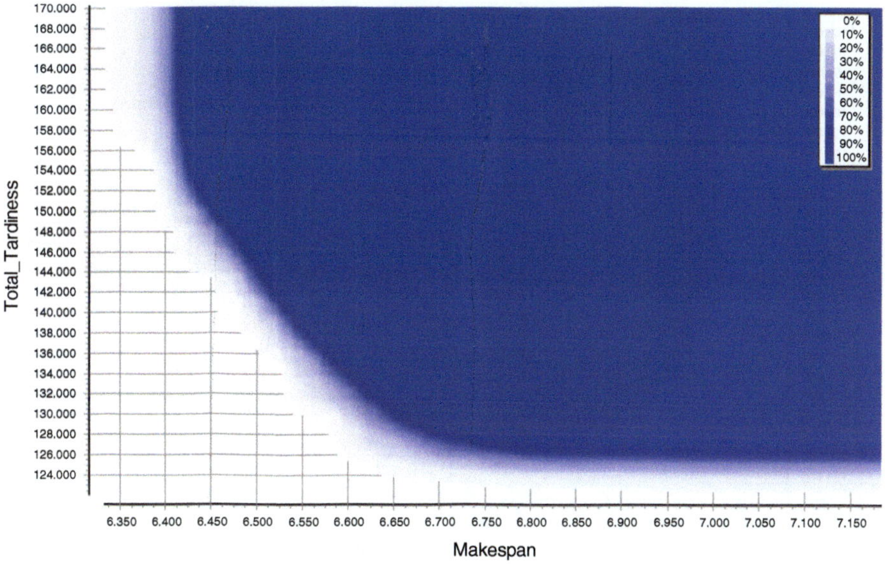

Fig. 10.5 Empirical attainment function after 100 runs of the algorithm MOSAII$_M$ of Varadharajan and Rajendran (2005) (modified by Minella et al. 2008) over instance Ta081 of Taillard (1993) with 100 jobs and 20 machines. Objectives of makespan and total tardiness

For instance, Fig. 10.6 shows a second EAF for the same instance as in Fig. 10.5 but in this case for the RIPG algorithm described in Sect. 10.5.2.1.

The Pareto front given by RIPG is closer to the axes and it contains solutions closer to the ideal point. Although both scales in Figs. 10.5 and 10.6 are identical, it is not easy to compare both plots. In these cases is where the Diff-EAFs are handy. We can subtract both figures and we plot a Diff-EAF with two gradients of colour (blue and red) to positive and negative values of Diff-EAF, respectively. This way, we are able to display the two EAFs as a plot which permits a quick and graphical identification of which algorithm is dominating each zone of the objective space. The intensity of the colour assigned for each algorithm shows the probability of dominating a point in the space of that algorithm over the other algorithm. Points of low colour intensity show lower differences between the algorithms. Notice that white or no colour indicates that algorithms cannot generate solutions dominating this region, or that both algorithms have the same probability of generating dominating points and hence the difference is zero. Figure 10.7 shows the corresponding Diff-EAF. The 'moustache' type of plot indicates that, for this instance, RIPG dominates MOSAII$_M$ over the whole objective space. Recall however, that EAFs need to be examined instance by instance.

Although several algorithms and tools are available so as to construct EAF plots, EAF are hard to code, and hard to manage. Another main drawback of both EAFs and Diff-EAFs is that one plot has to be generated for each instance and for each pair of algorithms. Furthermore, to the best of our knowledge, and at least at the time of the

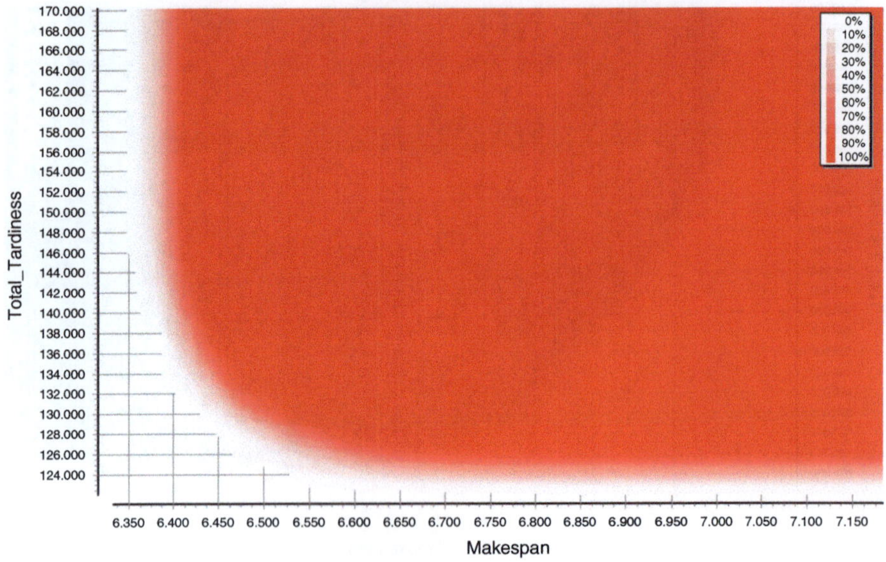

Fig. 10.6 Empirical Attainment Function (EAF) after 100 runs of the algorithm RIPG of Minella et al. (2011) over instance Ta081 of Taillard (1993) with 100 jobs and 20 machines. Objectives of makespan and total tardiness.

writing of this book, EAFs and Diff-EAFs have not been applied to multi-objective problems with more than two objectives.

10.7 Scheduling Interfering Jobs

So far, it has been assumed that the different objectives affect equally to all jobs to be scheduled. However, there are many situations in which several subsets of jobs can be identified among them, each set with different objectives. Particularly, within the rescheduling context, upon the arrival of new jobs to the shop floor, at least two sets of jobs naturally appear:

- Jobs already scheduled in an earlier scheduling decisions, but that have not been yet actually processed in the shop, and
- Arriving jobs which need to be scheduled.

One possibility is to schedule both sets anew, but in many situations it makes sense to employ different objectives for each set: Some performance measure is minimised for arriving jobs in order to obtain a short completion time for this set of jobs (such as for instance minimising their makespan or flowtime), while the objective for the existing jobs may be in line with either minimising the disruption from their initial schedule, or at least that this set of jobs does not exceed the already committed due dates.

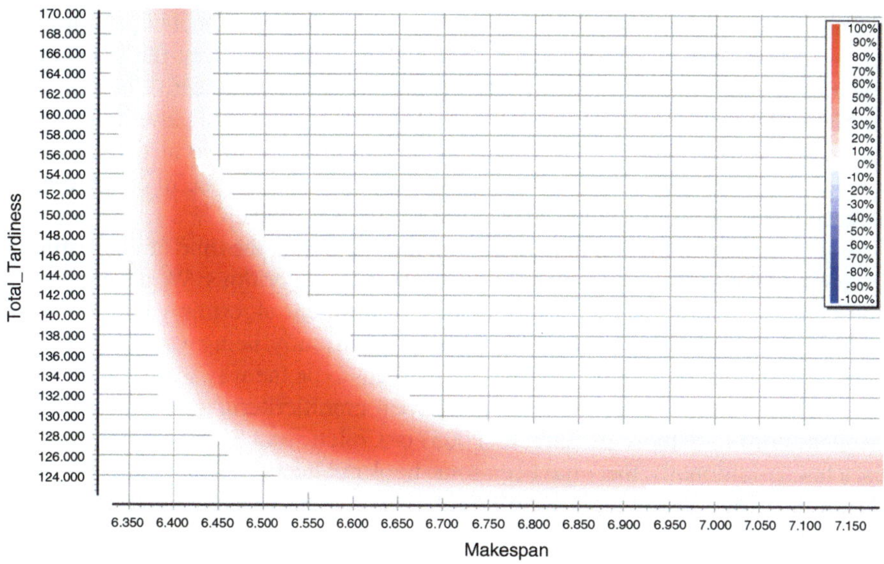

Fig. 10.7 Differential empirical attainment function (*Diff-EAF*) of the two EAFs given in Figs. 10.5 and 10.6 between algorithms MOSAII$_M$ and RIPG. Zones in blue indicate MOSAII$_M$ dominance whereas zones in red show RIPG dominance. Instance Ta081 of Taillard (1993) with 100 jobs and 20 machines. Objectives of makespan and total tardiness

So far, there is no unified name for these models (some of the most popular options are *interfering jobs*, *heterogenous-criteria scheduling* or *multi-agent scheduling*, which are, in fact, a general case of multicriteria scheduling models (note that all the previous multicriteria models are particular cases of the corresponding interfering jobs models.)

10.8 Conclusions and Further Readings

In this chapter we have given a brief introduction to multi-objective optimisation, techniques and quality assessment with special emphasis on scheduling problems. The initial motivation given in this text brings a clear conclusion: reality is multi-objective no matter how hard we try otherwise. We have presented some examples of multi-objective approaches classified mainly by how and when the decision-maker intervenes. The 'easiest' approaches are those in which the decision-maker provides information before the optimisation run. These are referred to as 'a priori' techniques. More difficult are the 'a posteriori' techniques that are basically represented by the Pareto optimisation approaches. These offer the decision-maker with a complete set of non-dominated solutions which represent the best trade-off solutions for the considered objectives.

Compared to single-objective scheduling, multi-objective techniques are less advanced and ready for action. This comes as no surprise, given the still limited penetration and usage of the many advanced existing single-objective optimisation methods. The difficulty and computational demands of multi-objective methods are also to blame here. Nevertheless, given the urgent need for these methods in practice, it is unquestionable that more research efforts must be done in order to bring advanced multi-objective algorithms to the production shops.

A very hot topic of research that is still going, at least at the time of the writing of this book, is the quality assessment of multi-objective algorithms. We have indicated how hard is to compare the outcomes of two different algorithms and the many methods that are being developed and that are currently being validated. Pareto-compliant performance indicators and EAFs seem to be the way to go.

A lot has been written about multi-objective (or multicriteria) optimisation. Some excellent books are those of Ehrgott (2005) or Collette and Siarry (2003), to name just a few. Additionally, there has been some book publications about multi-objective scheduling. The most notable contribution is the comprehensive book by T'Kindt and Billaut (2006) but there are also other less known works, as it is the book of Bagchi (1999).

Regarding the notation employed in the book, note that this is a minimum notation set as much more comprehensive listings and complete notations can be found in the books of Ehrgott (2005) and Collette and Siarry (2003) (to name just a few) and in the paper of Zitzler et al. (2003), later extended in Paquete (2005) and in Knowles et al. (2006). Furthermore, specific notation related to multi-objective scheduling is to be consulted in Vignier et al. (1999), T'Kindt and Billaut (2001), or in the excellent book of T'Kindt and Billaut (2006).

Regarding the types of models of multi-criteria scheduling, the list of references using a weighted combination of objectives in the multi-objective scheduling literature is endless. Some examples are Nagar et al. (1995); Sivrikaya-Şerifoğlu and Ulusoy (1998). Some of the earliest known papers about lexicographical approaches for scheduling problems are due to Emmons (1975). Other related references are Rajendran (1992); Gupta and Werner (1999), or Sarin and Hariharan (2000). The ε-constraint methodology has been employed by, for instance Daniels and Chambers (1990) or McCormick and Pinedo (1995). Some of the best well-known works about goal programming applied to scheduling problems are the papers of Selen and Hott (1986) and Wilson (1989). Regarding Pareto approaches, Daniels and Chambers (1990) was one of the first papers on the topic. More recent contributions are Varadharajan and Rajendran (2005), Geiger (2007) or Minella et al. (2011).

By far and long, most multi-objective metaheuristics refer to so-called evolutionary computation approach. In this regard, see the book by Bagchi (1999), or references like Deb (2001), Abraham et al. (2005), Coello et al. (2007), Coello (2004) or more recently, Knowles et al. (2008).

The issue of comparing multi-objective algorithms in a sound way is still a hot topic of research. Some important results are those of Zitzler et al. (2003), Paquete (2005), Knowles et al. (2006) or more recently, Zitzler et al. (2008). More specifically, a comprehensive study of the existing performance measures for Pareto optimisation problems can be found in Zitzler et al. (2008) following the previous studies of Zitzler et al. (2003) and Knowles et al. (2006). The hypervolume and the epsilon indicators were introduced in Zitzler and Thiele (1999) and in Zitzler et al. (2003). A fast algorithm for the calculation of the hypervolume indicator—otherwise a very time-consuming indicator— is provided in Deb (2001). The I_ε^1 version of this indicator is provided by Knowles et al. (2006). EAFs were first proposed by Grunert da Fonseca et al. (2001) and were later analysed in more detail by Zitzler et al. (2008). Differential EAF are described in López-Ibáñez et al. (2006) and López-Ibáñez et al. (2010). Finally, a classification of interfering jobs is Perez-Gonzalez and Framinan (2013).

References

Abraham, A., Jain, L., and Goldberg, R., editors (2005). Evolutionary Multiobjective Optimization: Theoretical Advances and Applications. Springer-Verlag, London.

Bagchi, T. P. (1999). Multiobjective Scheduling by Genetic Algorithms. Kluwer Academic Publishers, Dordrecht.

Coello, C. A. (2004). Applications of Multi-Objective Evolutionary Algorithms. World Scientific Publishing Company, Singapore.

Coello, C. A., Lamont, G. B., and van Veldhuizen, D. A. (2007). Evolutionary Algorithms for Solving Multi-Objective Problems. Springer, New York, second edition.

Collette, Y. and Siarry, P. (2003). Multiobjective Optimization: Principles and Case Studies. Springer-Verlag, Berlin-Heidelberg.

Daniels, R. L. and Chambers, R. J. (1990). Multiobjective flow-shop scheduling. Naval research logistics, 37(6):981–995.

Deb, K. (2001). Multi-Objective Optimization Using Evolutionary Algorithms. John Wiley & Sons, West Sussex, England.

Deb, K. (2002). A fast and elitist multiobjective genetic algorithm: NSGA-II. IEEE Transactions on Evolutionary Computation, 6(2):182–197.

Ehrgott, M. (2005). Multicriteria Optimization. Springer, Berlin-Heidelberg, second edition.

Emmons, H. (1975). A note on a scheduling problem with dual criteria. Naval Research Logistics Quarterly, 22(3):615–616.

Geiger, M. J. (2007). On operators and search space topology in multi-objective flow shop scheduling. European Journal of Operational Research, 181(1):195–206.

Grunert da Fonseca, V., Fonseca, C. M., and Hall, A. O. (2001). Inferential performance assessment of stochastic optimisers and the attainment function. In Zitzler, E., Deb, K., Thiele, L., Coello Coello, C. A., and Corne, D., editors, Proceedings of the First International Conference on Evolutionary Multi-Criterion Optimization, volume 1993 of Lecture Notes in Computer Science, pages 213–225, Berlin. Springer.

Gupta, J. N. D. and Werner, F. (1999). On the solution of 2-machine flow and open shop scheduling problems with secondary criteria. In 15th ISPE/IEE International Conference on CAD/CAM, Robotic, and Factories of the Future, Aguas de Lindoia, Sao Paulo, Brasil. Springer-Verlag.

Knowles, J., Corne, D., and Deb, K. (2008). Multiobjective Problem Solving From Nature: From Concepts to Applications. Springer-Verlag, Berlin-Heidelberg.

Knowles, J., Thiele, L., and Zitzler, E. (2006). A tutorial on the performance assessment of stochastic multiobjective optimizers. Technical Report 214, Computer Engineering and Networks Laboratory (TIK), ETH Zurich, Switzerland. revised version.

López-Ibáñez, M., Paquete, L. F., and Stützle, T. (2006). Hybrid population-based algorithms for the bi-objective quadratic assignment problem. Journal of Mathematical Modelling and Algorithms, 5(1):111–137.

López-Ibáñez, M., Stützle, T., and Paquete, L. F. (2010). Graphical tools for the analysis of bi-objective optimization algorithms. In Proceedings of the 12th annual conference on Genetic and evolutionary computation (GECCO 2010), pages 1959–1962, New York, NY, USA. ACM.

Mc Cormick, S. T. and Pinedo, M. (1995). Scheduling n independant jobs on m uniform machines with both flowtime and makespan objectives: a parametric analysis. ORSA Journal on Computing, 7(1):63–77.

Minella, G., Ruiz, R., and Ciavotta, M. (2008). A review and evaluation of multi-objective algorithms for the flowshop scheduling problem. INFORMS Journal on Computing, 20(3):451–471.

Minella, G., Ruiz, R., Ciavotta, M. (2011). Restarted iterated pareto greedy algorithm for multi-objective flowshop scheduling problems. Computers & Operations Research, 38(11):1521–1533

Nagar, A., Heragu, S. S., and Haddock, J. (1995). A branch-and-bound approach for a two-machine flowshop scheduling problem. Journal of the Operational Research Society, 46(6):721–734.

Paquete, L. F. (2005). Stochastic Local Search Algorithms for Multiobjective Combinatorial Optimization: Method and Analysis. PhD thesis, Computer Science Department. Darmstadt University of Technology. Darmstadt, Germany.

Perez-Gonzalez, P. Framinan, J. M. (2014). A Common Framework and Taxonomy for Multicriteria Scheduling Problems with interfering and competing jobs: Multi-agent scheduling problems, European Journal of Operational Research, 235(1):1–16.

Rajendran, C. (1992). Two-stage flowshopnn scheduling problem with bicriteria. Journal of the Operational Research Society, 43(9):871–884.

Ruiz, R. and Stützle, T. (2007). A simple and effective iterated greedy algorithm for the permutation flowshop scheduling problem. European Journal of Operational Research, 177(3):2033–2049.

Sarin, S. C. and Hariharan, R. (2000). A two machine bicriteria scheduling problem. International Journal of Production Economics, 65(2):125–139.

Selen, W. J. and Hott, D. D. (1986). A mixed-integer goal-programming formulation of the standard flowshop scheduling problem. Journal of the Operational Research Society, 37(12):1121–1128.

Sivrikaya-Şerifoğlu, F. and Ulusoy, G. (1998). A bicriteria two-machine permutation flowshopn problem. European Journal of Operational Research, 107(2):414–430.

Srinivas, N. and Deb, K. (1994). Multiobjective optimization using nondominated sorting in genetic algorithms. Evolutionary Computation, 2(3):221–248.

Taillard, E. (1993). Benchmarks for basic scheduling problems. European Journal of Operational Research, 64(2):278–285.

T'Kindt, V. and Billaut, J.-C. (2001). Multicriteria schduling problems: A survey. RAIRO Recherche operationnelle - Operations Research, 35(2):143–163.

T'Kindt, V. and Billaut, J.-C. (2006). Multicriteria Scheduling: Theory, Models and Algorithms. Springer, New York, second edition.

Varadharajan, T. and Rajendran, C. (2005). A multi-objective simulated-annealing algorithm for scheduling in flowshopns to minimize the makespan and total flowtime of jobs. European Journal of Operational Research, 167(3):772–795.

Vignier, A., Billaut, J.-C., and Proust, C. (1999). Les problèmes d'ordonnancement de type flow-shop hybride: État de l'art. RAIRO Recherche opérationnelle, 33(2):117–183. In French.

Wilson, J. M. (1989). Alternative formulations of a flowshopn scheduling problem. Journal of the Operational Research Society, 40(4):395–399.

Zitzler, E., Knowles, J., and Thiele, L. (2008). Quality assessment of pareto set approximations. In Multiobjective Optimization: Interactive and Evolutionary Approaches, pages 373–404, Berlin, Heidelberg. Springer-Verlag.

Zitzler, E. and Thiele, L. (1999). Multiobjective evolutionary algorithms: A comparative case study and the strength pareto approach. IEEE Transactions on Evolutionary Computation, 3(4): 257–271.

Zitzler, E., Thiele, L., Laumanns, M., Fonseca, C. M., and da Fonseca, V. G. (2003). Performance assessment of multiobjective optimizers: an analysis and review. IEEE, Transactions on Evolutionary Computation, 7(2):117–132.

Part IV
Scheduling Tools

The Part IV of the book includes Chaps. 11 and 12, and deals with software tools to embed models and methods. We start by discussing in detail the requirements and main features of modern scheduling tools and describe the software architecture of a generic scheduling tool. Advanced design issues, such as integration with other decisions systems in the company, data requirements and accuracy, robustness and user interaction are also treated in this part of the book.

Chapter 11
Overview of Scheduling Tools

11.1 Introduction

This is the first chapter of a series in the book in which we address a number of issues related to software tool building containing the models and solution procedures introduced in the previous parts of the book so that the piece of software may support the scheduling decision process. This topic is not as usual as some of the aspects of manufacturing scheduling discussed in the precedent chapters, and in some cases, it is plainly ignored. However, as we will see, the use of manufacturing scheduling tools is key to the real-world applicability of the models and methods presented earlier.

More specifically in this chapter, we

- discuss the convenience of performing manufacturing scheduling decisions with the aid of a software tool, i.e. a Decision Support System (DSS) (Sect. 11.2),
- summarise the (otherwise rather brief) story of the scheduling tools in the last few decades (Sect. 11.3),
- analyse a manufacturing scheduling tool under the perspective of a business information system, and acknowledge the design of the architecture of such a system as the main generic block of a scheduling tool (Sect. 11.4),
- present earlier approaches to the architectures of scheduling tools (Sect. 11.5),
- identify the different generic requirements of the scheduling tool (Sect. 11.6) and
- discuss an architecture of a scheduling tool fulfilling these requirements (Sect. 11.7).

11.2 Scheduling Tools

Scheduling tools constitute a topic scarcely treated in the scheduling literature, at least as compared to the attention devoted to scheduling methods and models. One reason usually stated is that it is a difficult topic, as the usability of scheduling tools is, as many reports attest, rather limited. In addition, discussing the design and implementation of a software tool represents a hands-on approach that may be perceived

J. M. Framinan et al., *Manufacturing Scheduling Systems*,
DOI: 10.1007/978-1-4471-6272-8_11, © Springer-Verlag London 2014

by the scheduling academic community as only marginally connected to their work, lacking the abstract elegance of dealing with models and procedures. Another reason may lie in the fact that one great motivation for the academic community is to disseminate the results of their research, usually by publishing these results in highly specialised journals, and it is difficult to publish descriptions of implementations of scheduling tools in such journals. Finally, the development of such tool requires at least some degree of expertise in a number of different disciplines, most notably software engineering and business information systems in addition to operations research and production management.

A look into the scarce literature reveals that most of the works fall into one of the following two extremes: on one hand, we have a rather naive approach considering (either explicitly or implicitly) that the scheduling models and solution procedures can be implemented into a software tool in a more or less automatic fashion, so this part can be skipped in most books and papers, and on the other hand, there is a pessimistic stream of papers blaming the entirety of the scheduling field for the lack of practical applications. Perhaps ironically, most of these works accept this situation as a given and offer no new insights for a more practical approach.

In this book, we adopt an intermediate position: on one hand, the development and implementation of a business software tool is far from being a straightforward task, as it is continuously experienced by practitioners in companies. On the other hand, while it is clear that there is a gap between the state of the theoretical knowledge and their application in practice, this problem is commonplace in operations research and management science as it has been attested since the times of Ackoff. In addition, failure rates and underutilisation of scheduling tools is not very far from those of other specific business information systems, even if addressing scheduling decisions is probably one of the less standardised decisions in manufacturing companies, as numerous references of case studies may attest.

Finally, it has to be taken into account that scheduling is a business function that companies *do* accomplish in one way or another, and that, in this regard, a computerised tool offers room for improving the way in which this function is carried out, even if it does not incorporate state-of-the-art models and procedures. As many reports state, simply the fact of replacing the physical Gantt charts by a computerised version may represent a sufficient improvement for the schedulers (McKay and Black 2007).

Therefore, in this book, by scheduling tool we mean a software application that can be used by the decision-makers in order to get support for scheduling decisions. This tool should be built upon scheduling models (to capture the essential parts of the real-life decision problem to be analysed) and scheduling methods (to provide solutions for the models).

Although, in principle, there would be no need of a software tool, the complexity of most practical scheduling problems and the recurring nature of the decision problem make (almost) necessary such tool. While one approach could be to give complete computer control of the scheduling function, a rather different idea seems to be more realistic: The human scheduler and the computer should be considered part of a team in which one complements the other. As expressed in Morton and Pentico

(1993), humans have a broad base of common sense, are intuitive and have an ability to summarise knowledge. However, human expertise is difficult to teach to others and is frequently overloaded by computation. Besides, there is a variability in the response that may be expected from a human. On the other hand, computerised systems possess very fast computation abilities and can be designed to cope with an extraordinary number of different variables. In contrast, they cannot easily learn from experience and definitely do not have common sense.

Therefore, the (so far) most successful approach to combine the strengths and weaknesses of both human and computer is that the latter offers interactive support to the decisions that should be done by the human. By this approach, the computation abilities of the computer are employed to iteratively present the human (the ultimate decision-maker) with useful information that helps him/her to identify problems and to make decisions. This type of tool is named Decision Support System or DSS. Note that the term DSS encompasses a large variety of (mostly computer-based) tools, including the so-called 'expert systems', which include a sophisticated database (named knowledge base) in which a number of rules (usually derived from the human behaviour) are obtained. The majority of the manufacturing scheduling systems described can be considered as DSS. However, there is some controversy on the suitability of this type of systems for scheduling, as some authors claim that very few of these tools have been successfully implemented (Wiers 1997).

11.3 Review of Scheduling Tools

In this section, we will conduct an extremely brief review of the main characteristics of the scheduling tools employed in the last few decades. The field starts with the early attempts of interactive scheduling tools in the 1960s and 1970s, which are summarised in Godin (1978). It is to note that most of the systems discussed in this reference were prototypes and that none of them was successful enough to lead to the development of commercial software. All these tools shared a number of features and difficulties (see, e.g. Morton and Pentico 1993 for a discussion). We have grouped them into the following categories:

- Scheduling state-of-the-art. While there is a general consensus that interactive scheduling systems could provide a response to scheduling decision-making, the state of the art in the scheduling field is not ready for solving large combinatorial problems. The research was very disjoint and the authors do not communicate with each other.
- Technology. Computer technology was still quite expensive and not affordable for all companies, interactive systems were still in its infancy, and software houses avoided the area due to the needs of a customised deployment and expensive user training.
- Humans. First, schedulers do not understand their own job and were not motivated by customer goals. Second, decision-makers did not fully understand the

advantages of computers which—up to this point—were only employed for electronic data processing. A key role is given to the graphical representation of the schedules.

In the 1980s and early 1990s, a number of scheduling tools were presented, such as ISIS (Fox 1994), OPIS (Smith 1994), SONIA (Le Pape 1994) or Micro-Boss (Sadeh 1994). As in the previous group of tools, it is possible to establish a number of common ideas:

- Scheduling state-of-the-art. The focus of these tools is on finding feasible solutions within the constraints imposed by the manufacturing system, which lead to a line of research attempting to capture the knowledge domain and to categorise the types of constraints. Also, opportunistic scheduling (see, e.g. Smith 1994; Sadeh 1994 or Le Pape 1994) appears as a concept of modifying the schedules, providing a number of strategies to repair existing schedules in view of the events in the shop floor. Finally, most of the approaches to model and solve the problem fall within the field of Artificial Intelligence (AI), employing techniques such as constraint propagation, beam search, etc.
- Technology. As computer technology becomes more affordable, the tools are built to be run on workstations or personal computers with graphical capabilities. Interactive scheduling manipulation is available in several of these tools (see, e.g. Sadeh 1994).
- Humans. It is to note that many of these systems were validated with real-life data, but only a few of them were put into production. Maybe this fact can explain the relatively little attention paid to the humans in these works, although some of these tools can be classified as expert systems.

In the 1990s and 2000s, a large number of successful experiences are described, particularly in the European market, although it is true that many of these experiences refer to the deployment of these tools in a highly automated manufacturing plant. Perhaps the most well known is the Leitstand system, which is a generic name for several commercial software (see Adelsberger and Kanet 1991). A good summary of some of the tools that appeared from the 1990s is provided in Pinedo (2009). The characteristics of most of these scheduling tools can be classified as follows:

- Scheduling state-of-the-art. There are plenty of accurate methods to solve a large number of scheduling problems. However, just a few of these methods are embedded in most commercial programs. Although the principles of these methods are widely known, the details of the implementation are often hidden to the decision-makers, thus acting as a black box.
- Technology. Cheap technology is available, and these tools range from PC-based software to sophisticated client/server software. Connectivity and interoperability is an issue in these tools, usually integrating—at least on a theoretical level—several manufacturing or business functions. However, this movement often leads to feeding the scheduling tool with data of poor quality.
- Humans. Decision-makers continue not to trust on computers for the decision-making process, and there is no sufficient evidence that the implementation and

customisation effort of a scheduling tool pays off. The implementation cases described stress the need of a proper training and adequate consideration of the usability of the tool to ensure its success.

As it can be seen from this brief review, some of the problems and difficulties have remained throughout the decades. While the available technology has changed radically, human problems remain and there is no general perspective of what a scheduling tool should accomplish or how it should be evaluated. In this part of the book, we will attempt to address these issues. In the remainder of this chapter, we will analyse the requirements of a scheduling tool in order to provide the building blocks (or, more specifically, the architecture) of a scheduling tool. Some detailed design concepts will be addressed in Chap. 12.

11.4 A Business Information Systems Approach to Scheduling Tools

A computerised scheduling tool can be considered a particular case of business information systems, i.e. information systems supporting business functions. Usually, business information systems can be divided into packaged (standard) software, and customised software (see e.g. Kirchmer 1999). The technological peculiarities of different production environments make it difficult to come up with a general purpose scheduling approach (Brandimarte et al. 2000), and quite often the code developed for the customisation of a packaged scheduling software turns to be more than half the code of the final version (Pinedo 2009). As a result, the development of a generic scheduling tool that can be widely installed has eluded the many vendors who have tried (McKay and Buzacott 2000). Therefore, without loss of generality, we would assume that a computerised manufacturing scheduling tool falls into the category of customised software. Nevertheless, most of the subsequent discussion could also apply to the development of standard scheduling tools, or to its customisation for a specific company.

Given the increasing complexity of software, a full branch of computer science—called software engineering—has been developed to provide methodologies to efficiently develop and maintain software. Software engineering has been particularly successful in providing methodologies to develop information systems. Although these methodologies are very different, the development of a customised information system encompasses a number of activities, which are independent from the adopted software development process (Kurbel 2008). These activities are:

- Requirement analysis (or requirements engineering). Requirements analysis consists in determining the needs of the information system to be developed. This activity usually involves a series of interviews with all parts involved in the life-cycle of the resulting system (the so-called *stakeholders*) in order to determine what the system should do. Note that this activity does not properly provide a technical answer in terms of how the information system should look like, but rather establishes a number of questions that should be answered while designing the tool.

- System design. The design of the system provides a conceptual solution to the requirements that have been identified while carrying out the requirements analysis. Conceptual design means that no computer code is developed, but rather a number of documents or drawings from which the development of a software can be done. Usually, system design is broken down into (1) the description of the software components of the system and their relationships (what is called the *architecture* of the system) and (2) the detailed design of these software elements (Jacobson et al. 1999). This division into two parts will be very important for us, since there we will place the frontier between the design of a generic scheduling tool and the design of a specific scheduling tool.
- System implementation. The implementation of the system refers to transforming the conceptual solution from the previous activity into a piece of software. This involves the translation of the design into source code able to implement the requirements set at the beginning of the process.
- System testing. This activity consists of carrying out a validation and verification of the implemented tool.

Note that these activities are not usually carried out in a sequential manner, but using different software development models based on the concepts of incremental (newer versions of the piece of software contain more functionalities than the previous versions) and iterative (the functionalities of the newer version of the piece of software are more refined than in the previous versions) improvement. The main problem for conducting a business software development process in a sequential fashion is the fact that the requirements analysis cannot be completed and 'frozen' in advance, so it is more efficient to perform a number of iterations in which the different stakeholders will acquire a complete view of the requirements of the system. In contrast, it is clear that a sequential approach may shorten (if properly managed) the development effort. These aspects will be further analysed in Chap. 14.

Although customised manufacturing scheduling tools are, by definition, different for each company, it is clear that some activities in the development process may be common to all companies, as they refer to high-level descriptions of the purpose of the system. This is illustrated in Fig. 11.1, where generic (common) and specific activities are depicted according to the development scheme described above. While there are a number of company-dependent purposes for which a scheduling tool can be implemented (see e.g. Aytug et al. 2005 for a classification of the different purposes of scheduling), most scheduling tools share a number of requirements, as all of them must have a number of common functionalities. Since these requirements are reflected into components of the architecture of the system, it is clear that some parts of this architecture may also be common to most manufacturing scheduling tools. The reuse of an efficient, validated architecture instead of designing a new tool from scratch has a number of advantages, i.e.:

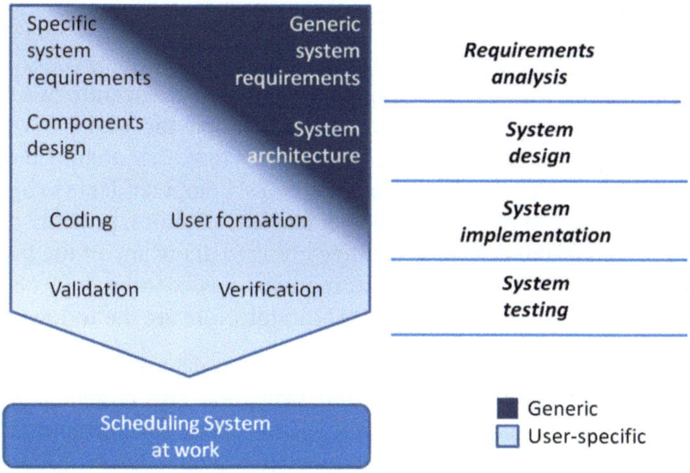

Fig. 11.1 Activities in the development of a manufacturing scheduling tool

- It shortens the development cycle of the tool, as it saves time (and money) that otherwise should have been allocated to the activities of requirements analysis and design.
- It ensures that the main functionalities of a scheduling system are adequately covered, thus the architecture acts both as a checklist and as a design guide for the developers. Note that failure to properly discover the functionalities of a system may not be discovered until later stages and may result in an expensive redesign and recoding.
- It allows the reutilisation for future systems of part of the code developed, provided that the architecture is described in terms of blocks or function-specific modules. This strategy is adopted in reuse-oriented process models for software development.
- It serves to scale the effort by prioritising certain blocks in the development process, thus allowing a gradual development of the system, keeping enhancements of the system for a later stage.

In the following sections, we will focus onto identifying a generic architecture for a scheduling tool. To do so, we first review a seminal work dealing with architectures of scheduling tools (Pinedo and Yen 1997) and discuss their advantages and limitations. Building upon this early work, we will identify the main (generic) requirements to be covered by a manufacturing scheduling tool. Finally, these requirements are discussed and grouped together in order to propose an integrated modular architecture implementing them.

11.5 Early Design of Architecture of Scheduling Tools

One of the pioneering works in describing a conceptual architecture of a scheduling tool is Pinedo and Yen (1997). This architecture adopts the well-known three-tier architecture for software development. Usually, in a three-tier architecture, the user interface, functional process logic or *business rules*, computer data storage and data access are developed and maintained as independent modules, possibly on separate platforms. The aim of the three-tier architecture is to allow any of the three tiers to be upgraded or replaced independently as requirements or technology change.

The main parts of Pinedo and Yen (1997) architecture are the following:

- User Interface. This module handles the input from the user and the presentation of the information offered by the schedule generator. This module is paid a great importance within the architecture, as it is considered that, without an excellent user interface, the system would be too unwieldy to use (Pinedo 2009).
- Schedule Generator. The schedule generator consists of the following modules:
  - Preprocessor. In the preprocessor, the input data of the instance to be scheduled are analysed. Statistical analysis is performed in order to determine due date tightness, congestion factors, setup time factors and other statistics of interest.
  - Algorithm Library. The algorithm library consists of a number of algorithmic procedures, ranging from simple sorting criteria to complex algorithms.
  - Algorithm Generator. The algorithm generator allows the user to construct a new algorithm by combining the existing ones in the Algorithm Library. Additionally, the user may want to apply several available algorithms to a particular problem.
  - Scheduler and Dispatcher. This module generates all schedules and sequences for the problem instance according to the algorithms selected from the Algorithm Library. Two main functions are identified in this module:

    Simulator
    Performance measure

- Database. In this module, the data required by the system are maintained. The database module may be able to manipulate the data, perform various forms of statistical analysis and allow the decision- maker to see the data in graphical form.

Among these three modules, the schedule generator is described in greater detail, and an object-oriented approach is employed to present a number of objects and methods containing their basic functionality. This architecture has been implemented and it is currently supported by a computer application (i.e. CUISE and Lekin, see Pinedo 2012). Therefore, at least our starting point is an architecture that has proved able to stand real-life scheduling. In addition, most of the modules described in the architecture are present (either implicitly or explicitly) in most architectures underlying scheduling tools, such as T'Kindt et al. (2005). In addition, this proposal has the advantage of employing a formal language (object-orientation) to describe the architecture of the system and the relations among the different modules, thus moving towards a greater formalisation of the design of scheduling tools.

Despite these facts, this seminal work in the field of scheduling tools includes a number of issues in this architecture that we retain worth to be discussed before proceeding further. These are the following:

- A very general critique of the architectural design refers to the main modules in the architecture: although there is a database module, it is not clear which kind of data it should contain. Initially, one may think of production process data (such as the routing schema or the processing time of the machines, etc.) as well as job-related data. However, on one hand, in most real-life systems, a big part of these data—particularly job-related data— are already stored in an information system (such as an ERP system). The interoperability of the scheduling tool with the rest of business information system is not explicitly considered in the presentation of the architecture.
- Another issue refers to the adopted model. As mentioned before, it follows the three-tier architecture adopted in many information systems. Although this model is quite useful for software reutilisation in many developments, we hesitate that it can be easily adopted for a scheduling tool. First, the user interface (the presentation tier) should not only interact with the schedule generator, but also with the database, as it is clear that many data should be entered into the system by the user in a (possibly) graphical way. Therefore, the database should lie at the heart of the system, and the rest of the modules of the tool should have direct access to it.
- The algorithm library itself could be part of the database, particularly if the implementation follows the pattern adopted by some commercial systems (i.e. in SAP R/3 the functionalities of the system are stored in the database in the format of an interpreted-like language that is loaded and executed in memory once it is required).
- One may assume that at least part of the schedule generator must be included in the User Interface, as it is very difficult to separate the presentation of a schedule from the business logic that should be included in order to allow the user to manipulate the solutions. This fact also speaks for the aforementioned difficulty of the three-tier architecture adopted in the design.
- The architecture proposed adopts basically a technical, problem-solving approach, and perhaps overlooks the cyclical nature of the scheduling process. As such, functionalities such as the analysis and management of different scenarios, self-learning issues, etc., are not explicitly considered in the description.
- Finally, it is not clear whether reactive scheduling can be carried out with these modules, or it is considered out of the scope of the tool. This is a very important issue, as we shall see in the next sections.

In summary, we believe that Pinedo and Yen's architecture should be understood as a basic, high-level description of the main functionalities of a scheduling tool. However, in order to produce a usable decision support system, this architecture should be enhanced by addressing the issues mentioned before. In order to systematically capture the needs of such software, in the next section we review the main requirements of a scheduling tool.

11.6 Requirements of a Scheduling Tool

As mentioned before, the first step for developing a software tool consists of analysing its requirements. In software engineering, it is usual to distinguish between *functional requirements* and *non-functional requirements*. The former specify what the system should do, i.e. what functions it should provide to its users. Non-functional requirements are general criteria that the system should meet apart from solving specific application problems. Such criteria often refer to software quality and performance of the system. Some examples are maintainability, reliability, scalability, robustness, etc., and collectively define a number of characteristics of a good piece of software. As such, it is hard to see how these can be distinguished from other business decision support software, and thus there is nothing inherently specific to scheduling tools. Therefore, we will omit this type of requirements in this chapter and refer the interested reader to some of the many good books on software engineering, such as Kurbel (2008). In contrast, we will concentrate on the functional requirements, particularly on those generic requirements that allow us to identify an architecture of the systems when we break down into the design phase. These are grouped into several categories and described in more detail in the next sections. In addition, it is also very common in software engineering to prioritise the different functionalities of the system to be developed. Since the development process rarely goes smoothly, it is usual that some functionalities have to be limited or removed from the system due to the lack of time or resources to accomplish all of them. In such cases, a categorisation of the functionalities may help to know which are the ones regarded as critical, and which can be removed without greatly impacting the product. In our case, we do not dare to make a strict categorisation, but at the end of each section we provide some hints on how 'basic' or 'advanced' these categories are. A summary of these categories and their related functionalities is shown in Fig. 11.2.

11.6.1 Scope of the System

The scope of a manufacturing scheduling system refers to the set of business functions targeted by the system within the context of supporting decisions for production management. With this regard, there are two issues that should be discussed: The first one is to define the set of business functions that define scheduling, and second, whether there are additional production management decisions—usually considering to be outside of the scheduling function—that are so intertwined with scheduling that should be included into a software tool.

With respect to the scheduling function, generally speaking, there are two views or levels, depending on the time horizon (Aytug et al. 1994):

- A higher level that uses the output of production planning to set up the dates for the beginning of each job on each machine. This level is often referred as *release scheduling*.

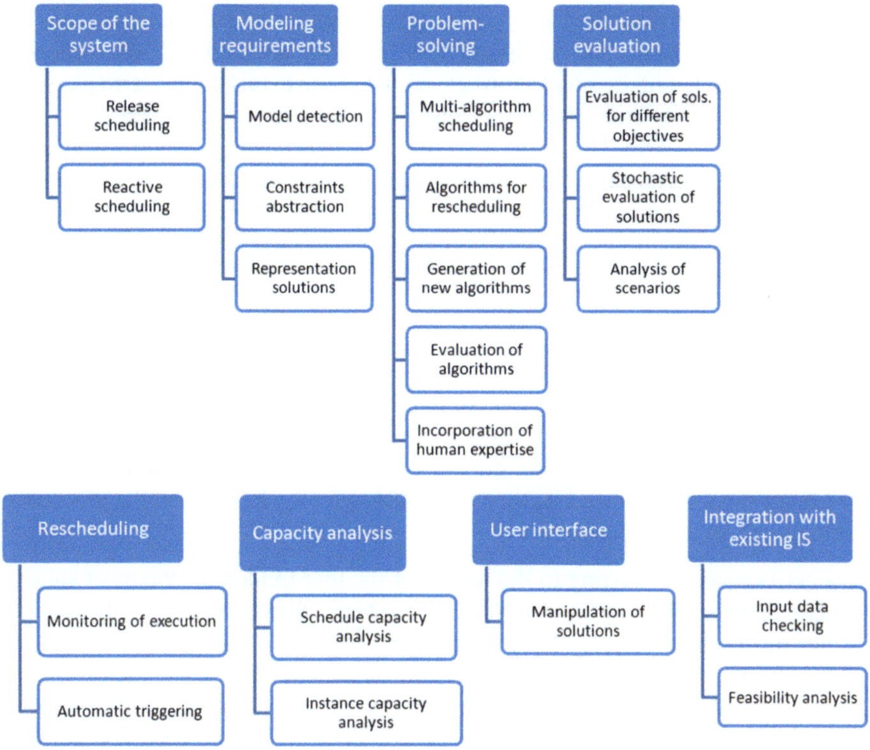

Fig. 11.2 Summary of functionalities identified

- A lower level which is involved with real-time item movement planning. This level is usually denoted as *reactive scheduling*.

There is a consensus in that these two levels should be adequately covered by the scheduling system, which means that a scheduling system's architecture should integrate functionalities regarding the monitoring and execution of the planned schedules. This aspect is what McKay and Wiers (1999) label as 'sustained control', meaning that the schedulers should be able to monitor the progress of production and to solve problems if the actual situation deviates from the scheduled situation. In other words, there is no *scheduling problem* to be solved, but a *scheduling decision process* which may (or may not, depending on the situation) involve solving decision (optimisation) problems. Therefore, in order to build a tool that supports the scheduling process, the lower level of scheduling (reactive scheduling) should be contemplated in the software.

In addition, it has been already mentioned that there are a number of manufacturing activities that are related to scheduling. Perhaps the most obvious is production planning, which covers production at a multi-week level and makes up the input data for release scheduling. Despite this clear connection, planning and scheduling activities are usually handled independently, except for a few attempts (see, e.g. Huang et al. 1995; Kanet and Adelsberger 1987; Zhang and Mallur 1994). Indeed, it has been argued that many small shops have little scope to make long-term plans and that systemic complexities also force scheduling to be directed to very short-time horizons (Higgins 1996). Therefore, we will not include planning functionalities into our proposal, although the architecture should contain the necessary coordination between planning and scheduling.

11.6.2 Modelling Requirements

Modelling requirements refers to a set of requirements related to the modelisation of the shop floor to be entered in the system and its corresponding schedules. Several functionalities regarding this area have been proposed in the literature. Although all of them seem to be desirable, their usefulness greatly depends on the theoretical developments in the topic.

- Model Detection, referring to the ability of the tool to identify the most suitable (theoretical) scheduling model from the instance data provided to the system. Since most scheduling research refers to specific models (i.e. a combination shop floor setting, constraints and objectives) due to the complexity and inefficiency of general scheduling model representation and solution procedures, model detection may be an important feature of scheduling tools, as it may help selecting the most suitable algorithms to solve the problem or drive the development of new algorithms. This functionality is present in the e-OCEA architecture (T'Kindt et al. 2005). In another contribution, Lamatsch et al. (1988) develop a tool that uses a metric to match scheduling problems. It is to note that, in the references mentioned, Model Detection mainly refers to matching the layout with that of a theoretical model, whereas little is discussed about the constraints in the scheduling process, which may be much more difficult to be identified. It is likely that, in order to better match the real situation, this functionality identifies a relatively complex model that a human could easily simplify by taking out of the model non-relevant elements. However, it is clear that it may be useful for thc human Decision-Maker in order to guide him/her in identifying the underlying scheduling model. This functionality can be seen as an evolved functionality from the functionality 'Schedule Capacity Instance' which is presented in Sect. 11.6.6 (see in this section a discussion of the relationships between both functionalities).
- Constraints Abstraction. This refers to the ability of the tool to work with a simplified subset of shop floor constraints during the obtention of the solutions for the scheduling problem, so constraints with a little impact in the evaluation of the solution are ignored. This aspect, mentioned in Fox (1990), may be regarded as a multi-layer model since the representation of the constraints in the scheduling tool is already an abstraction of the real constraints in the shop floor. Note that several surveys on scheduling practices (see, e.g. McKay et al. 1998) reveal that

hundreds of constraints can be identified in the shop, and that some of them are paramount while others can be safely ignored. Therefore, the system should use (at least) two layers—or views—of a model: An aggregated view employed during the search of solutions, and a detailed one employed to present the solutions to the Decision-Maker. The suitability of the approach depends on ensuring that the constraints that have been ignored do not have a major effect in the feasibility of the so-obtained solutions. This can only be verified by a thorough validation process during the deployment of the tool.
- Representation of the Solutions. The schedules provided by the solution procedures have to be transformed into starting and finishing times for each job. This functionality is mentioned by Pinedo and Yen (1997) as 'simulator', and follows from the fact –already mentioned– that the tool should work with a simplified subset of shop floor constraints. Obviously, this level of abstraction with respect to the constraints greatly influences the complexity of this module, as it would possibly must 'rework' the schedules provided by the scheduling generator to make them feasible, and, if this process is not properly bounded, it may greatly reduce the efficiency of the scheduling engine as there may not be a correlation among good schedules (as provided by the schedule generator) and good solutions (as finally presented to the user), therefore rendering useless any optimisation effort.

If the three functionalities have to be classified or prioritised, it is clear that the two latter represent basic aspects of a scheduling tool, whereas the first one (Model Detection) constitutes an advanced feature of the system. In addition, its usefulness would depend on how well covered are the rest of functionalities (e.g. if the algorithm library does not contain algorithms for a large number of scheduling models, its nearly automatic identification would not be very useful).

11.6.3 Problem-Solving Functionalities

This section refers to the functionalities related to the solution procedures to be integrated in the system and how to use them. The following functionalities could be identified:

- Multi-algorithm scheduling. This functionality refers to the separation of model and solution procedures, so different solution procedures can be applied to a given problem instance. This is usually denoted as *algorithm library* by several authors (see, e.g. Pinedo and Yen 1997). Ideally, the algorithm library should contain all algorithms available in the system for all scheduling problems that the system is able to deal with. Here, we can distinguish between application-specific algorithms and generic algorithms. Application-specific algorithms are able to solve a particular scheduling problem, while generic algorithms may be employed for most (or all) the scheduling problems. Both types of algorithms have advantages and disadvantages (McKay et al. 2002). Generic algorithms have to be customised to each problem, requiring compromises in terms of representation, objective specification and schedule generation, so the scheduler may encounter potential difficulties in specifying the problem and in obtaining acceptable results. In contrast, application-specific algorithms can be prohibitively expensive to develop in some cases and can be very hard to change and/or enhance if the situation changes. Even in the case of most generic algorithms, it is clear that not all of them can be used for all models, therefore a database matching each model and the corresponding solution procedure(s) should be kept in the software tool. This functionality could be linked to that of the evaluation of the algorithm.

- Algorithms for rescheduling. In many occasions, it is not possible to completely rebuild existing schedules, but it is preferable to perform a reschedule in which a subset of jobs change their schedule. In order to perform the rescheduling process, specific algorithms (usually considering stability-related objectives such as minimisation of the changes of the existing schedule) should be available in the system. This functionality is mentioned to have to be covered by a manufacturing scheduling tool by several authors, including Błażewicz et al. (2001), Collinot et al. (1988), Lee et al. (1997), Fox (1994), Kempf (1994), Smith (1994), Hadavi et al. (1990), Sauer and Bruns (1997), Prietula et al. (1994) and it follows more or less straightforward from the scope of the scheduling tool. Clearly, these algorithms for rescheduling could be seen as a part of the algorithm library described in the previous item.
- Generation of new algorithms. In order to make a dynamic and efficient use of the algorithm library, the tool should allow for the integration of new algorithms in an easy manner. While this aspect is frequently mentioned in the literature, the proposed solutions range from a semi-automation of the process of generating new algorithms (see T'Kindt et al. 2005, where a specific meta-language for describing scheduling algorithms is suggested), the description of an Application Program Interface (API) to embed the algorithms into the existing system (see the LEKIN system by Pinedo 2012, where several general purpose languages can be employed to be linked with the system), or the inclusion in the algorithm library of 'chunks' of heuristics that can be combined to obtain new heuristics (see Sauer 1993). In addition, the generation of new algorithms should balance the performance and generalisation of these new algorithms (which may be undoubtedly higher if a low-level integration via API is performed) against the time required to integrate them into the system.
- Evaluation of algorithms. This functionality refers to the fact that the tool must be able to choose/suggest the most suitable algorithm required for any specific problem to be solved. This feature may be required by at least two reasons. First, it may happen that the specific problem instance to be solved does not match any of the problems that can be solved by the algorithms included in the library. In such case, the tool can propose/choose employing algorithms that solve problems closer to the original. Second, it may be that more than one algorithm could be employed for the problem corresponding to the specific instance. Among these, the system should propose/choose the one(s) most suitable for the specific instance.

  Several proposals have been done to capture the knowledge about the suitability of the algorithms. Sauer and Apelrath (1997) suggest using the knowledge generated from solving actual instances of the problem (i.e. past scheduling problems), while T'Kindt et al. (2005) propose a mechanism to generate benchmark instances derived from actual instances of the problem to be solved that are subsequently employed to test the different algorithms. Finally, Gupta et al. (2000) propose the application of neural networks to select the best heuristic algorithm to solve a given scheduling problem. In their paper, the objective is to choose the heuristic with higher statistical quality of the results.

- Incorporation of human expertise. This refers to the ability of the system to incorporate the knowledge of the human schedulers into the tool, and it has given rise to a relatively robust literature on scheduling expert systems. A list of the potential benefits of this functionality can be found, e.g. in Reinschmidt et al. (1990). However, there is some controversy on the usefulness of the approach, since many researchers are sceptical that true human experts exist in scheduling (Steffen 1986). This is motivated by the belief that most real-life scheduling environments are beyond the cognitive capability of most schedulers (Fox 1990). Indeed, some experiments have been carried out (Baek et al. 1999) in order to analyse the performance of a human scheduler who must develop a schedule from scratch against that of a software tool. In these experiments it is shown that, if the problem is complex, the performance of the solution obtained by the human scheduler from an initial schedule provided by the software tool is better than that obtained from scratch by the human scheduler. Besides, even if human experts exist, some authors consider that the expert system would merely automate their (good or bad) decisions (Kanet and Adelsberger 1987). As a consequence, the need/suitability of this functionality is highly controversial. Since we are describing a general architecture of a scheduling system, we believe that this functionality should be considered, and let its implementation to the particular cases that may arise.

Regarding the categorisation of this set of functionalities, the ability to support a group of algorithms for scheduling and for rescheduling must be considered a basic functionality of a scheduling tool. Without these features, a scheduling tool is so severely limited that hardly serves to represent and manipulate the schedules proposed by the decision-maker. The ability to easily incorporate new algorithms may be regarded as an advanced feature, as we suspect that for many implementations, if new, specific algorithms are needed, then this task should be accomplished by the development team and not for the company's staff. If this is the case, the advantages of having a tool that helps building new algorithms in a quick manner can be limited by the knowledge of the software possessed by the developers. Regarding the incorporation of the human expertise, this topic is very debatable and there are many voices against it that it cannot be considered a basic functionality. Finally, in our view, the state of the art with respect to the evaluation of solutions is rather limited for this functionality to have a great impact, although its usefulness would clearly depend on how repetitive the data are.

11.6.4 Solution-Evaluation Functionalities

This set of functionalities refers to how the solutions found by the problem-solving functionalities described in Sect. 11.6.3 are evaluated. The following functionalities have been identified from the literature review:

- Evaluation of solutions for different objectives. The tool must allow the evaluation of a solution with respect to several objectives, including a combination of different objectives. This functionality is mentioned by Pinedo and Yen (1997) and Sauer (1993). In practice, implementing this functionality in the tool translates—at least—into two issues:
  - A large number of objective functions should be implemented in the tool. This would range from the rather typical literature-based scheduling criteria, to cost-related objective functions. Note also that it is likely that the decision-maker would like to see each objective function in this functionality to be also included as an objective function for the algorithms in the algorithm library described in a previous section. In practice, this would mean that at least some general purpose algorithms should be implemented by the tool, in order to cover these objectives for which efficient specific algorithms are not known, or they are simply not worth the effort for its development. A further refinement would be to allow the user to include new objectives, much in the manner of the 'generation of new algorithms' functionality.
  - A smart way to present the evaluation of the algorithms to the user should be implemented, as it is likely that the user would like to compare different solutions for different objectives. If the objectives are numerous, it would be hard for the user to make sense of the different results, so the tool could ideally detect the objectives with relevant differences among the solutions under comparison and hide those for which these differences are not significant. Another option would be to make a hierarchy of the different objectives and show them in—at least—two different levels. An alternative to limit this problem is to select a minimal set of objectives, as in practice most of them would be related. This aspect is discussed in Chap. 12.

- Stochastic evaluation of solutions. The solutions obtained by the problem-solving functionalities are usually based on deterministic assumptions, therefore it would be of interest to evaluate the performance of the solutions under the stochastic conditions of the real shop floor. This functionality is suggested by Błażewicz et al. (2001), who propose the use of queuing theory or simulation to accomplish this evaluation. The combination of simulation and scheduling is discussed in detail in Fowler et al. (2006). Note, however, that this is only one of the approaches that can be employed to address variability in scheduling, which include the following (see e.g. Black et al. 2006 for a summary of the different approaches):

- Using different methods for modelling uncertainty, such as, e.g. fuzzy logic,
- Researching issues associated with measuring robustness in the face of uncertainty,
- Developing different techniques for constructing schedules (e.g., rescheduling and rolling horizons), using real-time information for rescheduling or schedule repair and
- Inserting idle or slack time, in a general manner or placing slack at tactical locations in the schedule to allow for expansion and schedule adjustment caused by variability.

- Analysis of scenarios. The system must allow the management of different solutions for what-if analysis. This functionality is particularly useful if the user is allowed to manually modify the schedules proposed by the system (see Sect. 11.6.7 for a discussion). It is mentioned in McKay and Wiers (2003).

With respect to the categorisation of this set of functionalities, the evaluation of solutions for different objectives and the analysis of scenarios have to be clearly labelled as basic functionalities. In contrast, our view on the stochastic evaluation of solutions indicates that, although it seems potentially interesting to combine the optimisation results with a simulation, we are sceptical with respect to its usefulness on a general basis. On one hand, detailed simulation models are extremely expensive to construct and to maintain, and stochastic models based, e.g. on queuing theory may not provide an accurate capacity check and thus may mislead the scheduler in his/her decision. On the other hand, some of the methods later mentioned can be used in a more or less sophisticated manner to perform an analysis of the effect of the variability in the shop floor without expensive developments. Therefore, this functionality cannot be regarded as basic.

11.6.5 Rescheduling Functionalities

This set of functionalities refers to the ability of the tool to perform rescheduling. Two different aspects can be considered:

- Monitoring of execution. While one option for the user is to decide whether to reschedule or not in view of the shop floor information found outside the scheduling system, the system could monitor the execution of planned schedules in order to measure the deviation between the planned and the actual situations and trigger some alarm under certain conditions. This functionality is mentioned in Błażewicz et al. (2001), Collinot et al. (1988), Smith (1994) and Numao and Morishita (1989).
- Automatic triggering of scheduling/rescheduling functions. A step beyond is to allow the tool to decide, in view of the deviation from the plans, whether the current jobs have to be rescheduled, or a full schedule is to be developed. Although this functionality is found by several authors (see, e.g. Błażewicz et al. 2001; Collinot et al. 1988; Smith 1994), we think that this could not be a good option except for

fully automated systems, as we believe that the approach of a scheduling tool is that of a decision support system and not that of an automatic control system. Indeed, post-evaluation case studies such as the one by Lin et al. (2007) indicate that the schedulers want to be in full control of the scheduling tool. In contrast, it would be interesting if the tool includes the capability of learning under which conditions the human scheduler is conducting a rescheduling of the existing solution, and perform an automatic triggering of the alarms. Even if this learning process is conducted using relatively simple procedures, we believe that this would greatly help the scheduler in making more consistent rescheduling decisions, as the system would indicate that, in the past, a reschedule was triggered under similar conditions. This would make the scheduler reflect and modify his/her procedures, or at least to make explicit constraints or objectives not initially contemplated.

In summary, monitoring of execution should be regarded as a basic functionality, helping the scheduler to be in control of the system, whereas the automatic triggering of scheduling and rescheduling functions can be labelled as advanced.

11.6.6 Capacity Analysis Functionalities

This set of functionalities refers to the ability of the tool to detect critical points and slacks in the system, and consequently to help the scheduler to focus his/her attention in the most critical parts of the process. These functionalities can be grouped as follows:

- Schedule capacity analysis. By using this functionality, the system must be able to detect potential bottlenecks and under-loaded resources according to the planned schedule (or even according to the actual schedule, if the system allows monitoring of execution). This functionality may help the schedulers to focus their attention during the execution of the schedule into the critical sections of the shop floor, and to identify under-utilised resources to be loaded during the building of the next schedule, or transferred to the critical sections. This functionality is mentioned by Collinot et al. (1988), Kempf (1994), Prietula et al. (1994).
- Instance capacity analysis. By using this functionality, the system may detect potential bottlenecks and under-loaded resources *before* a solution for the problem instance is provided by the system. Note that this functionality is different from that of schedule capacity analysis, as there is no planned schedule. This functionality is called *preprocessor* by Pinedo and Yen (1997), and it can be linked with the Model Detection functionality described in Sect. 11.6.2, since it can be regarded as a wizard to help identifying a suitable model for the shop floor (i.e. a preprocessor that identifies that there is a bottleneck machine regardless the specific job schedule may serve to point to the single-machine scheduling model as the most suitable model to schedule jobs).

In our view, both functionalities are somewhat advanced, as they may not be strictly needed for some scenarios. Among these two, schedule capacity analysis is less advanced and requires less sophistication that the instance capacity analysis. Instance capacity analysis may be quite complicated to implement, as the state of the art regarding this issue is practically inexistent and many practical problems for which there is no previous guidance may arise during the process.

11.6.7 User Interface

There are a number of functionalities regarding the user interface that are required for a generic business information system, mostly referring to the ability of the system to provide an interactive way to enter input data into the system, or to extract output data (typically reports) from the system. Here, we focus onto those specific features of a manufacturing scheduling system, mostly referring to a user interface which acts both as input and output of the system:

- Manipulation of schedules. Most authors highlight the need of human interaction (see e.g. the works by McKay et al. 1995a or Trentesaux et al. 1998) meaning the manipulation by the human scheduler of the solutions offered by the system, an aspect mentioned by many authors, among them Collinot et al. (1988), Numao and Morishita (1989), Prietula et al. (1994), Sauer and Bruns (1997), Sauer (1993), Sauer and Apelrath (1997), T'Kindt et al. (2005) and McKay and Buzacott (2000). Note that this functionality should include mechanisms for repairing the possible infeasibility of the solutions generated by the human scheduler. Regarding the presentation of the solutions, while Gantt charts are mostly used as the primary mean for this interaction, Higgins (1996), questions its adequacy and suggests employing the so-called Job Screens and machine-loading boards as a starting point. In Chap. 12, this discussion will be rejoined when introducing some different alternatives to represent schedules.
- Comparison of manual versus automatic solutions. It is clear that the manipulation of the resulting solutions may result into a notable worsening of the original objective function. While this is not necessarily a problem (for instance, it may be the case where the original solution did not contemplate some complicated constraints that the scheduler had in mind), the tool must provide reports indicating the deviation of the so-modified schedules with respect to the ones suggested by the algorithms. The purpose of this subfunctionality is two-fold: In the short term, it helps the scheduler to balance the importance of these 'implicit' objectives/constraints with respect to the overall performance of the schedule. In the long term, the continuous monitoring of these deviations may help to notice important objective/constraints that were not initially taken into account but that have a large influence in the final results.

From these two functionalities, it is clear that the first one is a basic requirement of a scheduling tool. Even if a fine implementation of this functionality can be costly, without at least some basic manipulation features, the tool gives little space for user interaction and its value and usability decreases enormously. The second functionality is somewhat more advanced, although paradoxically it is easier to implement that the first one.

11.6.8 Integration with Existing Information Systems

Integration in the organisational environment means that the scheduling tool should be able to import the data required and to export the results obtained from/to the rest of the systems. This is a critical aspect in the development of the tool, as several surveys (see e.g. McKay et al. 1998) identify up to 25 different sources of data related to scheduling. A number of functionalities can be mentioned regarding this aspect:

- Input data checking. The system must provide functionalities to check the consistency and the comprehensiveness of the data automatically entered into the system. Significant discrepancies between historical and new data could be checked so that the system may trigger alerts to the Decision-Maker to allow/discard these changes.
- Feasibility analysis. The system can check the existence of all resources needed for scheduling the jobs, namely the availability of raw materials, state of the machines, etc. This functionality is mentioned in Błażewicz et al. (2001) and Hadavi et al. (1990), and can constitute an output of the scheduling tool towards different information systems, such as the MRP.

An important issue in order to accomplish a seamless integration of the scheduling system is to define a format or language that allows the precise specification of these data for a wider number of scheduling problems. Given the diversity of scheduling problems, this is not an easy task. So far, some research has been carried out on languages for describing scheduling problems (see, e.g. Yen 1997 or Zentner et al. 1998). However, keeping in mind the aforementioned need of integration of scheduling activities with existing manufacturing information systems, the development of a standard language is a must. In this line, perhaps scheduling systems can benefit from the XML standard and, more specifically, from the Business Process Modelling Initiative (BPMI) (BPMG.org 2004). This initiative is seeking to establish an XML-based language and notation (BPML and BPMN, respectively) for the specification of business process models.

Regarding the prioritisation of the different functionalities, since input data are critical for the scheduling system and may come from many different sources of different quality, such checking functionality is considered to be of great importance (McKay and Wiers 2003). Indeed, the lack of quality in the input data is mentioned as a reason to implement data filtering utilities and spreadsheets as an aside product of different scheduling tools (Berglund and Karltun 2007). The only exception in the importance of this functionality would be if the scheduling tool is completely autonomous with respect to the data, which may not be the most usual case. In contrast, feasibility analysis can be regarded as a more advanced functionality.

11.6.9 Summary of Revised Functionalities

This section summarises the contributions discussed in the early sections regarding the functionalities that have been identified. The main outcome is presented in Table 11.1, where the functionalities, the main references dealing with them and their character (basic or advanced) are presented.

11.7 An Integrated Modular Architecture for Scheduling Systems

As mentioned in Sect. 11.5, there are several references addressing a high-level description of the main building blocks of the architecture of a scheduling system. These blocks can be grouped into three subsystems, which constitute the standard blocks of a decision support system, i.e. a data base management system, a model-based management system and a user interface (see e.g. Pinedo and Yen 1997 or Ecker et al. 1997). We propose an extended modular architecture that contains a compromise between specificity and generality by adding important missing elements from existing proposed architectures, such as an explicit mechanism for handling the communication among the modules. This architecture also captures the reviewed functionalities from the previous section while giving enough details so as to enable an effective design and implementation of a scheduling system. More specifically, the scope of the proposed system entails both production scheduling and shop floor control, in accordance with the requirements identified in Sect. 11.6.1.

The modules identified in the architecture are the following:

- Business Logic Unit/Data Abstraction Management Module. This module acts as a broker to guarantee the required abstraction when accessing to the data in the tool.
- Database Management Module. This module stores the data in the system and provides a mechanism to interface with existing Business Information Systems in the company.
- User Interface Module. This module provides the required interface between the user and the scheduling tool.
- Schedule Generator Module. This module contains all functionalities related to providing schedules to the user.

A deployment diagram based on the Unified Modelling Language (UML) representing these modules is given in Fig. 11.3. The modules and subcomponents are briefly detailed below and would be extensively discussed in Chap. 12.

The proposed architecture is highly modular at all possible levels. All parts communicate through Advanced Programming Interfaces (APIs) and through the Business Logic Unit/Data Abstraction Module (BLU/DAM) in order to isolate the different modules. Otherwise, small changes in one module could have a serious—and

Table 11.1 Summary of functionalities identified

Type of functionality	Functionality	Main references
Basic	Production planning	Huang et al. (1995), Kanet and Adelsberger (1987), Zhang and Mallur (1994)
Basic	Shop floor control of schedules	McKay and Wiers (1999)
Advanced	Model detection	T'Kindt et al. (2005)
Basic	Constraints abstraction	Fox (1994)
Basic	Representation of the solutions	Pinedo and Yen (1997)
Basic	Rescheduling	Błażewicz et al. (2001), Collinot et al. (1988), Fox (1994), Kempf (1994), Smith (1994), Hadavi et al. (1990), Sauer and Bruns (1997), Prietula et al. (1994)
Basic	Multi-algorithm scheduling	Pinedo and Yen (1997), Błażewicz et al. (2001), Kempf (1994), Smith (1994), Sauer and Bruns (1997), Sauer (1993), Sauer and Apelrath (1997), T'Kindt et al. (2005), McKay et al. (2002)
Advanced	Generation of new algorithms	Pinedo and Yen (1997), T'Kindt et al. (2005)
Advanced	Evaluation of algorithms	Sauer (1993), T'Kindt et al. (2005), Sauer and Apelrath (1997)
Advanced	Incorporation of human expertise	Reinschmidt et al. (1990), Steffen (1986), Kanet and Adelsberger (1987), Fox (1990)
Basic	Evaluation of solutions for different objectives	Pinedo and Yen (1997), Sauer (1993), Sauer and Apelrath (1997)
Advanced	Stochastic evaluation of solutions	Błażewicz et al. (2001)
Basic	Analysis of scenarios	McKay and Wiers (2003)
Basic	Monitoring of execution	Błażewicz et al. (2001), Collinot et al. (1988), Smith (1994), Hadavi et al. (1990)
Advanced	Automatic triggering of rescheduling/scheduling functions	Błażewicz et al. (2001), Collinot et al. (1988), Smith (1994)
Basic	Schedule capacity analysis	Błażewicz et al. (2001), Collinot et al. (1988), Kempf (1994), Prietula et al. (1994)
Advanced	Instance capacity analysis	Pinedo and Yen (1997)
Basic	Interactive scheduling	Collinot et al. (1988), Numao and Morishita (1989), Prietula et al. (1994), Sauer and Bruns (1997), Sauer (1993), Sauer and Apelrath (1997), T'Kindt et al. (2005), McKay and Buzacott (2000), Higgins (1996)
Basic	Input data checking	McKay and Wiers (2003)
Advanced	Feasibility Analysis	Błażewicz et al. (2001), Hadavi et al. (1990)

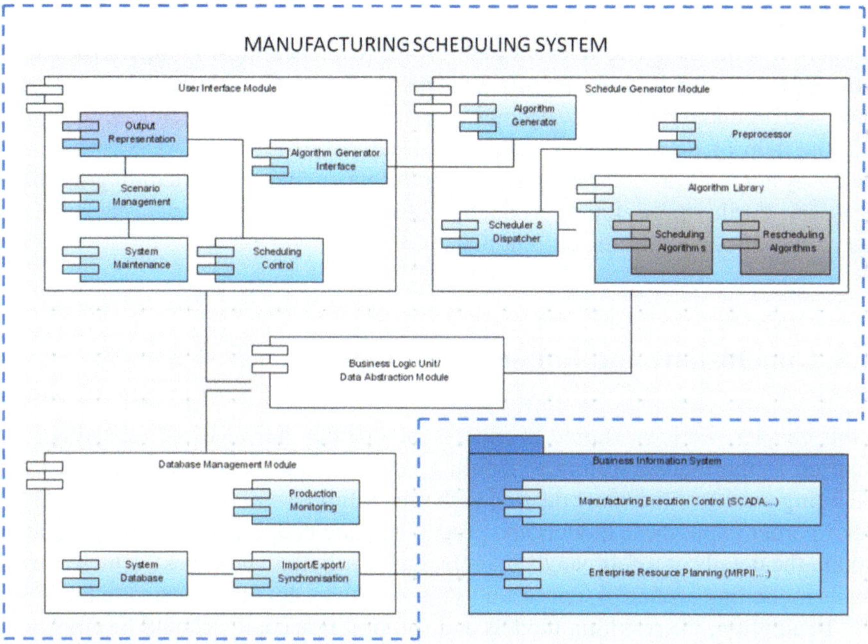

Fig. 11.3 Extended modular architecture

sometimes difficult to predict or to spot—impact on other modules. The user interacts with the graphical User Interface, which should provide an abstraction layer over the database and scheduling algorithms. The user interface must be personalised for each scheduling setting as each production problem is unique (Pinedo 2012). Interconnection with the company's existing systems is carried out by the Database Management Module, which is in turn only accessed by the BLU/DAM, providing a high level of abstraction. All algorithms and routines related to scheduling are also isolated from the system and included in separated modules again only accessible by the BLU/DAM.

One typical use case for a scheduling system based on the previous architecture is the following: Information about the production floor is initially loaded into the application. From that point, the current production plan is created or loaded from the ERP or existing business information systems. The user selects the available resources, calendar constraints and all other significant aspects of the problem, along with the optimisation objective. Once all data are loaded, the user selects the running time of the optimisation method (when possible) and generates a production sequence. Special importance is given to the graphical representation of such sequence since all vital information must be easily and readily available. Once again, this graphical representation contains a good deal of personalisation.

Changes to several previous data elements should be allowed at some step, like on-the-fly changes of lot sizes, due dates, addition and/or cancellation of lots, etc.

Usually, after a few iterations, the human scheduler will be satisfied with a final production sequence. The relevant data of this sequence must be again passed on to the ERP and all necessary reports must be generated. Once production starts, the optimised production sequence becomes the 'current plan' that is therefore controlled with the help of Supervisory Control and Data Acquisition (SCADA) production control systems. If production events occur, resequencing or rescheduling might be necessary. Furthermore, future production plans are scheduled as per request of the user as a continuation of the current plan.

11.8 Conclusions and Further Readings

In this chapter, we have started to address an area that has received relatively little attention in the manufacturing scheduling field, at least as compared to the overwhelming wealth of literature dealing with scheduling models and procedures. However, in order to put these models and procedures into practice, they must adequately capture the details of a company's manufacturing scheduling process, which usually entails modelling scenarios, constraints and objectives not often treated in the literature. In addition, the resulting models and solution procedures should be embedded into a manufacturing scheduling system that feeds them with reliable and timely data and provides an efficient way to interact with the users. This gives rise to a question on which the main blocks or components of a manufacturing scheduling system are and how they are related. Despite the specificity of scheduling systems, a number of features or functionalities can be identified, thus constituting the architecture of a manufacturing scheduling tool. In order to analyse these functionalities, we have reviewed and classified existing contributions on the architecture of manufacturing scheduling systems, as well as works dealing with the description of a manufacturing scheduling system's requirements. From this analysis, we have proposed an integrated, modular architecture of scheduling systems covering all the so-identified functionalities. We do not claim that our proposal is the only way (or even the best way) to represent the architecture of a scheduling system, although it has been validated both through literature analysis and through industrial practice. By presenting it, we hope to enrich the rather scarce literature on the topic and thus help bridging the widely recognised gap between the development of scheduling models and procedures, and their implementation in a real-life industrial settings.

We believe that the opinions raised in this chapter may be highly influenced by the experiences of the authors, and that there may be complex, highly constrained, non-repetitive scenarios in which the human expertise may outperform an automated manufacturing system, whereas an automated system could be preferable in other scenarios with a large number of different orders to be scheduled over a number of different machines on a repetitive basis. In any case, in our view, the automated system should serve to support the decisions of the human scheduler, and not to replace him/her. In the proposed architecture, this idea is explicitly supported by stressing the need that the tool should allow for the manual modification of the schedules, to manage different scenarios possibly with different objectives/constraints, or to include rescheduling algorithms, among other features.

So far a very high-level analysis of the scheduling tool is presented, which is a level of detail sufficient to be presented as a common architecture, but it just describe what to do and not how to do it. A refinement of this system analysis will be done in the next chapter, where some important issues that raise when moving from the architecture design to the development of a scheduling tool may arise.

Regarding further readings, this chapter assembles and extends the ideas in Framinan and Ruiz (2010). The classical paper for the definition of an architecture of a scheduling system is Pinedo and Yen (1997), while many of the discussion on the functionalities of a scheduling tool are present in several articles authored or co-authored by Kenneth McKay, such as McKay et al. (1995a,b), McKay and Black (2007).

References

Adelsberger, H. and Kanet, J. (1991). The leitstand-a new tool for computer-integrated manufacturing. *Production and Inventory Management Journal*, 32(1):43–48.

Aytug, H., Bhattacharyya, S., Koehler, G. J., and Snowdon, J. L. (1994). A review of machine learning in scheduling. *IIE Transactions on Engineering management*, 41(2):165–171.

Aytug, H., Lawley, M. A., McKay, K., Mohan, S., and Uzsoy, R. (2005). Executing production schedules in the face of uncertainties: A review and some future directions. *European Journal of Operational Research*, 161(1):86110.

Baek, D., Oh, S., and Yoon, W. (1999). A visualized human-computer interactive approach to job shop scheduling. *International Journal of Computer Integrated Manufacturing*, 12(1):75–83.

Berglund, M. and Karltun, J. (2007). Human, technological and organizational aspects influencing the production scheduling process. *International Journal of Production Economics*, 110(1–2):160–174.

Black, G., McKay, K., and Morton, T. (2006). Aversion scheduling in the presence of risky jobs. *European Journal of Operational Research*, 175(1):338–361.

Błażewicz, J., Ecker, K., Pesch, E., Schmidt, G., and Węglarz, J. (2001). *Scheduling Computer and Manufacturing Processes*. Springer, Berlin, second edition.

BPMG.org (2004). Business process modelling tools. http://www.bmpg.org.

Brandimarte, P., Rigodanza, M., and Roero, L. (2000). Conceptual modelling of an object-oriented scheduling architecture based on the shifting bottleneck procedure. *IIE Transactions*, 32(10): 921–929.

Collinot, A., Le Pape, C., and Pinoteau, G. (1988). Sonia: A knowledge-based scheduling system. *Artificial Intelligence in Engineering*, 3(2):86–94.

Ecker, K., Gupta, J., and Schmidt, G. (1997). A framework for decision support systems for scheduling problems. *European Journal of Operational Research*, 101(3):452–462.

Fowler, J., Mnch, L., and Rose, O. (2006). *Handbook of Production Scheduling*, chapter Scheduling and Simulation: The Role of Simulation in Scheduling, pages 109–133. Springer.

Fox, M. S. (1990). Constraint guided scheduling: A short history of research at CMU. *Computers in Industry*, 14(1–3):79–88.

Fox, M. S. (1994). *Intelligent Scheduling*, chapter ISIS: A Retrospective, pages 3–28. Morgan Kaufmann.

Framinan, J. and Ruiz, R. (2010). Architecture of manufacturing scheduling systems: Literature review and an integrated proposal. *European Journal of Operational Research*, 205(2):237–246.

Godin, V. (1978). Interactive scheduling: Historical survey and state of the art. *AIIE Transactions*, 10:331–337.

Gupta, J. N. D., Sexton, R. S., and Tunc, E. A. (2000). Selecting scheduling heuristics using neural networks. *INFORMS Journal on Computing*, 12(2):150–162.

Hadavi, K., Shahraray, M., and Voigt, K. (1990). Reds-a dynamic planning, scheduling, and control system for manufacturing. *Journal of Manufacturing Systems*, 9(4):332–344.

Higgins, P. G. (1996). Interaction in hybrid intelligent scheduling. *International Journal of Human Factors in Manufacturing*, 6(3):185–203.

Huang, S. H., Zhang, H.-C., and Smith, M. L. (1995). A progressive approach for the integration of process planning and scheduling. *IIE Transactions*, 27(4):456–464.

Jacobson, I., Booch, G., and Rumbaugh, J. (1999). *The Unified Software Development Process*. Addison-Wesley Professional.

Kanet, J. J. and Adelsberger, H. H. (1987). Expert systems in production scheduling. *European Journal of Operational Research*, 29(1):51–59.

Kempf, K. (1994). *Intelligent Scheduling*, chapter Intelligent scheduling semiconductor wafer fabrication, pages 517–544. Morgan Kaufmann.

Kirchmer, M. (1999). *Business Process Oriented Implementation of Standard Software*. Springer, Berlin, second edition.

Kurbel, K. (2008). *The making of Information Systems*. Springer.

Lamatsch, A., Morlock, M., Neumann, K., and Rubach, K. (1988). Schedule - an expert-like system for machine scheduling. *Annals of Operations Research*, 16(1–4):425–438.

Le Pape, C. (1994). *Intelligent Scheduling*, chapter Scheduling as intelligent control of decision-making and constraint propagation, pages 67–98. Morgan Kaufmann.

Lee, C.-Y., Lei, L., and Pinedo, M. (1997). Current trends in deterministic scheduling. *Annals of Operations Research*, 70:1–41.

Lin, C., Hwang, S., and Wang, E. (2007). A reappraisal on advanced planning and scheduling systems. *Industrial Management and Data Systems*, 107(8):1212–1226.

McKay, K. and Buzacott, J. (2000). Application of computerized production control systems in job shop environments. *Computers in Industry*, 42(2):79–97.

McKay, K., Safayeni, F., and Buzacott, J. (1988). Job-shop scheduling theory: What is relevant? *Interfaces*, 18(4):84–90.

McKay, K. and Wiers, V. (2003). Integrated decision support for planning, scheduling, and dispatching tasks in a focused factory. *Computers in Industry*, 50(1):5–14.

McKay, K. N. and Black, G. W. (2007). The evolution of a production planning system: A 10-year case study. *Computers in Industry*, 58(8–9):756–771.

McKay, K. N., Pinedo, M. L., and Webster, S. (2002). Practice-focused research issues for scheduling systems. *Production and Operations Management*, 11(2):249–258.

McKay, K. N., Safayeni, F. R., and Buzacott, J. A. (1995a). "Common sense" realities of planning and scheduling in printed circuit board production. *International Journal of Production Research*, 33(6):1587–1603.

McKay, K. N., Safayeni, F. R., and Buzacott, J. A. (1995b). Review of hierarchical production planning and its applicability for modern manufacturing. *Production Planning and Control*, 6(5): 384–394.

McKay, K. N. and Wiers, V. C. S. (1999). Unifying the theory and practice of production scheduling. *Journal of Manufacturing Systems*, 18(4):241–255.

Morton, T. E. and Pentico, D. W. (1993). *Heuristic Scheduling Sysmtems With Applications to Production Systems and Project Management*. Wiley Series in Engineering & Technology Management. John Wiley & Sons, Hoboken.

Numao, M. and Morishita, S. (1989). A scheduling environment for steel-making processes. *Proceedings of the 5th Conference on Artificial Intelligence Applications*, pages 279–286.

Pinedo, M. L. (2009). *Planning and Scheduling in Manufacturing and Services*. Springer, New York, second edition.

Pinedo, M. L. (2012). *Scheduling: Theory, Algorithms, and Systems*. Springer, New York, fourth edition.

Pinedo, M. L. and Yen, B. P.-C. (1997). On the design and development of object-oriented scheduling systems. *Annals of Operations Research*, 70(1):359–378.

Prietula, M., Hsu, W., Ow, P., and G.L., T. (1994). *Intelligent Scheduling*, chapter MacMerl: Mixed-Initiative Scheduling with Coincident Problem Spaces, pages 655–682. Morgan Kaufmann.

Reinschmidt, K. F., Slate, J. H., and Finn, G. A. (1990). Expert systems for plant scheduling using linear programming. In *Proceedings of the 4th International Conference on Expert Systems in Production and Operations Management*, Head Island, USA.

Sadeh, N. (1994). *Intelligent Scheduling*, chapter Micro-opportunistic scheduling: The Micro-Boss Factory Scheduler, pages 99–135. Morgan Kaufmann.

Sauer, J. (1993). Meta-scheduling using dynamic scheduling knowledge. In Dorn, J. and Froeschl, K., editors, *Scheduling of Production Processes*, pages 151–162, Upper Saddle River, N.J. Ellis Horwood.

Sauer, J. and Apelrath, H.-J. (1997). Knowledge-based design of scheduling systems. In Nahavandi, S. and Saadat, M., editors, *Proceedings of WMC97. International Symposium on Manufacturing Systems*, Auckland.

Sauer, J. and Bruns, R. (1997). Knowledge-based scheduling systems in industry and medicine. *IEEE Expert*, January-February:24–31.

Smith, S. (1994). *Intelligent Scheduling*, chapter OPIS: A Methodology and Architecture, pages 29–66. Morgan Kaufmann.

Steffen, M. S. (1986). A survey of artificial intelligence-based scheduling systems. In *Proceedings of the Fall Industrial Engineering Conference*, volume Proceedings of the Fall Industrial Engineering Conference.

T'Kindt, V., Billaut, J.-C., Bouquard, J.-L., Lenté, C., Martineau, P., Néron, E., Proust, C., and Tacquard, C. (2005). The e-OCEA project: towards an internet decision system for scheduling problems. *Decision Support Systems*, 40(2):329–337.

Trentesaux, D., Moray, N., and Tahon, C. (1998). Integration of the human operator into responsive discrete production management systems. *European Journal of Operational Research*, 109(2):342–361.

Wiers, V. (1997). *Human-Computer Interaction in Production Scheduling: Analysis and Design of Decision Support Systems for Production Scheduling Tasks*. Technische Universitiet Eindhoven.

Yen, B. P.-C. (1997). Scheduling description languages. Technical report, Dept. of Industrial Engineering and Engineering Management, Hong Kong University of Science and Technology, Hong Kong.

Zentner, M. G., Elkamel, A., Pekny, J. F., and Reklaitis, G. V. (1998). A language for describing process scheduling problems. *Computers and Chemical Engineering*, 22(1–2):125–145.

Zhang, H.-C. and Mallur, S. (1994). An integrated model of process planning and production scheduling. *International Journal of Computer Integrated Manufacturing*, 7(6):356–364.

Chapter 12
Advanced Design of Scheduling Tools

12.1 Introduction

In this chapter we move from the architecture discussed in Chap. 11 to discussing some design issues of a manufacturing scheduling tool by entering into the details of the different modules identified. In doing so, we also move from the generic (i.e. company-independent) to the specific (company-dependent), according to the different activities in the development of a manufacturing scheduling tool presented in Fig. 11.1. Therefore, it is not possible to give specific rules on how to implement the functionalities of the architecture, as this would largely depend on the specific tool to be designed. Instead, in this chapter we will go along the four modules of the architecture in order to discuss the main issues while keeping the discussion as much generic as possible.

More specifically, in this chapter we present the following modules:

- Business Logic Unit/Data Abstraction Module (Sect. 12.2),
- Database Management Module (Sect. 12.3),
- User Interface Module (Sect. 12.4), and
- Schedule Generator Module (Sect. 12.5).

12.2 Business Logic Unit/Data Abstraction Module

This module mimics modern computer languages and development paradigms which insist on abstraction, modularity and encapsulation. Databases evolve over time, since new requirements often imply new fields and tables. The same can be said about the user interface. If a scheduling tool is built in a monolithic way, a small change in the database could have a wave effect on many other modules. This would result in a software that is both very rigid and difficult to maintain. To ensure complete modularity and a high degree of independence, we propose the usage of elaborated class hierarchies that allow inter-module communication with published

J. M. Framinan et al., *Manufacturing Scheduling Systems*,
DOI: 10.1007/978-1-4471-6272-8_12, © Springer-Verlag London 2014

interfaces (like those available, e.g. in the C# programming language). Notice that this architecture is very different from standard approaches where data is simply retrieved from databases and stored in memory structures. A distinct feature we propose is *data polymorphism* with which the same data entity might have different representations depending on the moment and intended use. Picture for example the entity that modelises a given quantity of a product to be produced. First of all, this might be referred to as a job, task, lot, production order and many other different names, depending on the company. Second, the data required to work with such an entity depends on the module. When visualising the production plan, detailed information like family, colour or characteristics of the product to produce are required. In addition, internal data employed by the tool (such as the status of a job) have to be handled. Finally, different data (and data formats) may have to be provided to the different information systems interoperating with the scheduling tool. However, all this information is not relevant for generating schedules, as typically only the code of the job and processing time might be all that is needed. As a result, the type of interface published by a data class or entity to each module differs according to the intended use.

With this architecture, the BLU/DAM acts as a hub between all modules, ensuring maximum expandability, flexibility and reusability. With the proposed architecture, it should be easy to provide a Web-based user interface as a Web-based IU could be easily plugged in to the BLU/DAM without knowing any detail of the other modules. Although the BLU/DAM does not directly incorporate any functionality, it includes the functionalities related to constraint abstraction discussed in Sect. 11.6.2.

12.3 Database Management Module

This module contains three submodules, namely System Database, Production Monitoring and Import/Export/Synchronisation, which we discuss in the following items:

- System database. Among the three modules, this one is the most standard, and it contains all necessary persistent data in the tool. Modern relational databases are perfectly capable for this task. However, large databases require some extra efforts in order to assure a responsive software. Existing architectures, like the aforementioned by Pinedo and Yen (1997) fail to address important aspects of the database management which can be of paramount importance for an effective and real implementation of scheduling tools. Some important data will have to be retrieved from existing information systems and ERP software. Usually, this information is not complete, specially if the company is implementing a third-party production planning/scheduling software. As a result, some additional data will have to be introduced into the System DB, which in turn adds the necessity of keeping consistency of this data with the ERP system. Keeping this consistency is extremely complex and requires an specific module to import, export and to synchronise the Business Information System with the System DB. This module can incorporate the Schedule and Instance Capacity Analysis functionalities discussed in Sect. 11.6.6.

The challenges presented in this module should not be underestimated. Although there are clear concepts of jobs, machines and so forth, the implementation of the functionalities described in the architecture render these concepts more complex in practice. For instance, the analysis of scenarios requires to store all relevant data related to a single schedule, as the Decision-Maker may want to 'play' with different processing times of certain machine (maybe trying to make a guess of the worst and best cases available), or different sets of jobs (for instance when deciding if certain job is to be scheduled, or not). As a result, storing a scenario may consist in virtually storing certain layout (together with the characteristics of the machines), a given set of jobs (which may be different for each scenario, or at least with different features), and different schedules (or solutions) for each scenario. Keeping all this information in the database forces to a careful design of a data model far away from a trivial task.

- Production Monitoring. This submodule is concerned with production control. As we will see later on, controlling the current production programme requires of special features. SCADA systems and in general, the business information systems of the company contain manufacturing execution control software that generate real-time information. This information is urgent and must be dealt with rapidly.
- Import/Export/Synchronise. This module should carry out lengthy operations in batch mode in order to keep up to date the System DB. As a result, an ad hoc module must be written in order to load real-time production control data. This data is then interfaced to the BLU/DAM and the necessary events are fired in order to notify all other modules that real-time data has been received. Together with the BLU/DAM, this module can incorporate the functionalities to perform the monitoring of execution (*Production Monitoring* submodule) as required in the analysis of functionalities carried out in Sect. 11.6.5.

12.4 User Interface Module

Although this module might seem straightforward, it is indeed one of the most complex modules in the architecture. Effective, complete and easy to work with visualisation is a hard task that requires extensive coding. This comes as no surprise, since different scheduling problems require not only different solution procedures but also equally varied and complex solution representations.

When building this module, the following design principles should be taken into account:

1. The tool should be flexible enough to allow the Decision-Maker to follow his/her preferred logical process when building a schedule. There are numerous studies describing the unstructured nature of the decision process adopted by the human schedulers (see, e.g. MacCarthy and Wilson 2001), therefore it does not seem too practical to force the schedulers to adopt a particular one, even in the case that the schedulers themselves have actively participated in the design of the tool.

2. The tool should allow the Decision-Maker to focus in a specific view or step of the scheduling process, starting from a general view and moving (if required) to a more specific one. While it may be very tempting that the decision-maker can modify input data while analysing schedules, we believe that this should be done via different interfaces, so the user does not loose track of what he/she is doing.
3. The tool should not overload the decision-maker with data/information outside the context in which the information is originally presented. Although this is a difficult decision on which information should be displayed and which not, we believe that this could be done by a careful selection of the information and by drill-down menus that, in any case, may allow the scheduler to customise the information that should be depicted.
4. The tool should be able to capture all relevant scenarios for the decision-making process, i.e. the Decision-Maker might wish to try different alternatives of performing the scheduling tasks, including different sets of jobs, modes of operations for the machines, shifts, etc. For each of these scenarios, different schedules may be stored (some of them generated by the Schedule Generator submodule, while others are manually built/modified by the Decision-Maker). All this information should be available in an structured manner so the task of the Decision-Maker is facilitated.

These principles would naturally lead to the separation of this module into different interfaces depending on the view of the scheduling process addressed. While the connection among these interfaces should be enforced, the logical distinction may allow the scheduler to make sure about the step in the process in which he/she is currently involved. Consequently, up to five submodules can be identified inside the user interface. These are the following:

- Output Representation. This submodule implements a set of functionalities devoted to present the information to the Decision-Maker. As we shall discuss, this submodule can be employed also as an input from the Decision-Maker, if the manipulation of the schedules is allowed.
- Scenario Management. This submodule serves to implement a number of functionalities related to handling different scheduling scenarios. By scheduling scenarios we mean different schedules or even different instances created from the original schedule provided by the Schedule Generator module due to the manual intervention of the schedule. As we shall discuss, this submodule is instrumental in providing the tool with the ability to perform a what-if analysis.
- System Maintenance. This submodule refers to the part of the user interface devoted to ease the creation and maintenance of the data required by the scheduling tool, including a number of functionalities so that the consistency and comprehensiveness of the data entered are checked, and alarms are triggered if discrepancies between past and actual data are observed.
- Scheduling Control. This submodule deals with the functionalities related to visualise (possibly) in real-time the state of the shop floor. This submodule is closely linked to the Production Monitoring submodule described in Sect. 12.3.

- Algorithm Generator Interface. This submodule may allow the user to edit new algorithms in an easy manner.

These submodules are described in detail in the following sections.

12.4.1 Output Representation

Once a solution for a given scheduling problem or production plan has been provided by the Schedule Generator Module, it has to be visualised. In addition, in order to build an interactive tool, the same interface used to visualise the solution can be employed to allow its modification. In addition, it has been argued (Pinedo 2009) that Output Representation interfaces should allow (at least up to a certain extent) for the modification of the input data, as some of these data (for instance, due dates) often have to be changed during a scheduling session.

Several interfaces for the presentation and manipulation of solutions can be considered. These are:

- Gantt Charts. The Gantt Charts are the most popular form of schedule manipulation and have been already presented throughout this book. The typical form of the Gantt chart is an horizontal bar chart in which the x-axis represents time and the y-axis represents the different machines or stages in the manufacturing process. Jobs or tasks are depicted as rectangles whose basis correspond to the processing time of the tasks. The identification of the different jobs in a Gantt chart is done via labelling these rectangles and/or depicting them with different colours or patterns. This format can be also employed to depict some other characteristics of the jobs and/or the resulting schedule, such as the tardiness or earliness of some jobs. In addition, some additional features related to the properties of the jobs, machines or schedules can be represented in a Gantt Chart, such as job earliest starting times, due dates and deadlines (that can be shown as vertical lines), frozen jobs (that may be rendered in a different colour/pattern), machine unavailabilities (that may be pictured as shadowed), etc. Gantt Charts are a well known and (relatively) easy-to-understand visualisation tool and therefore they constitute the primary form of showing aggregate information in an scheduling tool (Wiers 1997).
  Regarding the manipulation of the schedule shown in the Gantt chart by the scheduler, several options are possible (see, e.g. Speranza and Woerlee 1991, or Baek et al. 1999), i.e.:
  - Add/Delete. The news are manually entered into or are removed from the existing schedule. This leads to the necessity of showing the Decision-Maker a zone in the interface where non-scheduled jobs are shown (or, alternative, where jobs removed from the schedule can be stored).
  - Interchange. The relative sequence of two or more jobs is interchanged. This usually leads to a limited change in the schedule (at least as compared to other options).

  - Split/Join. A job is splitted into two or more tasks. Conversely, two or more tasks are merged into a single job. The implications of this movement are not straightforward, as the processing times of the machines for the joint/splitted jobs are different from those in the initial schedule. Therefore, it would be advisable to invoke the Schedule Generator once again in order to detect new (possibly improved) schedules.
  - Lock. A job is locked in the Gantt chart, i.e. their starting and finishing times are fixed. An alternative would be to lock the relative position of a job in the machine, so that the starting and finishing times can be changed, but not the order of the job with respect to the rest of the jobs in this machine.

  Note that some of these options can be done by clicking, dragging and dropping the individual jobs. However, providing the interface with these capabilities is far from being a trivial task, since after changing the position of a job on a machine, the rest of the jobs have to be pushed backward or forward to maintain feasibility. The effect of this movement on the rest of machines is referred as *cascading* or *propagation* (see, e.g. Pinedo 2009) and it may lead to substantial changes with respect to the schedule provided by the application. Therefore, an option that should be available for the decision-maker is whether to invoke or not a rescheduling procedure, possibly after locking some of the jobs.

  In addition to a detailed representation of the individual jobs, an aggregated view of the whole schedule (possibly representing shifts or sets of related jobs, and groups of machines or stages) would be interesting for the Decision-Maker. This aggregated view should be static, i.e. it should not allow the manipulation of the individual schedules. However, it has the advantage of showing a global view of the process (particularly needed when the number of jobs/machines in the plant is rather large) that can be useful when assessing the impact of the local changes in the schedule.

  After allowing all this manipulation, the Decision-Maker may want to save the schedule, or to 'undo' certain modifications. The first option is discussed when describing the functionalities of Scenario Management in Sect. 12.4.2. With respect to 'undoing', we believe that this option should be available (in order to recover from mistakes when manipulating schedules), but only for a limited number of moves, as it is easy that the Decision-Maker loose the track of what he/she is doing. It is preferable, thus, that the tool offers to save the actual schedule as an scenario every time a short number of manipulations has been performed.

- Job Screens. A Job screen (see, e.g. Higgins 1996) is an ordered list of the jobs that should be processed on each machine. Another name for job screens is Dispatch lists (see Pinedo 2009). A number of characteristics of the jobs, such as the processing times, their corresponding weights and/or priorities, the starting, finishing times and deadlines, etc. can be shown. The job screens provide a representation of the schedule that is focused on each manufacturing resource, and therefore seems to be quite suitable for the manipulation of the schedule. For instance, given the list of the jobs to be processed in front of a machine, the scheduler can alter this order according to constraints or considerations that have not been captured by the model such as certain set-ups, different age/characteristics of the machines or the different skills of the workers in a shift. Nevertheless, it has the disadvantage that, in principle, it only considers local information, so the effects of this (local) manipulation in the full schedule are not shown. In our previous example, it may be that a reasonable change in the order of the processing of the jobs in an early stage have a negative effect when evaluating the schedule in all machines. Therefore, it seems reasonable to accompany the representation of the job screens with global information of the schedule (such as the aggregated Gantt Chart discussed in the early item), so changes in the objective functions are also visible. Local changes in this global information view can be visualised by rendering the changed zone in a different colour and by showing both aggregate Gantt Charts (before and after the local changes).

Although it is not usually shown in Job screens from several tools (such as LEKIN, see Pinedo 2009), additional information related to each machine could be shown. Particularly, information regarding machine utilisation, total set-up time, etc., could be also very useful for the decision-maker to focus onto a particular machine.

Therefore, our suggestion for this type of interface would be to implement a two-level interface. A first-level graph (sketched in Fig. 12.1) would display each one of the machines in the manufacturing process, together with some summary information related to each one of the machines. By doing so, the scheduler can visualise the whole process and focus onto the overloaded resources (or, in contrasts, in those idle). This first level would be primarily devoted to show the output, and little manipulation or no manipulation at all should be provided. A second level can be accessed by the user by clicking on each one of the resources, and would provide the information corresponding to the jobs in this machines. Alternatively, an integrated Job Screen such as the one shown in Fig. 12.2 can be employed.

- Capacity buckets. In this type of interface (see e.g. Pinedo 2009), the time axis (usually horizontal) is partitioned into a number of time slots or buckets that may correspond to either days, weeks or any other time unit. For each machine (usually depicted in the y-axis), the corresponding capacity is shown. The idea is similar to the one of the machine load boards (see, e.g. Higgins 1996). An example of capacity buckets is shown in Fig. 12.3. This type of interface may help the scheduler to identify which machines in the plant are operating near their capacity (in the same manner as the first level of the Jobs Screens proposed before), but also during which time interval. It is primarily an output interface, and interactivity is, in principle, not required. However, a second level of information could be providing by allowing the scheduler to click on a particular bucket of an specific machine in order to obtain the list of the jobs to be processed, or possibly, a single-machine Gantt chart in order to gain some insight on the causes of the congestion (or idleness) of the machines.
- Throughput diagram. These type of diagrams usually represents the evolution of the value of a cumulative performance measure (usually represented in the y-axis) against the time (usually depicted in the x-axis). The performance measure may be (Pinedo 2009), the total amount of orders received, orders produced or

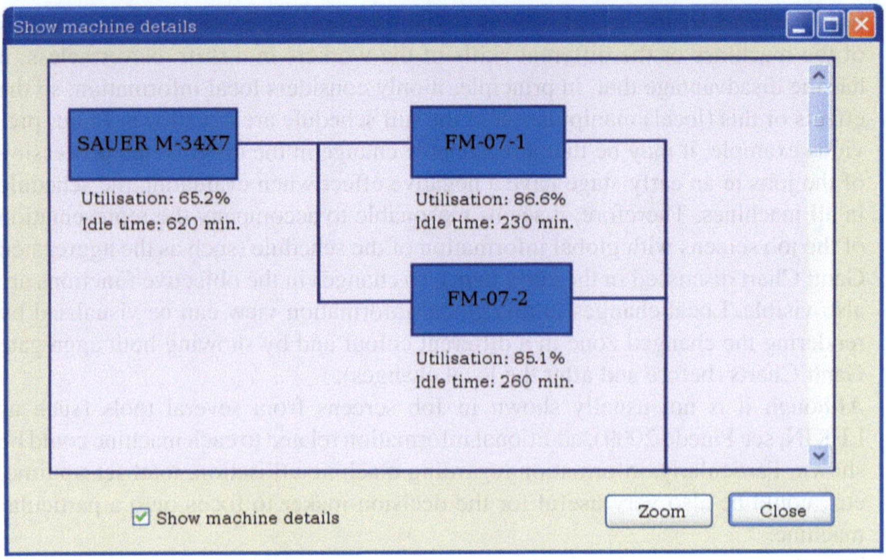

Fig. 12.1 First level of the proposed job screen interface

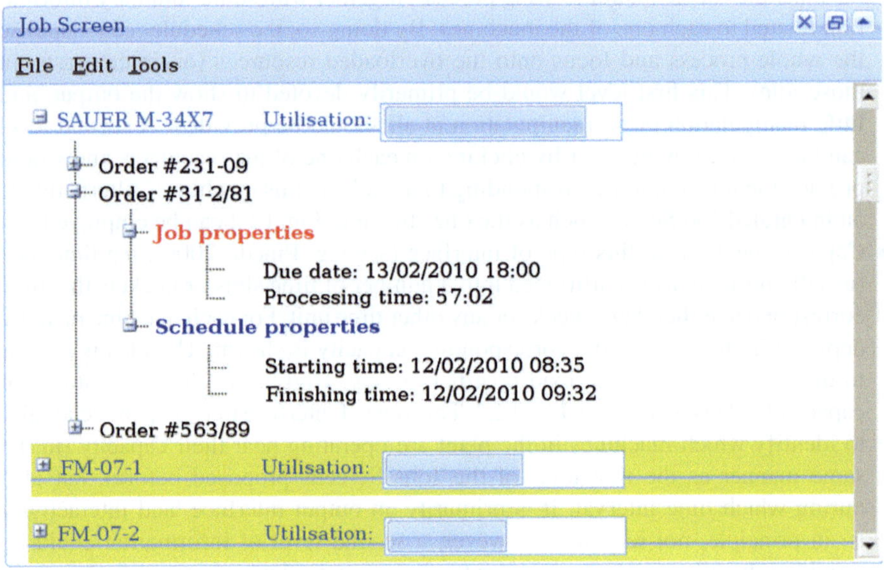

Fig. 12.2 Integrated job screen interface

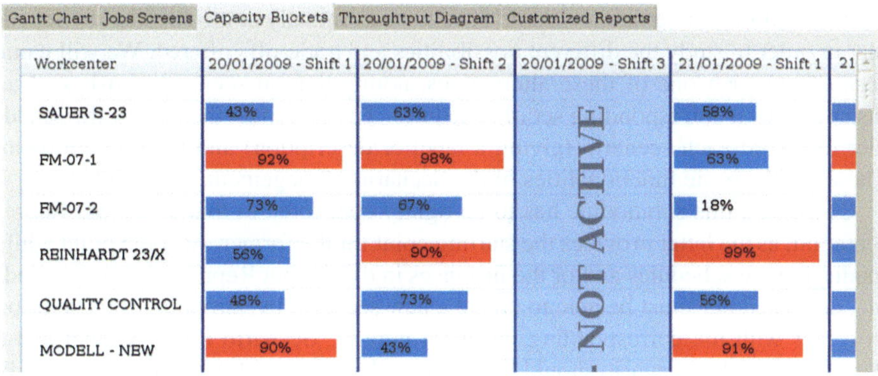

Fig. 12.3 Example of capacity buckets

orders shipped. Note that the difference between the curves of orders produced and shipped amounts for the work-in-process. In addition, the expected amount of orders can be also depicted, so this diagram allows visualising the deviations from the planned schedules, particularly if the data shown in the diagram are captured in real-time. Throughput diagrams are devised for representation of the schedules and not for their manipulation, therefore there is no interactivity linked to this type of diagrams. For a detailed explanation on the uses of throughput diagrams, we refer the reader to Wiendahl and Tonshoff (1988).

Alternative names for these diagrams are IOP (Input-Output Planning) (Wiers and Van Der Schaaf 1997), and are available in many standard software packages for production planning and control (Wortman et al. 1996).

In addition to the types of diagrams described before, a number of reports containing statistic pages and performance criteria sheets must be provided, among many other possible data visualisations. It is tempting to overload the decision-maker with reports and screens very rich in information, but in this case we believe that it is preferable to concentrate in only one of the different views or perspectives of the scheduling process. Most of these reports should be developed ad hoc, as they must comply with company-specific forms and data.

12.4.2 Scenario Management

As soon as the scheduler manually modifies the schedule proposed by the tool via some of the interfaces described in the previous section, a new *schedule* is generated. Analogously, if the scheduler modifies some input data (e.g. in order to anticipate a delay in the arrival of raw materials required to start certain job, its corresponding release times are changed), a new *instance* is generated, and several schedules can be offered by the tool, or be obtained after a manual modification by the scheduler.

Each one of these alternatives would have to be eventually confronted by the scheduler in order to study the different possibilities and trade-offs offered. We will name *scenario* to each one of these alternatives, noting that an scenario consists of an instance plus a corresponding set of schedules. The tool must then allow the scheduler to record each scenario (giving a label or description) and to compare them. These are the basic functionalities of the Scenario Management submodule.

Obviously, this submodule has to be tightly connected to that of Output Representation, as the latter provides the starting point for the former. At some point while modifying the schedules and/or the instances in the Output Representation submodule, the scheduler must be able to create a new scenario by 'dropping' this solution (together with the corresponding instance) into an 'Scenario List'. A label and a description of this scenario should be given, so that the scheduler can easily identify it at a later stage. Once the initial selection of scenarios has finished, the user must be able to see the main performance measures of the selected scenarios and to see the full details of the scenario, perhaps by double-clicking on it. In order to make this process systematic, our opinion is that the scheduler should not be able to modify the scenario from this submodule. However, he/she can transfer the scenario back to the Output Representation module (by 'copying' the scenario) and then make all necessary manual modifications and save it as a new scenario in the 'Scenario List'.

The differences among the scenarios in the Scenario List should be clearly visible, both by forcing the Decision-Maker to introduce different, meaningful labels for each scenario, and by allowing the Decision-Maker to see a summary of the main characteristics of each scenario (related both to input and output data).

The submodule may allow to discard scenarios that have been initially selected, but given the complexity of generating an, in principle valid, scenario, we believe that the scenarios should not be completely removed unless explicitly stated by the decision-maker. To do so, a 'folder' of discarded scenarios should be kept apart from those under scrutiny.

We believe that this submodule may greatly enhance the decision support capabilities of a scheduling tool, as the analysis of scenarios is pointed out by McKay and Wiers (2003) as an important element in a scheduling tool. Let us not forget that once a scheduling tool is available to the user, more and more strategic decisions can be taken. For example, questions like what would be the result with an additional machine? What if I cancel the night shift? can be answered with a Scenario Management submodule.

12.4.3 System Maintenance

Of course, all the persistent information in the System Database must be maintained, from the configuration of the shop and existing machines to the detailed production characteristics of each product. As simple as this might seem, the sheer amount of production data often results in a cluttered user interface. A number of interfaces can be built to present and edit these data, such as the plant layout interface and the jobs interface.

The plant layout interface may depict graphically the work centres and machines in the plant. If possible, it should also represent the routing between the work centres, although for many shop floors this may not be possible or may result in a very confusing representation. In such cases, it is preferable to depict just the work centres together with their description and characteristics. An important part is to include (either in this interface or in a separate one) an entry point of the resource calendar, as it is likely that the machines are not always available.

The jobs interface would serve to enter all data related to the jobs and to their relationship with the machines. Note that some of these data must be particularly awkward to be entered, such as, e.g. those related to the sequence-dependent setups. Special care should be put to devise simple screens for this type of interface, as it relates to data that are frequently modified. Nevertheless, some of these data would not change, therefore past configurations should be stored and the scheduler should be asked whether these data must be pre-loaded in order to ease the task of entering new information, or not.

In order to be usable, maintenance screens should be easy to work with. We advocate the usage of simple data wizards and automation tools. For example, when purchasing an additional parallel machine, it is surely easier to copy an existing one and later specifying the small differences than to introduce a new machine from scratch. The same can be said regarding entering new jobs into the system.

Finally, note that even in the case of a scheduling tool seamlessly connected to other information systems providing all sources of data, we believe that this submodule should be kept, so that the tool is able to generate new scenarios for the Scenario Management submodule.

12.4.4 Scheduling Control

A separate interface should be devised for handling in real-time the data coming from the BLU/DAM module. Regarding the Scheduling Control submodule, it is to note that real-time data coming from the BLU/DAM must be handled independently. Should the current schedule receive an event from the production floor, this event must generate a visual warning to the user. Once this warning is accounted for, a detailed report of changes to the current schedule should be provided. Some of this information is needed in order to cope with functionalities mentioned in Sect. 11.6.8. More specifically, input data checking and feasibility analysis must be confronted upon each new scenario.

12.4.5 Algorithm Generator Interface

Last but not least, the scheduling tool should allow for the creation of new algorithms. In order to hide the small details of the algorithms to the user, a friendly interface based on wizards is preferable, which is foreseen in our proposal to be the goal of the Algorithm Generator Interface submodule. The functionality of new algorithms was put forward in Sect. 11.6.3.

12.5 Schedule Generator Module

This module contains four main submodules, i.e. Algorithm Library (which in turn may be divided into Scheduling Algorithms and Rescheduling Algorithms), Algorithm Generator, Scheduler and Dispatcher, and Preprocessor. In order to be adaptable, the optimisation algorithm must work on an abstraction of the scheduling model, or better, on an abstraction of the representation of a possible solution. For example, a typical permutation flow shop solution for makespan criterion can be represented with a simple permutation of the jobs. The detailed schedule with all the start and finish times of every task is many times of little relevance for the optimisation methods. As mentioned before, these abstractions are provided by the BLU/DAM. Notice that once the solution to a problem is returned to the BLU/DAM, a complete and much more detailed schedule might be built by the user interface module.

What should be mentioned as a part of the architecture is the separation of all non-essential constraints from the schedule generation process. Many constraints can be accounted for when the detailed schedule is constructed. Of course some level of accuracy will be lost in the process but it has to be carefully studied whether this decreased accuracy is compensated by a more general and adaptable optimisation method. In order to reflect this, we separate the Algorithm Library submodule from the Scheduler and Dispatcher submodule. A given—and more general—algorithm produces an optimised production schedule where all major constraints have been considered. In a second step, the Scheduler and Dispatcher submodule produces an even more detailed schedule where minor constraints are also considered. By operating in this way, the scheduling tool might accommodate new constraints and new situations in a more flexible and agile way.

Similarly, the number and variety of optimisation criteria or performance measures found in practice is extremely large. As a result, for the calculation of the performance measure for a given solution, a two-stage approach is preferable. In the first stage, a detailed schedule is derived from the abstract representation of the solution. All relevant data, and most importantly, completion times of all tasks, are calculated. In a second step, specific performance criteria are derived from the completion times calculated in the first stage. This two-phase approach allows for a large flexibility and independence when defining new and ad hoc criteria. With the previous premises, a given scheduling algorithm does not need to know in detail neither the specific constraints and small details of the underlying model nor the particular scheduling criterion.

Notice also that the Algorithm Library should contain not only full-scale optimisation algorithms but also some smaller schedule repair and rescheduling algorithms that might be invoked by the user interface module as a result of changes in a given solution or as the response to events in the production floor. Examples of these specific approaches are summarised, e.g. in Li et al. (2000).

We would like to point out an important issue here. As of today, and according to our experience, most companies are using extremely simple scheduling methods like list algorithms, if at all. Therefore, when implementing and deploying a new scheduling tool, the focus should not be on the accuracy and effectiveness of the optimisation algorithms as these are likely to outperform human schedulers in many scenarios. Conversely, special care should be put on assuring the success of this scheduling software as a whole by providing flexibility and expandability. Among the algorithms, there is a broad distinction between specific algorithms and generic algorithms. Specific algorithms are designed for solving a particular model, while a generic algorithm may be applied to a wide range of scheduling problems. In practice, most commercial scheduling tools employ general-purpose algorithms, such as genetic algorithms or tabu search (e.g. SAP-APO employs genetic algorithms). As mentioned, both types of algorithms have advantages and disadvantages. However, there are some reasons for not overlooking specific algorithms:

- Exact algorithms. With the increasing power of computers, more and more problem instances may be solved by exact algorithms (i.e. enumeration methods). Typical generic algorithms for exact methods are based in integer programming models that are relaxed into continuous models in order to obtain bounds to perform a branch and bound method. However, it is known that, for most scheduling problems, such bounds are extremely bad and therefore are not efficient in the bounding procedure. A promising avenue in this regard is to employ the so-called hybrid methods (i.e. combining exact methods and metaheuristics, see Jourdan et al. 2009 for a review).
- Unfeasible solutions. Most real-life scheduling models must deal with unfeasible solutions—in fact, the aim of scheduling is, in most cases, simply obtaining a feasible solution for the problem (Hopp and Spearman 1996). While the performance of metaheuristics is extremely good for problems where the codification of the solutions does not lead to unfeasible solutions, this is not true for the case where such algorithms should handle unfeasible solutions. In contrast, specific algorithms for the model under consideration may yield very good (and feasible) solutions in a relatively short time period.
- 'In-house' algorithms. It may be that the schedulers have devised efficient solution procedures for the specific scheduling problem based on their experience. Although some of these algorithms may be naïve, they can cope well with specific performance criteria, or perform efficiently for certain specific constraints.
- Complementarities between generic and specific algorithms. For many cases, the performance of the generic algorithm heavily depends on the initial solution, usually provided by one or several specific algorithm(s).

The Algorithm Library should give some hints regarding the selection of the algorithm. On one hand, it may be that the specific problem under consideration cannot be fit into the models that can be solved by the algorithms contained in the library. In this case, the tool should suggest/propose the usage of algorithms included in the library that can solve problems which are similar to the one under consideration.

On the other hand, it may happen that more than one algorithm can be employed to solve a particular problem. From all these, the tool should suggest/propose those more suitable for each case.

Regarding the capacity of a manufacturing scheduling tool to be enhanced with new algorithms, McKay et al. (2002) note that few available tools contain libraries of configurable algorithms that plug and play. It has to be noted that the requirement is not only that completely new algorithms can be included in the library, but also that existing algorithms can be tailored and that certain parts of these algorithms can be integrated into new ones. With respect to the extensibility of scheduling tools, also McKay et al. (2002) state that few of them allow the enhancement or combination of existing algorithms in a simple manner. For most commercial software either these algorithms cannot be enhanced or the procedure for their enhancement is extremely complex (Stadtler and Kilger 2002).

The blocks described in this section serve to implement many functionalities discussed in an earlier chapter. First, model detection functionalities (Sect. 11.6.2) can be implemented in the Preprocessor submodule although this requires intervention from the user interface module. Regarding to problem-solving functionalities (Sect. 11.6.3), rescheduling algorithms are included into the Algorithm Library submodule together with scheduling algorithms (thus allowing multi-algorithm scheduling). An Algorithm Generator submodule is connected to the Algorithm Generator Interface (already discussed in Sect. 12.4) to facilitate the generation of new algorithms, which again requires a tight integration with the user interface. The Algorithm Library can also contain functionalities for the evaluation of algorithms. The only functionality presented in Sect. 11.6.3 and not included in our proposal is the incorporation of the human expertise, as their advantages are a subject of controversy. However, their integration in the proposed architecture is relatively straightforward. Finally, functionalities related to the evaluation of solutions for several objectives (discussed in Sect. 11.6.4) are covered by the Scheduler and Dispatcher submodule. Cleanly modularising this part of the architecture is a challenging task as it is tempting to incorporate algorithm logic in other modules. However, abstraction and modularisation is a must in order to avoid re-implementation of other modules at each implementation of a scheduling tool.

12.6 Conclusions and Further Readings

In this chapter we have moved from the high-level, conceptual issues regarding the development of a manufacturing scheduling tool described in Chap. 11 to detailed design issues. This has resulted in an outline of the main contents of the modules identified in the architecture in Chap. 11.

Summing up, the modules described in this section address a number of functionalities discussed in Chap. 11: In addition to the representation of the solutions (see Sect. 11.6.2) which is covered by the *Output Representation* submodule and the capability that the scheduler manipulates the solutions found by the tool (discussed in Sect. 12.4), the proposed architecture setup the basis for an intuitive generation of new algorithms (see Sect. 11.6.3) by using the *Algorithm Generation Interface* submodule, and eases the analysis of scenarios (see Sect. 11.6.4) by employing a submodule for *Scenario Management*. Finally, functionalities referred to reactive scheduling (see Sect. 11.6.5) are foreseen by the submodule *Scheduling Control*.

There are a number of issues that can be concluded from this exposition. The first one is the careful choice of the technology (computer, databases, programming languages, etc.) to be employed in the manufacturing scheduling tool. Note that many of the features described in this chapter (and particularly those referring to the user interface) cannot be implemented, e.g. on a light client or Web-based user interfaces.

Another important conclusion is to realise the enormous amount of 'overhead' work related to the development of a manufacturing scheduling tool beyond the effort to build appropriate models and solutions procedures. Even for a tool designed to support a single scheduling model and its corresponding solution procedure, in order to make the software usable, a sheer number of functions for entering, registering, tracking and checking data have to be written. This may be another reason explaining both the difficulty to move from the optimisation-based scheduling to the decision-making scheduling, as well as the high failure rate of some implementations of scheduling tools if these supporting, yet vital, functions are not carefully addressed.

Despite the applicability of the issues treated in the chapter regarding the implementation of a scheduling tool, the existing literature is quite scarce. This chapter has been elaborated around the ideas in Framinan and Ruiz (2010), using a lot of material from Higgins (1996), who discussed the main problem related to the interaction of the user in scheduling software. Several articles cited in the document (such as e.g. Wiers 1997, or Pinedo 2009) constitute a good point to extend some of the topics treated here. Finally, additional information on throughput diagrams can be found in Wiendahl and Tonshoff (1988).

References

Baek, D., Oh, S., and Yoon, W. (1999). A visualized human-computer interactive approach to job shop scheduling. *International Journal of Computer Integrated Manufacturing*, 12(1):75–83.

Framinan, J. and Ruiz, R. (2010). Architecture of manufacturing scheduling systems: Literature review and an integrated proposal. *European Journal of Operational Research*, 205(2):237–246.

Higgins, P. G. (1996). Interaction in hybrid intelligent scheduling. *International Journal of Human Factors in Manufacturing*, 6(3):185–203.

Hopp, W. J. and Spearman, M. L. (1996). *Factory physics. Foundations of manufacturing management.* Irwin, New York, USA.

Jourdan, L., Basseur, M., and Talbi, E.-G. (2009). Hybridizing exact methods and metaheuristics: A taxonomy. *European Journal of Operational Research*, 199(3):620–629.

Li, H., Li, Z., Li, L., and Hu, B. (2000). A production rescheduling expert simulation system. *European Journal of Operational Research*, 124(2):283–293.

MacCarthy, B. L. and Wilson, J. R., editors (2001). *Human performance in Planning and Scheduling.* Taylor & Francis.

McKay, K. and Wiers, V. (2003). Integrated decision support for planning, scheduling, and dispatching tasks in a focused factory. *Computers in Industry*, 50(1):5–14.

McKay, K. N., Pinedo, M. L., and Webster, S. (2002). Practice-focused research issues for scheduling systems. *Production and Operations Management*, 11(2):249–258.

Pinedo, M. (2009). *Planning and Scheduling in Manufacturing and Services*. Springer, New York, second edition.

Pinedo, M. L. and Yen, B. P.-C. (1997). On the design and development of object-oriented scheduling systems. *Annals of Operations Research*, 70(1):359–378.

Speranza, M. and Woerlee, A. (1991). A decision support system for operational production scheduling. *European Journal of Operational Research*, 55(3):329–343.

Stadtler, H. and Kilger, C. (2002). *Supply chain management and advanced planning*. Springer, Heildelberg.

Wiendahl, H.-P. and Tonshoff, K. (1988). The throughput diagram - an universal model for the illustration, control and supervision of logistic processes. *CIRP Annals - Manufacturing Technology*, 37(1):465–468.

Wiers, V. (1997). *Human-Computer Interaction in Production Scheduling: Analysis and Design of Decision Support Systems for Production Scheduling Tasks*. Technische Universitiet Eindhoven.

Wiers, V. and Van Der Schaaf, T. (1997). A framework for decision support in production scheduling tasks. *Prod Plann Control*, 8(6):533–544.

Wortman, J., Euwe, M., Taal, M., and Wiers, V. (1996). A review of capacity planning techniques within standard software packages. *Production Planning & Control*, 7(2):117-128.

Part V
Scheduling Systems

The Part V of the book rejoins the elements described in the previous parts and focuses on scheduling system as a whole. We first discuss in Chap. 13 the organisational perspective in scheduling, paying attention to how humans conduct the scheduling process. In Chap. 14, we present a roadmap for developing a scheduling system, addressing all relevant issues throughout its life cycle and pointing to the corresponding section of the book where these are treated. This part concludes with Chap. 15, in which a real-life case study is described and analysed.

Chapter 13
Overview of Scheduling Systems

13.1 Introduction

In this chapter, we approach for the first time the integration of the three elements discussed in precedent chapters (i.e. models, methods and tools) in a manufacturing scheduling system. To do so, we need to gain an in-depth understanding on how the manufacturing scheduling process is carried out and how the three elements can contribute to improve this process.

More specifically in this chapter, we

- discuss the integration of the three elements within a system (Sect. 13.2),
- present the organisational perspective in scheduling (Sect. 13.3),
- investigate the main tasks carried out by human schedulers (Sect. 13.4) and how these are carried out (Sect. 13.5),
- identify the main features of human scheduling in view of the outcome of the previous item (Sect. 13.6) and
- discuss the integration of the human element in the scheduling system (Sect. 13.7).

13.2 The Integration of Models, Methods and Tools

In the previous chapters, most of the different elements constituting a scheduling system have been discussed. Given the width and complexity of some of these issues, some aspects would have been just briefly mentioned while a more extensive discussion would have been possible for another. It is now the point where all these elements are assembled together to make a scheduling system.

Recall that in Chap. 1 we define a scheduling system as a combination of *models*, *methods* and *tools* together with the human schedulers involved in the process. Issues regarding scheduling models have been discussed in Chaps. 3–6, while the main aspects for scheduling methods are addressed from Chaps. 7–10.

J. M. Framinan et al., *Manufacturing Scheduling Systems*,
DOI: 10.1007/978-1-4471-6272-8_13, © Springer-Verlag London 2014

Chapters 11 and 12 discuss the main issues related to the development of a manufacturing scheduling tool. In the next three chapters, we relink these different aspects related to a scheduling system. Particularly, we will focus onto the integration of the manufacturing scheduling tool (with its models and methods) into an organisation. In order to conduct such integration in an effective manner, the tool should first reflect the needs and expectations of the company, and second, the so-developed tool should be deployed in an effective manner.

The first of these issues (reflecting the needs and expectations of the company in which the tool is installed) is closely related to the understanding of how scheduling tasks are performed in companies. Since these tasks are, as we shall see later, generally performed by specific staff (schedulers), this lead us to the need of understanding the nature of human scheduling, their main activities and features, and how these could be integrated into the scheduling tool. All these aspects constitute the content of this chapter.

The second issue (how to implement and deploy the tool into a company) obviously encompasses many company-specific aspects, although there may be a number of lessons that can be regarded as common for most scheduling systems and that can be thus extracted from an analysis of past implementations. Therefore, guidelines to effectively conduct the deployment of manufacturing scheduling systems could be developed from the analysis of evidences supported by the experiences described in the literature. In Chap. 14, we discuss and classify these experiences and extract such guidelines. Finally, a case study that can be employed to illustrate the process of deploying a manufacturing scheduling system is provided in Chap. 15.

13.3 The Organisational Perspective in Scheduling

At this point, it must be sufficiently clear that manufacturing scheduling is a multi-facet complex problem, which leads to different views and perspectives, each one focusing onto specific issues within the scheduling process. These different perspectives are usually summarised into three aspects: problem-solving, decision making and organisational (Herrmann 2006). The first two aspects can be seen as a 'technical' view of manufacturing scheduling (MacCarthy and Wilson 2001), as opposed to a 'organisational' (or human) view of the topic.

Both views should not be seen as mutually exclusive, but complementary. Although the issue of which view should be predominant is clearly company-specific, there is evidence that both organisational support (to reduce the complexity of the scheduling and improve communication) and technical support (to provide better data integration and scheduling functionalities) are required (Vernon 2001). The limitations of treating scheduling as essentially mathematical (i.e. technical) problems fully specified and then solved for feasibility or optimality have been frequently noted (see, e.g. MacCarthy and Wilson 2001; MacCarthy 2006; Shobrys and White 2000; Buxey 1989). In this regard, it is important that both users and developers of the

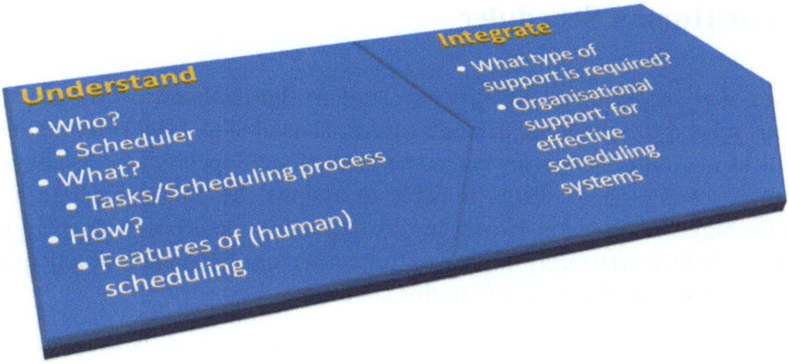

Fig. 13.1 Supporting the organisational view

manufacturing scheduling system understand that this system is neither a database or a piece of software, nor an optimisation procedure (Herrmann 2006).

The organisational or human issues in companies are really complex matters whose detailed analysis would fall beyond the scope of this book. Therefore, here we would restrict ourselves to understanding the role of the human scheduler and the nature of the scheduling practice as a basis to develop more effective scheduling systems (Crawford 2001) as well as to investigate how these could be integrated into existing manufacturing scheduling systems. Indeed, some authors point out at the lack of fit between the human element and the formal or systemised element found in the scheduling methods as one of the causes for the gap between the theory and practice of scheduling (McKay and Wiers 1999, 2006).

Therefore, to properly address the organisational perspective of manufacturing scheduling, at least the following issues should be considered:

- What is the 'human' approach to scheduling, i.e. how is the scheduling task performed in companies?
- How the features identified in the previous issue can be integrated into a scheduling system? More specifically, how these features can be taken into account when designing the scheduling tool, models and procedures?
- How the resulting scheduling tool can be effectively deployed into an organisation?

While the last issue will be the subject of the next chapter, here we will focus onto the first two. An outline to the rationale to be followed in this chapter is depicted in Fig. 13.1. Clearly, answering the first one involves analysing *who* performs scheduling in companies along with the activities assigned, *what* is the process followed to accomplish these activities and *how* the process is accomplished. These three aspects are studied in the next sections: Sect. 13.4 is devoted to describing the main tasks allocated to the humans in charge of the scheduling process, whereas in Sect. 13.5 the way in which these tasks are accomplished is presented. Finally, in Sect. 13.6 the main features of humans when performing scheduling are discussed.

13.4 The Human Scheduler

First, we start by trying to identify the staff in the company who accomplishes decisions related to scheduling. In one survey, Halsall et al. (1994) found that 64 % of small-to-medium enterprises had a member of staff specifically responsible for production scheduling. It seems sensible to assume that this percentage is much higher for bigger companies. However, this scheduler role is not exclusive for many companies: Indeed very rarely the title of 'scheduler' can be found in practice (Vernon 2001). The tasks usually allocated to this staff are:

- Assigning jobs to machines. Although (in theory) this seems to be the primary focus of schedulers, field studies show that the time to build and manipulate schedules only represents 10–20 % of their time (Fox and Smith 1984; Grant 1986).
- Dealing with problems affecting to the schedules. This is, indeed, the most cited task in many case studies (see, e.g. Crawford 2001). This task is usually addressed by dynamic changes into resources, processes, quantities, materials, date, operators, etc. (McKay and Wiers 2006). An important characteristic of this task is the fact that, as we will discuss later, is usually performed in an urgent fashion.
- Anticipating and avoiding future scheduling problems. To do so, they spend a lot of time in checking and correcting inconsistencies between different sources, and in gathering new information (Webster 2001). In addition, schedulers may change the current schedule in order to avoid problems in the future (Crawford 2001): An example may be the case of a scheduler who knows that there will not be raw materials required for some specific orders, and then it does not release these particular orders in the shop floor in order to avoid accumulating inventory and/or creating future problems (Fig. 13.2).

13.5 The Scheduling Process

Regarding to how these tasks are carried out, McKay and Buzacott (2000) found that a common scheduling strategy is adopted—explicitly or implicitly—across different companies. This strategy (or *process*, in the following) consists of seven steps:

1. Situation assessment. The current shop situation is analysed and the differences between the actual situation and the expected situation is determined. This usually involves a walk about and taking note of the most relevant aspects, the status of the different orders, and how well the process is doing. Note that, in assessing the situation, the scheduler adopts much of the role of 'information investigator' (Crawford 2001). Typical examples about the information required may be the changes occurred in the daily shipping requirements, the jobs processed in the last hours, or the current inventory levels (McKay and Wiers 2006).
   By accomplishing this step, the problem definition is refreshed and updated, and it emerges a new view of what the problems are.

Fig. 13.2 Scheduling tasks (adapted from McKay and Buzacott 2000)

2. Crisis identification. The next step is to identify the special problems or 'crises' in the shop. These are likely to be the most constrained or the most important activities in the shop, that are prioritised in order to be treated as soon as possible before reflecting on other problems in the shop. Typically, crises are jobs running behind their processing estimates, jobs where the wrong material was used, jobs where the yield is lower than required, jobs where the setups were substandard, jobs which made the wrong parts, jobs where the inventory records were corrected and significant losses occurred, etc.
   Interestingly, part of this step is either to classify a problem as a crisis to be solved, or simply as a potential issue that has to be monitored. Evidence exists that expert schedulers know that responding too quickly to disturbance may be detrimental for the schedule, as sequentially occurring disturbances can cancel a previous disturbance or compound it (Vernon 2001).
3. Immediate resequencing and task reallocation. Using the information about the status of the shop (obtained from Step 1) and the priority of the different issues (gathered in Step 2), a number of changes are devised to correct the deviations

from the expected plans. This is done in an urgent manner with the purpose of altering the trends in the plan running currently in the shop and may entail a broad spectrum of activities, focusing onto the jobs (jobs expedition, job preemption, cancellation of existing jobs, changing the order of the current jobs, etc.), onto the resources (subcontracting, overtime, moving operators from one resource to another, etc.) or even onto the processes (alternative processes can be employed for current jobs, or designed for the new jobs).
Note that, in this step, the crises are addressed by resequencing and reallocation, meaning that there is no intention to provide a new plan for those jobs not directly affected by the crises to be solved (hot jobs). Indeed, by altering the plan for the hot jobs, the plan may become now infeasible for other jobs in the immediate future (McKay and Wiers 2006). McKay and Wiers (2006) also state that, while a decision-support system can be of help for this step, the ability to make and negotiate dynamic changes to the problem definition are currently in the realm of human capabilities.

4. Complete scenario update. The overall scenario (that is, not only the hot jobs that have been altered in Step 3) is updated. Therefore, the picture of the remaining jobs emerges so the scheduler knows the current tasks and allocations as a starting point. As mentioned in the previous step, the reallocation may have caused new problems in the future jobs, so these new work requirements have to be considered from this new starting point.
5. Future problem identification. Once the urgent problems are considered to be fixed, the scheduler has a look at a longer time horizon in order to identify problems with the schedule or sequence outside the immediate dispatching horizon. Some of them can be considered to be second-order effects or issues that break the feasibility or desirability of the schedule (McKay and Wiers 2006), but some other problems in the future can be caused externally to the shop (see, e.g. the case described by Crawford 2001).
At this point, McKay and Buzacott (2000) identify two variants in the process. In a completely manual system or one with a traditional MRP mechanism, the scheduler will look through the list of upcoming work and isolate work needing special attention. In most cases, it is possible to obtain a list of planned work by part or by work centre and restrict the time horizon for data selection. If the scheduler has the benefit of a scheduling decision-support system, the work may be displayed visually on the Gantt chart using simple rules such as earliest due date. The basic information is the same as if a resource report was used, e.g. sequence of work orders for the resource.
In both cases, the schedulers scan the list looking for jobs that stick out. Since planning is an iterative process and continues day to day, the completely new work is usually towards the end of the time horizon and rolls in. The majority of work within the time horizon would have been looked at already in a past cycle and corrective action taken. They will look for new work within the time zone, but in most cases, the decisions modified one day will hold for the next.
6. Constraint relaxation and future problem resolution. Following and during their reflection in phase five, they identify hot jobs and processes from the future and try

to place them on the schedule that they think is best. This might involve changing some other decisions or it might not. During this process, they identify possible risks and try to set up the schedule to minimise or avoid the impacts. These future problems will be attempted to be solved with sequencing or allocation strategies (McKay and Wiers 2006). However, if the problem cannot be addressed, the situation will be dealt with by relaxing constraints. For example, a part might not have been made for 6 months, and a test batch is needed before the full batch is run. Or, certain resources are being overloaded and an alternative resource is made available for a limited time and assigned the work. Once the scheduler relax the constraint(s) that make the problem unfeasible, he/she assumes that the problem is fixed.
An important activity within this step is the development of contingency plans, which is considered to be a key aspect of the scheduler's behaviour (see, e.g. McKay et al. 1995b or Vernon 2001). From the viewpoint of the development of a manufacturing scheduling software tool, the need of creating, storing and manipulating these contingency plans (i.e. alternative schedules) has to be considered. Note that the 'analysis of scenarios' functionality described in Chap. 11 constitutes an implementation of this requirement.
7. Scheduling by rote. Finally, the routine work (i.e. jobs that have not been identified as critical in the previous steps) is scheduled mechanically. The scheduler tries to balance due dates and capacity restraints while trying to satisfy Industrial Engineering requirements for preferred resources, etc. During this phase, he/she tries to resolve conflicts as they appear by, e.g. looking at alternative resources, batch splitting, etc.

From the description of this process, which is considered to be commonplace for many schedulers, a number of clues can be obtained on how a decision-support system can be of the best aid for the scheduling process. In addition, the process itself may lead to certain features typical of human scheduling. These will be discussed in the next section.

13.6 Features of Human Scheduling

Although it is difficult to establish a set of characteristics that can be applicable to humans in general, and, particularly, to how humans perform an specific task, a review of the literature may serve to identify a number of common features:

- Humans tend to have a 'myopic view' of the scheduling process, as human schedulers do not look very much into the future (Sanderson and Moray 1990). Indeed, it is often argued that they plan their scheduling behaviour only 1/2 to 1 hour ahead (Beishon 1974; Crawford and Wiers 2001). This feature is clearly a reflection of the process described in the previous section, although it is debatable whether the origins of this feature are due to a cognitive limitation of humans when addressing scheduling problems, or to the changing and problematic environment in which they perform their work. Nevertheless, this widely studied and accepted feature is to be taken into consideration when accounting for human performance in scheduling, and also when designing a scheduling software tool.

- Human schedulers reduce the problem space quickly (McKay et al. 1992). In some cases, they (implicitly) adopt a Drum-Buffer-Rope approach (Goldratt and Cox 1992) and focus onto the bottleneck of the plant (Webster 2001). They also may create different sets of jobs according to their status in order to deal with smaller (and more manageable) subsets.
  This can be seen as an advantage for many situations where finding good schedules (or even formalising the scheduling process) is difficult, but on the other hand this can lead to an oversimplification which creates infeasible/suboptimal schedules (Dessouky et al. 1995). In addition, the task of reducing the complexity of the problem may represent a great percentage of schedulers' time, as studies suggest that they spend 80–90 % of their time determining the constraints that will affect scheduling decisions (see, e.g. Fox and Smith 1984; Grant 1986), which means that only a few time is spent on the generation and modification of schedules.
- There is a large variability in (human) scheduling practice, as field studies show that the specific decisions taken by the humans depend on the particular scheduler (Norman and Naveed 1990; Wiers 1996). As a consequence, for most cases, the developers of a specific manufacturing scheduling system in a company would not have at their disposal explicit descriptions of scheduling procedures can be easily transferred to a software tool. On the contrary, such descriptions (or 'best practices') would emerge (and possibly negotiated) during the requirements elicitation phase. This aspect would be elaborated further in Sect. 13.7.
- Human schedulers use their own sources of information. Instead of dealing with the enormous amount of information required to perform scheduling decisions (or maybe due to this fact), schedulers establish and use their own 'information networks' from which they collect the necessary information (McKay et al. 1995a). Part of this information comes from verbal interactions in the shop floor (Vera et al., 1993), or include 'informal' sources of information (Vernon, 2001). Other sources of information include interaction with other functions within the manufacturing company that fall outside of the scheduling area (Crawford 2001).
  As they collect a big deal of information, one may think that schedulers also constitute a source of information for different (related) functions in the company. In this regard, some authors (Crawford 2001) consider the schedulers an 'information hub', both using and disseminating the information. True as it is, in some case studies it is found that schedulers may hide information to other upper-level planners, due to the fact that in some cases he/she adopts different objectives than those stated by these upper levels (Vernon 2001).
  It has also to be noted that not all the information gathered by the human schedulers is strictly required, as some of the information sources are used for verification, leading to a duplication of the effort in searching the information (Vernon 2001).
- Human schedulers tend to use untaught and non-routine heuristics in many circumstances, but particularly in conditions that they consider exceptional.

- Human scheduling is (specific) goal-driven. For most cases, rather than generating the schedule by a specified means or procedures, schedulers generate their schedules to meet specific goals (Crawford and Wiers 2001). Some field studies (Vernon 2001) describe these goals as related to production performance rather than to customer's satisfaction, which not always induce a behaviour in the scheduler that is congruent with management expectations.
- The role of humans when a decision-support system is employed is to take control of the DSS whenever there exist a disruption of the schedule (Nakamura and Salvendy 1994). Therefore, the major function of schedulers is dealing and preparing for contingencies (Thurley and Hamblin 1962).

13.7 Integrating the Human Piece into the Scheduling System

In this section, we summarise the different features of human schedulers and extract relevant conclusions within the context of a scheduling system.

13.7.1 Humans on Command

Obvious as it may seems, the scheduling system should place humans at the top of the decision-making system. Therefore, a scheduling tool must truly support and do not replace human decision-making (Crawford and Wiers 2001). This also should have been made clear from the scheduling process described in Sect. 13.4, where some of the steps cannot simply be accomplished by a software tool. The variety in the sources from which the human schedulers obtain information (some of them informal and thus out of the scope of the company's information systems), the unstability of the environment and thus the ability to react to unexpected events are at least two reasons for this.

This view of the scheduling system is what is called 'interactive' or 'hybrid' approach (Vernon 2001), where the assumption is that this type of systems outperforms both fully automated and fully manual approaches (see, e.g. Haider et al. 1981; Nakamura and Salvendy 1994; Higgins 1996; Baek et al. 1999).

13.7.2 Acquiring Knowledge from Schedulers

13.7.2.1 Devising 'Scheduling Best Practices'

It has been discussed earlier in the chapter that there is a large variability in scheduling practice and that many scheduling heuristics are untaught and non-routine. The implications of these facts are that, on one hand, scheduling performance may largely depend not only on the specific scheduler, but on the particular disposition, mood or awareness of the same scheduler. The issue becomes more complicated when different schedulers with different degrees of knowledge, experience, motivation, etc. have to operate in the same shop floor.

On the other hand, it is likely that these heuristics and scheduling practices are not explicit, not to mention documented in a formal way that allows their reuse and improvement. Therefore, particular (i.e. individual) scheduling practices should be investigated, formalised and documented. If possible, a collection of scheduling 'best practices' should be extracted in order both to reduce the variability of the scheduling process and to collectively improve scheduling performance.

A special remark should be done regarding the objectives to be pursued in the scheduling activity, as these have to become a necessary part of the collection of best practices. In order to make them aligned with the expectations of the company's management, the management goals should be integrated, for which the final word of these best practices should be given to the top management and not confined to the staff in charge of manufacturing.

As the untaught and non-routine heuristics tend to be more used in exceptional conditions, a procedure for registering 'new' heuristics that have been used can be implemented.

13.7.2.2 Using Schedulers to Build the Scheduling Decision-Support System

As discussed earlier, human schedulers reduce the problem space. This characteristic of human scheduling is more useful in the design phase of a scheduling system (where schedulers may help to remove constraints/objectives that do not have an impact on the performance of the system) that during the operation phase (where schedulers may overlook important improvements as they oversimplify the problem). As a consequence, human schedulers may be of great help when designing a scheduling system, but, when finished, this system should provide them with all schedule information, including all constraints and objectives.

In addition, it has been mentioned that schedulers build their own networks of information, which make them to be 'information hubs' (Crawford 2001). This knowledge is relevant not only for designing decision-support systems, but to gain an in-depth understanding of the shop floor.

13.7.3 Training and Evaluation

13.7.3.1 Theoretical Support

It has been shown that a factor affecting scheduling success includes scheduler's cognitive abilities and how they use theoretical scheduling techniques (Stoop and Wiers 1996). In addition, there is a wealth of literature stating that schedulers tend to use primarily heuristics and procedures that they are able to understand (Baek et al. 1999; Wiers 2001; Freed et al. 2007, among others). Therefore, emphasis must be placed in enhancing the knowledge of schedulers by training and upgrading.

13.7.3.2 Objectives

It has been shown in the previous section that, in many cases, schedulers generate their schedules in order to meet specific goal, and that schedulers have a myopic view of the process. All this naturally leads to the fact that their suggested scheduling objectives may change over time, and that they are short-range objectives. Typical examples are to achieve a certain output during a shift (see, e.g. the case study by Vernon 2001), which may lead to speed up 'easy' jobs by delaying others that are more important, but also more complex or time-consuming.

Therefore, an effort to agree scheduling objectives within the company, and to make them explicit has to be carried out in order to be prevented against local, short-time objectives that may not be beneficial for the company on the long run.

13.7.3.3 Understanding the Manufacturing Process

The literature points the need that scheduling is to be carried out by personnel with first-hand and intimate knowledge of the shop floor, given the complexity in scheduling created by demands from customers due to product customisation (see Thurley and Hamblin 1962; Crawford and Wiers 2001 among others). Therefore, training of schedulers must not focus solely on the scheduling process in which they are primarily actors, but on the whole process of the company. Note that the purpose here is not that the schedulers gain a deep knowledge of the rest of the activities of the company, but that they understand their work and the implications on their own task. Some authors (e.g. MacCarthy 2006) state that ignoring the higher level functions within which lower level allocation decisions are defined and constrained and ignoring the dynamic rolling pipeline of planned production, and the problems and opportunities it provides, constitutes a major drawback.

13.7.3.4 Practice

There are many field studies that (perhaps unsurprisingly) found relevant differences in performance between expert schedulers and novice schedulers (Randel and Pugh 1996). One of the reasons is the fact that expert schedulers are more efficient at searching for the relevant information when building the schedules (Vera et al. 1993), which is an expertise that can be acquired mainly through practice.

13.7.3.5 Evaluating Schedulers' Performance

An important issue raises when evaluating the performance of the schedulers. As we have discussed previously, the environment in which scheduling decisions are taken may be highly unstable and the human scheduler has to react against a number of unexpected events. In this situation, the measures of the scheduler's performance should foster (whenever possible) a proactive, problem-solving approach rather than zero-risk, low-profile solutions that result in a systematic delay of the schedules upon any unexpected event.

On the other hand, schedulers should be measured against targets that are within their domain. In Crawford (2001), it is described an example, arguing that the failure of suppliers to deliver materials is partly the responsibility of the scheduler, as he should monitor ongoing deliver. However, this may be translating a suppliers' failure into a scheduler's failure, and therefore the supplier should be measured as a separate factor.

13.7.4 Smoothing the Environment

The environment in which schedulers must take their decisions is almost unanimously described as dynamic, complex and relatively unstable. It has to be noted as well that, according to the findings in the previous section, the main way for the human scheduler to reduce the complexity of the environment is via (over)simplification. Therefore, efforts should be carried out in order to avoid the potential under-performance due to this fact. Interestingly, some researchers have found that the parameters influencing human scheduling behaviour are not plant specific (Crawford and Wiers 2001); therefore, while there remains for the scheduler the need of an intimate knowledge of the specific shop floor in which he/she is to perform his/her activities, it is possible to devise a set of common guidelines to ease schedulers' activities.

Taking the environment as a whole, one (possibly obvious) solution that is mentioned in the literature to reduce the level of uncertainty is to try to automate most of the manufacturing environment, although there is some instability that is inherent to this environment. In order to minimise the level of risk and uncertainty that a human scheduler has to deal with, it has been proposed that a higher level solution needs to be formulated. For instance, material, capacity and personnel shortages can be reduced at a higher planning level. Basically, this can be done in three ways:

- Securing against uncertainty by building up inventories of material, as well as human and machine slacks. This is the traditional form of dealing with uncertainty, but it is far from being efficient or even feasible in today's ever-competitive environments.
- Taking into account scheduling-specific considerations at a planning level. In practice, this leads to the monolithic approach of production planning that has been widely regarded as unfeasible in many production environments.

- Providing an adequate coordination between the scheduling function and the higher level planning functions. This is probably the best approach, and it is has been discussed in Chap. 3. On the other hand, the issues mentioned in the previous section regarding the tendency of human schedulers to hide information should be taken into account.

Another approach is to break down the causes of instability and try addressing them, if possible, one by one. Instability refers to changes in the production process, and in the materials and personnel employed, and it is usually due to internal and to external uncertainty (Halsall et al. 1994, but see Hopp and Spearman 1996 for a slightly different classification with the names of internal and external variability).

Regarding internal uncertainty, companies appear to have some degree of certainty about their stages in the production process, but the processing times and set up times of these stages are less well known (Crawford and Wiers 2001). In some cases, the uncertainty of the processing times may be inherent to the manufacturing process (see, e.g. the tile production process described in Chap. 15), so there is little to do. For the rest of the cases, the lack of reliable data (regarding processing times) or documented, standard procedures (regarding setup times) can be fixed by proper actions. Another source of instability is the breakdown of machines. Although there is little that can be done for the scheduler, at least from a short-term perspective, one way to handle this type of instability is to provide the scheduler with tools to assess the criticality of the schedules with respect to the specific resources, so he/she can think of alternative schedules in case a breakdown happens, or at least cap the overall effect of the breakdown in the system.

Regarding external uncertainty, most problems have come from the fact that either the manufacturing process must start before the exact requirements of the customer are known, or these requirements are subject to frequent changes. Note that this type of uncertainty may be in reality a sort of 'internal' uncertainty, as in some companies the necessary coordination between Sales and Manufacturing is missing. Whenever this is the case, appropriate actions should be taken.

Addressing uncertainty coming from the customer is, however, a different thing. Nevertheless, for certain products, customer's demand is sensitive to price and lead times, so the company can quantify the losses due to this source of uncertainty and offer discounts based on demand stability.

A final source of external uncertainty is the lack of reliability (both in time, quantity or quality) of the suppliers. Some companies have embarked in just-in-time programmes with their suppliers to reduce this uncertainty, while others have simply translated them the responsibility of managing the supplies (together with the costs associated).

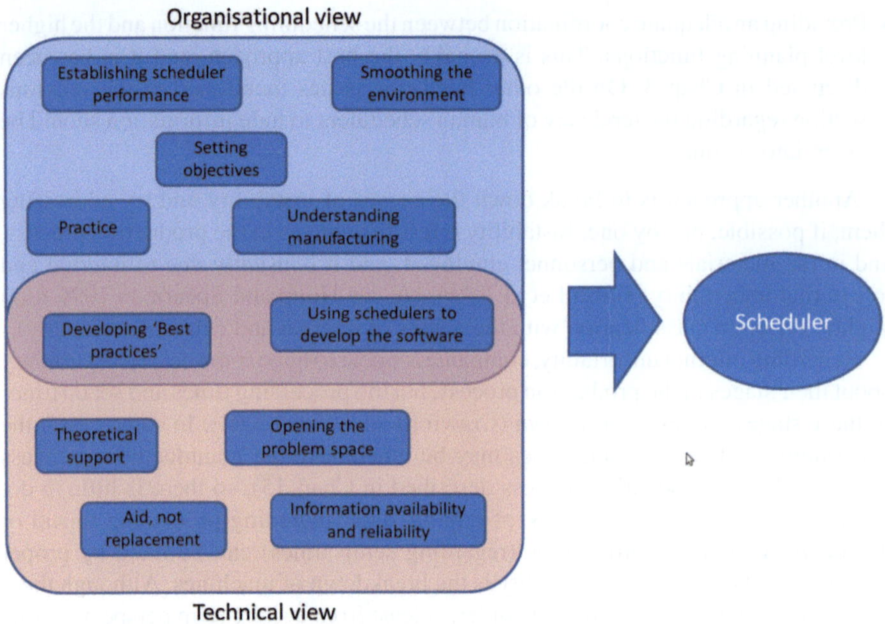

Fig. 13.3 Types of support required in a scheduling system

13.8 Conclusions and Further Readings

In this part of the book, we have started to address the integration of manufacturing scheduling models, procedures and tools into an organisation in order to constitute an scheduling *system*. To do so, we have focused in this chapter in understanding the nature of human scheduling, their main activities and features. From this analysis, it turns out that most of the tasks carried out by the human scheduler are reactive (i.e. avoiding disturbances in schedules) rather than predictive (developing schedules). This clearly conditions how human schedulers perform their tasks, and give a number of valuable clues for better placing them into a scheduling system (Fig. 13.3).

One of the—in our opinion—best sources regarding the issues discussed in this chapter are in the book *Human performance in planning and scheduling* edited by McCarthy and Wilson. Several chapters in this book have been cited throughout the chapter, including Crawford (2001); Vernon (2001); Crawford and Wiers (2001). This book is one of the few works devoted to this topic, which is both bad news (as an important issue in manufacturing scheduling is misrepresented) and good news for the scholar (as new research avenues with relevant practical implications are open).

References

Baek, D., Oh, S., and Yoon, W. (1999). A visualized human-computer interactive approach to job shop scheduling. *International Journal of Computer Integrated Manufacturing*, 12(1):75–83.

Beishon, R. (1974). *The Human Operator in Process Control*, chapter An Analysis and Simulation of an Operator's Behaviour in Controlling Continuous Baking Ovens, pages 79–90. Taylor & Francis.

Buxey, G. (1989). Production scheduling: Practice and theory. *European Journal of Operational Research*, 39(1):17–31.

Crawford, S. (2001). *Human Performance in Planning and Scheduling*, chapter Making Sense of Scheduling: The Realities of Scheduling Practice in an Engineering Firm, pages 83–104. Taylor & Francis.

Crawford, S. and Wiers, V. (2001). From anecdotes to theory: reviewing the knowledge of the human factors in planning and scheduling. In MacCarthy, B. L. and Wilson, J. R., editors, *Human performance in planning and scheduling*, pages 15–44. Taylor & Francis.

Dessouky, M., Moray, N., and Kijowski, B. (1995). Taxonomy of scheduling systems as a basis for the study of strategic behavior. *Human Factors*, 37 (3):443–472.

Fox, M. S. and Smith, S. (1984). Isis: A knowledge-based system for factory scheduling. *Expert Systems Journal*, 1(1).

Freed, T., Doerr, K., and Chang, T. (2007). In-house development of scheduling decision support systems: Case study for scheduling semiconductor device test operations. *Int J Prod Res*, 45(21):5075–5093.

Goldratt, E. and Cox, J. (1992). *The Goal: A Process of Ongoing Improvement*. North River Press; 2 Revised edition (January 1992).

Grant, T. J. (1986). Lessons for O.R. from A.I.: A scheduling case study. *Journal of the Operational Research Society*, 37 (1):41–57.

Haider, S. W., Moodie, C. L., and Buck, J. R. (1981). An investigation of the advantages of using a man-computer interactive scheduling methodology for job shops. *International Journal of Production Research*, 19(4):381–392.

Halsall, D. N., Muhlemann, A. P., and Price, D. H. (1994). A review of production planning and scheduling in smaller manufacturing companies in the uk. *Production Planning & Control*, 5(5):485–.

Herrmann, J. (2006). *Handbook of Production Scheduling*, chapter Decision-Making Systems in Production Scheduling, pages 91–108. Springer.

Higgins, P. G. (1996). Interaction in hybrid intelligent scheduling. *International Journal of Human Factors in Manufacturing*, 6(3):185–203.

Hopp, W. J. and Spearman, M. L. (1996). Factory physics. *Foundations of manufacturing management*. Irwin, New York, USA.

MacCarthy, B. (2006). *Handbook of Production Scheduling*, chapter Organizational, systems and human issues un production planning, scheduling, and control, pages 59–90. Springer.

MacCarthy, B. L. and Wilson, J. R., editors (2001). *Human performance in Planning and Scheduling*. Taylor & Francis.

McKay, K. and Buzacott, J. (2000). Application of computerized production control systems in job shop environments. *Computers in Industry*, 42(2):79–97.

McKay, K., Buzacott, J., Charness, N., and Sayafeni, F. (1992). *Artificial Intelligence In Operational Research*, chapter The Scheduler's Predictive Expertise: An Interdisciplinary Perspective, pages 139–150. Macmillan.

McKay, K. and Wiers, V. (2006). *Handbook of Production Scheduling*, chapter The Human Factor in Planning and Scheduling, pages 23–57. Springer.

McKay, K. N., Safayeni, F. R., and Buzacott, J. A. (1995a). "Common sense" realities of planning and scheduling in printed circuit board production. *International Journal of Production Research*, 33(6):1587–1603.

McKay, K. N., Safayeni, F. R., and Buzacott, J. A. (1995b). Review of hierarchical production planning and its applicability for modern manufacturing. *Production Planning and Control*, 6(5):384–394.

McKay, K. N. and Wiers, V. C. S. (1999). Unifying the theory and practice of production scheduling. *Journal of Manufacturing Systems*, 18(4):241–255.

Nakamura, N. and Salvendy, G. (1994). *Design of Work and Development of Personnel in Advanced Manufacturing*, chapter Human Planner and Scheduler, pages 331–354. Wiley.

Norman, P. and Naveed, S. (1990). A comparison of expert system and human operator performance for cement kiln operation. *Journal of the Operational Research Society*, 41 (11):1007–1019.

Randel, J. and Pugh, L. (1996). Differences in expert and novice situation awareness in naturalistic decision making. *International Journal on Human Computer Science*, 45:579–597.

Sanderson, P. and Moray, N. (1990). *Ergonomics of Hybrid Automated Systems II*, chapter The Human Factors of Scheduling Behaviour, pages 399–406. Elsevier.

Shobrys, D. and White, O. (2000). Planning, scheduling and control systems: Why can they not work together. *Computers and Chemical Engineering*, 24(2–7):163–173.

Stoop, P. and Wiers, V. (1996). The complexity of scheduling in practice. *International Journal of Operations and Production Management*, 16(10):37–53.

Thurley, K. and Hamblin, A. (1962). The supervisor's role in production scheduling. *International Journal of Production Research*, 1:1–12.

Vera, A., Lewis, R., and Lerch, F. (1993). Situated decision-making and recognition based learning: Applying symbolic theories to interactive tasks. In *15th Annual Conference of the Cognitive Science Sociey, University of Colorado-Boulder*, pages 84–95.

Vernon, C. (2001). Lingering amongst the lingerie: An observation-based study into support for scheduling at a garment manufacturer. In MacCarthy, B. and Wilson, J., editors, *Human performance in planning and scheduling*, pages 135–163. Taylor & Francis.

Webster, S. (2001). A case study of scheduling practice at a machine tool manufacturer. In MacCarthy, B. and Wilson, J., editors, *Human performance in planning and scheduling*, pages 67–81. Taylor & Francis.

Wiers, V. (1996). A quantitative field study of the decision behaviour of four shopfloor schedulers. *Production Planning and Control*, 7(4):383–392.

Wiers, V. (2001). Design of knowledge-based scheduling system for a sheet material manufacturer. *Human Performance in Planning and Scheduling*, pages 201–215.

Chapter 14
Guidelines for Developing Scheduling Systems

14.1 Introduction

In this chapter, we keep addressing a number of issues related to the development of a manufacturing scheduling system. While in the previous chapters the functionalities of such system have been described in detail, here we focus on the process of making the system work in the shop floor. Different issues in this process (such as data, constraints, models, etc.) are discussed.

More specifically, in this chapter, we

- describe the different aspects of a scheduling system at work (Sect. 14.2) and give some general hints (see Sect. 14.3), and
- discuss the specific aspects related to
  - Layout modeling (Sect. 14.4),
  - Data required for the system (Sect. 14.5),
  - Objectives (Sect. 14.6),
  - Solution procedures (Sect. 14.7), and
  - Constraints (see Sect. 14.8).

14.2 Manufacturing Scheduling Systems at Work

Up to this chapter, we have paid attention to how to ensamble the different parts of a scheduling system in order to make it a useful support for scheduling decisions. In other words, we have focused on *what* a scheduling system should contain. In this chapter, we address the process of *how* to put this system at work in the company. This may be certainly be regarded as the 'last mile' to be covered, but it is crucial for the success of a scheduling system, and it is likely to be the root of some development failures. It is therefore, a critical issue that, unfortunately, has been seldom addressed in the scheduling literature.

J. M. Framinan et al., *Manufacturing Scheduling Systems*,
DOI: 10.1007/978-1-4471-6272-8_14, © Springer-Verlag London 2014

It is clear that the development of a scheduling system is a particular case of that of a decision support system. From this viewpoint, software development models employed for business information systems could (and should) be applied. In principle, none of these has to be preferred as compared to the others, but given the technical complexity of these systems and the critical nature of the interface with the human (the decision-maker), software development models intensive on prototyping and on fast iterations over versions of the software seem to be more suitable. In addition, the relatively small size of the scheduling system (at least as compared to other business information systems) would not favor the employment of heavily documented and rigid approaches that may increase the burden and costs of software development.

Along this line, our aim in this chapter is to elaborate a number of recommendations or guidelines in order to make the manufacturing system work. The goal is providing stakeholders with a set of policies that may help them in the development process. These policies are based on an analysis of the existing literature describing development cases of scheduling systems, on the requirements established by the architecture of a scheduling tool proposed in Chap. 11, and on our own hands-on experience as researchers and practitioners in the field. These guidelines are grouped into specific functions in an attempt to provide a more structured vision of the recommendations. Nevertheless, we believe that a further systematisation of these guidelines is company- and technology- dependent, and thus would be of little application within the scope of this book. Therefore, there is no intention of being comprehensive but informative and still, many of the hints (even if supported by literature and/or experience) may be debatable for particular developments.

The different aspects that would be taken into account are the following:

- General hints. Under this umbrella, we describe a number of hints (we do not dare to call them 'principles') that can be useful to drive the deployment effort. They relate to each other only in that they may potentially affect to all stages in the deployment process. These issues are discussed in Sect. 14.3.
- Layout modeling. Here we will discuss the main issues related to how to represent the company's shop floor layout into the scheduling tool, i.e. how to model the shop floor. The guidelines obtained regarding to this aspect are presented in Sect. 14.4.
- Data. Here the main issues related to data acquisition, manipulation, and maintenance are presented in Sect. 14.5.
- Objectives. In Sect. 14.6, a number of guidelines referring to selecting and prioritizing the objectives to be presented to the scheduler is discussed.
- Solution procedures. These guidelines refer to how to approach the design and implementation of the solution procedures included in the scheduling tool. This is discussed in Sect. 14.7.
- Constraints. Main issues regarding the modeling of the constraints into the manufacturing scheduling tool are discussed in Sect. 14.8.

Finally, note that in this chapter we will not make any distinction among customised and packaged scheduling tools. We believe that most of the guidelines apply to both types of software, and just a few of them (particularly those related to the design of solution procedures) may, in principle, be more useful for customised tools.

14.3 General Hints

As stated earlier, this collection of guidelines consists mainly on an underlying approach to address the different issues that may arise during the development process, rather than specific bullets that could be checked at the end of the process.

14.3.1 Incremental Deployment of the Scheduling System

As just any other business information system, scheduling systems are more suitable to be implemented in companies by employing an incremental approach where progress can be rapidly visualised rather than 'big-bang' approaches, where a final solution is first fully envisioned and only then implemented. Despite their potential shortcomings—being the most notable the increase in the deployment period—, the literature on implementation of business information systems is clearly stressing the suitability of an incremental approach. This recommendation is particularly strong in the case that the information systems contain company-specific features, which is precisely the case for most scheduling systems. Indeed, several cases in the literature (see e.g. Leachman et al. 1996; Wiers 2002) report the need of long, iterative approaches for the successful implementation of scheduling systems, as opposed to the classical waterfall approach. This rationale is also related to the best fit of a bottom-up approach (typical in incremental approaches) as compared to a top-down approach (typical for big-bang approaches). Therefore, even if the final goal would be to capture with maximum detail the complexity of the physical shop floor, a more simple model would suffice for start working with the scheduling system, and for making the first results of its adoption visible, which in turn would serve to ease the subsequent deployment steps (Missbauer et al. 2009). Note that the lack of sophistication in which the scheduling function is accomplished in many companies has made many authors to state that simply a visual representation of the Gantt schedules is sometimes regarded as a big advance with respect to the previous situation (see e.g. McKay and Black 2007) and thus can be used to make visible the advantages of a scheduling system.

14.3.2 A-B-C Analysis

Scheduling is about making and monitoring plans, not about capturing and responding to all possible exceptions and alternative scenarios, which in many real-life implementations are virtually infinite. There is a big temptation in stretching the limits of a tool which is potentially able to make plans with great detail. But with the ever increasing number of product variants manufactured by companies due to the rise of mass customisation, nearly all real-life shop floors would be open shops with very complex constraints in strict terms. However, the effort to model this complex shop floor has to be carefully weighted; for instance, it may turn out that, for most product variants, this complex setting results in a much simpler layout. Indeed, it is often claimed that many job-shops are flow shops for most of the jobs (see e.g. Storer et al. 1992; Knolmayer et al. 2002). Similar simplifications can be achieved by concentrating the scheduling activities on a few key operations (MacCarthy and Wilson 2001). On the other hand, in many factories producing a standard set of products there is a (usually small) percentage of jobs which are customised, or occasionally some works may need re-working due to quality defects. Unless the percentage of quality defects or the percentage of customised jobs is really high, it may be easy to implement simple shop floor directives, such as giving the highest priority to these jobs, so to avoid a more complex approach when modeling the shop. Some hints to handle both model and real-world complexity have already been given in earlier chapters.

There are additional reasons to perform such an attempt to simplify the resulting models, as the literature describing cases consistently state that not all details of the shop floor can be captured into the scheduling system (see e.g. Bensana et al. 1986; McKay et al. 1988). Indeed, in Kerr (1992) it is described that the attempt to carefully incorporating all constraints observed by the human scheduler was identified as one of the roots of the failure of the system.

Therefore, an instrument to obtain these reductions in the complexity of the original shop floor is to perform a Pareto (or A-B-C) analysis based either in the contribution of the jobs to the bottom line, the frequency of the different orders, or the usage of the manufacturing resources. As a result, it seems sensible to develop a system to schedule 'just' 80 % of the jobs rather than considering a much more complex modeling approach in order to handle the other remaining 20 %, particularly if we adopt the aforementioned incremental approach. It is clear that the jobs 'outside' the system would interfere and cause disruptions of the schedule made for those 'in' the system, but there are relatively easy mechanisms (such as booking resource capacity for the former) to minimise such disruptions.

14.3.3 Avoid Technology-Dominant Approaches

Whenever possible, the technicalities of the scheduling models, procedures, and tools should be hidden to the schedulers. Given the complex nature of the scheduling business function, it is unavoidable that the logic of the scheduling system is not understood by many users, and regarded as a black box (Laforge and Craighead 2000). On the other hand, IT specialists cannot fully grasp the dynamics of the company, so they need the input from these users, a factor which is believed to be a key for a successful deployment (Missbauer et al. 2009).

Better training may be needed in the implementation of scheduling packages, as they contain rather complex features and options for configuration (Laforge and

Craighead 2000). Obviously, even if we will recommend to keep these options at a minimum, training must be an important item in the deployment process. Note that this not only would affect to the quantity of the training (as measured in hours), but to the quality of the training. More than often, the user training on business information systems consists merely on showing which screens and buttons should be pressed to replace the old manual functions of the employees, but little is told about the radical change that represents moving from a manual procedure to an automated system. The typical case are the data fed into the system: as in a manual scheduling procedure the data are handled by several employees in the company, it is likely that typos and minor errors do not reach to the shop floor, as the human common sense would realise about these errors. In contrast, in a highly automated system, human intervention is minimised and thus the likelihood that errors in the input data translates into wrong production orders increases. The subtle but extremely important change in the role of the scheduler from an executioner to a supervisor must be stressed in the training so most of these errors can be prevented.

14.3.4 Modular Design and Implementation

While the integration of scheduling with related decisions (such as, e.g. lot sizing or material flow control) has attracted a wealth of interest in the last years, such complex models are difficult and expensive from the implementation viewpoint (see e.g. Higgins 1996). There are a number of reasons for this:

- Data requirements. It is obvious that an integrated system would require more data, which is not a trivial issue given the data-intensive nature of the different operations management functions (see e.g. Kathawala and Allen 1993). In addition, the required quality and granularity of these data may be substantially different, as for instance aggregated estimations of processing times may be more than sufficient for production planning while unacceptable for scheduling. To make an integrated system work, the only solution is to stick to the maximum quality (and quantity) of data required by each of the submodules composing the system, therefore increasing not only development expenses, but also maintenance costs.
- Number of stakeholders. Since a higher number of decision-makers have to be involved in the development process of an integrated system, agreeing on constraints and objectives (among other issues) becomes harder. In addition, given the hierarchical nature of some operations management functions (i.e. production planning and production scheduling), it is likely that higher level considerations override the desired functionality, goals and constraints of the scheduling system, which may result in less useful support for the scheduling function.
- Changing nature of the business. Given the dynamic nature of the companies, integrated models exhibit a shorter life cycle and may require substantial upgrading when manufacturing conditions change. A modular design and implementation would increase the reusability of those parts of the system not affected by the changes.

As a conclusion, it seems clear that—despite the potential advantages of a system virtually capable to handle all operations management-related decisions—simple, modular approaches may be better suited. The enormous advances in protocols for data exchange among different systems also speak for the convenience of a modular design.

14.4 Layout Modeling

Note that layout modeling represents the most persistent abstraction of the scheduling system from the physical factory, and therefore it is likely that this layout would remain stable for longer periods. It is also possibly one of the main issues affecting the validation in the shop floor of the results obtained by the scheduling system, as it would determine the models to be employed, the data required to feed the models, and the solution procedures to be employed for obtaining solutions for the scheduling process. Therefore, these decisions should be carefully regarded.

One advisable hint is to keep the focus on modeling relatively simple layouts—as much as the physical shop floor allows it—. Even if the physical layout of the factory may be very complex, there are a number of reasons indicating the suitability of focusing onto simpler structures, at least on the initial stages of the life cycle of the deployment of the scheduling system. There are a number of reasons for this hint, which are detailed in the following subsections.

14.4.1 Dynamic Nature of the Business

Whereas theoretical scheduling models assume a set of fixed resources, this is hardly the usual situation in practice. Quite often, additional resources can be moved, purchased, or removed, provoking changes in the physical layout that detailed, low-level models cannot easily accommodate (see e.g. Benavides and Carlos Prado 2002; Freed et al. 2007). Asides, the current shortening of product life cycles and their corresponding technological processes make such changes more likely. This means that detailed scheduling models should be updated, validated, fed with data and put into production at a pace that their maintenance cost could be simply so high that they may never pay off.

14.4.2 Preprocessing/What-If Analysis

As mentioned in earlier chapters, some form of preprocessing of the layout is usually required, which can be used both to set the scope of the scheduling decision problem and for what-if analysis. It is assumable that this preprocessor may contain a simplified model of the plant, including a simplified layout, as it is used prior to a detailed modeling.

14.5 Data

Here we discuss a number of issues related to data acquisition, manipulation, and maintenance. In several sections of the book, it has been mentioned that scheduling is a data-intensive process, and that the quality of the data greatly affects the performance of the scheduling models and procedures. Broadly speaking, two types of data are required:

- Model-based data. By model-based data we understand all data related to the decision model to be entered into the scheduling system. These typically include the number of machines to be employed at each stage, transportation and other constraints not depending on the specific schedule to be found, etc. These data are relatively stable (at least as compared to the instance-based data) and may be seen as meta-data for the instance-based data to be discussed in the next item.
- Instance-based data. These data typically include the processing times of the machines, the routing of the jobs, the cost/income associated to each job, the release and due dates of each job, their weight and/or priority, etc. These data vary from one decision to another.

It is clear that the nature of these two types is quite different. A number of issues affecting both types are discussed in the following subsections.

14.5.1 Development of a Database Interface

Usually, some of the data required will be already present in the ERPs or business information systems of the company. However, it is unlikely that all data needed will be present, so data collection becomes an important part in the deployment of a scheduling system. In fact, the data required for a scheduling system may well be dispersed into as much as more than 20 different sources (in this regard, see the case study by McKay et al. 1988).

For introducing and editing model-based data, it may be interesting to develop a graphical interface, but this may be extremely tedious and expensive to maintain for instance-based data. Therefore, time and resources for the development of database interfaces should be allocated in the deployment project. Also note that the construction of such interfaces is usually considered as one of the most time-consuming items for packaged software, and that many maintenance contracts of the business information system vendors and/or IT consultants stipulate that no third-party can access to these data (even if in read-only mode). For such cases, semi-automatic import/export utilities should be the only available option apart from contracting the vendors and consultants for the development of the interface.

14.5.2 Data Quality

Another source of concern is the quality of the data employed in the scheduling tool. There are case studies in which it is reported that less than 50 % of the required data were consistent (see Leachman et al. 1996). Poor data quality is frequently mentioned (see Berglund and Karltun 2007) as the cause that specific tools and spreadsheets have to be used to complement standard scheduling software tools. Indeed, as mentioned earlier, one of the task of the human scheduler is to filter the data to be employed in the scheduling process.

14.5.3 Data Abstraction

Already mentioned in a previous chapter, a good advice is to abstract internal data structures from the databases and data tables in order to speed up development and implantation. This is heavily related to the data-intensive nature of the scheduling system and may turn useless much of the developments with respect to models and procedures, as the users would distruss such system.

14.5.4 Maintenance of Data

Given how hard is to gather the data, an equally detailed system should be placed in order to update and modify existing data. SCADA systems integration is a promising method of automating data maintenance. Even modest improvements in the automation of data can greatly contribute to the accuracy of the data in the system.

14.5.5 Keeping Interfaces Simple

The ERP of the company hardly needs to know the full sequencing with all the start and finish times of the production sequence. Furthermore, providing the ERP with all this data could prove challenging. Probably, only the estimated finishing date for each product is needed. As a result, when connecting a scheduling system with existing ERP software, a good idea is to interface only the most basic information and extend the amount of communication as the needs arise.

14.5.6 Performance

Advanced models will require extensive datasets that have to be retrieved from the database as needed. Special care should be put in the performance of queries to the databases. Modern distributed tables and threaded queries are necessary.

14.6 Objectives

It is difficult to overestimate the importance of coming to terms with the scheduling objectives. An important part of the development of a scheduling system is the discovery of the objectives to be sought. Managers may know what is desirable for the company, but may not be able to formulate operating policies, at least at a scheduling level. Furthermore, the interaction between operating policies and objectives is often unclear (Bhattacharyya and Koehler 1998). A probably by-side effect of that is the fact that, at the first glance, virtually all objectives that one may pose could be of interest for the decision-maker. It has been often stated that single objective optimisation is controversial in practice, and that several objectives must be taken into account.

Nevertheless, our experience indicates that it is difficult to visualise and understand more than 2–3 objectives, as the decision maker can hardly make sense of the economic impact of more than 2–3 conflicting operational measures. Therefore, the main issue here is how to prioritise/select among these. There are several approaches to do it, which are discussed in the next subsections.

14.6.1 Prioritisation of Objectives

Scheduling encompasses rather short-term decisions, and it is rarely linked to strategic, long-term or medium-term issues. Therefore, considering objectives linked to long-term goals makes little sense for some situations. One typical example would be the maximisation of machine usage. While from a costs accounting perspective machine utilisation helps cutting the unit production costs, this objective could be rarely more critical than fulfilling due dates, which is a pure, short-term objective. Hence, there is no need to consider both objectives simultaneously. Note that we do not claim that machine utilisation (or other medium or long term objectives) are not important: we just stress that operational decisions should be made according to operational measures, and that it is the output of these operational decisions what should feed strategic models, which as a result would determine whether machine utilisation (or any other medium-long term measure) is sufficient to adjust capacity.

14.6.2 Taking Out Non-Conflicting Objectives

Some of the proposed objectives may not be conflicting among them, but are simply different ways to express the goals of the decision makers. Even if some objectives could be theoretically conflicting, it has to be checked whether this happens in the industrial setting where the model is to be deployed. A typical example are the number of late jobs and maximum tardiness. It seems clear that one may theoretically reduce the number of late jobs by systematically delaying a small set of jobs, which will in turn generate big tardiness values. However, this alternative cannot be accepted by many companies for which customer reliability is their key competitive advantage. If this results to be the only available option, the company would find a way to renegotiate their commitments and immediately will modify its order acceptance policy to ensure that this situation will not happen again. Therefore, one of the two indicators may suffice as a scheduling objective. In addition to the additional burden on evaluating a non-relevant objective, in a multi-objective optimisation context, it has the additional risk of generating biased solutions, thus not providing the decision-maker with an appropriate set of solutions.

14.6.3 Postpone the Selection of Objectives

Another approach could be to delay the decision on which objectives should be considered. This approach may make sense in some cases, as usually the objectives of scheduling are not an input of the development of the scheduling system, but an output. First of all, it has to be taken into account that, in many cases, the dominant schedule criterion is often scheduling feasibility (Graves 1981).

Even when the system is finally deployed, the objectives to be considered undergo frequent changes (Cowling 2003). Therefore, developing the model could serve to better understand the objectives of the scheduling model. In addition, for some scheduling situations, finding a feasible (and understandable) solution may suffice (see e.g. Cowling 2003; Gao and Tang 2008), at least during the first stages of the development, therefore the decision on the objectives may not be so critical at the early stages. While developing the scheduling model, the implicit insights of the decision-maker become explicit and he/she can state the objectives in a clearer way.

14.6.4 Transforming Objectives into Constraints

In practice, some objectives can be safely transferred into constraints. Our experience is that this may apply to many penalty or cost (profit) -related objectives, as it may not be so easy to derive an accurate penalty/cost (profit) function. A typical example are due dates: some companies accept the violation of committed due dates, and recognise that this should imply a sort of penalty which is approximately related to the number of violations (service level) and to the difference between the promised due dates and the completion times (average waiting time). However, establishing such relation in a meaningful function that adequately weights them and can be easily calculated for all jobs and customers is not so easy. Therefore, a simple alternative is to establish a maximum tardiness allowed and/or a maximum fraction of jobs late. Since most production departments include objectives or indicators linked to service level and lead times, obtaining these maximum levels of allowance is straightforward in many situations.

14.7 Solution Procedures

A great effort in the scheduling field has been devoted to solution procedures, at least from the most theoretical side. A general problem is that most existing solution procedures have been tightly linked to the models, therefore generally resulting in algorithms with low performance outside the original models for which they were conceived, if applicable at all. Since no algorithm can outperform the rest for all scheduling models, building specific algorithms may be justified from a theoretical viewpoint, but it should be also clear that no scheduling system can (1) store the myriad of specific algorithms and (2) select the most suitable one among them for any scheduling decision. In addition, we have already discussed the relatively short life-cycle of scheduling models in view of the extremely dynamic nature of manufacturing. As a consequence, the advantages of designing specific algorithm in terms of quality of the solutions should be balanced against their cost of development and deployment. In addition, we note the following issues:

14.7.1 Focus on Feasibility

It is quite necessary to balance the effort in developing the solution procedures against their advantages. As mentioned before, in some cases the problem is so restricted that there are few feasible solutions (Graves 1981; Benavides and Carlos Prado 2002; Cowling 2003). In these cases, most of the computational effort may be wasted in checking that no better feasible solutions are available. In addition, in some cases, the objective function is rather flat as compared to 'classical' objective functions. Again, in this cases most of the computational effort is spent on confirming optimal values. This speaks for the need of balancing the effort in developing the solution procedures, particularly taking into account the aforementioned changing nature of the objectives sought.

14.7.2 Take into Account the Available Decision Time

The algorithms employed have to make the best use of the available running (decision) time (Cowling 2003), so there is a need of developing solution procedures whose quality of solutions is scaled with the decision interval. As the decision interval is very context specific (it may not only vary from company to company, but also within a single company, depending on the shift, workload, etc.), it would be extremely interesting to build algorithms that their performance is (roughly) linear with time (i.e. they do not stall after a number of iterations or require very long decision intervals to obtain acceptable solutions).

Note that the focus towards this type of algorithms also implies the need of re-assessing the way some computational experiences of scheduling algorithms are carried out. Usually, a new solution procedure for a specific scheduling problem is labeled as 'efficient' problem because it yields better solutions than existing ones when all of them are allowed the same CPU time. However, the relative performance of this new solution procedure for different CPU times is unknown.

14.7.3 Providing a Set of Alternative Solutions

Even in the best-defined industrial scheduling problems, some aspects cannot be captured by the objectives or the constraints of the problem, even if they remain important in decision-maker's eyes. Therefore, it is interesting providing him/her with a set of 'good' alternative solutions. This can be done by using the support of different algorithms for the problem (see e.g. Tang and Wang 2008), for which it is clear that the algorithms should be fast enough to fit within the decision interval.

An alternative option is to provide the scheduler with at least all solutions with equal objective values found by the algorithm. This would clearly help him/her to safely pick the one fitting best into these 'implicit goals', and at the same time, it would allow him/her to make these goals more explicit so they can later be incorporated into the system.

14.7.4 Transparency of Solution Procedures

An special effort has to be made to develop a transparent system who can be understood by the users (see Wiers 1997, 2001; Freed et al. 2007). It does not seem reasonable to believe that decision-makers with years of practice in an specific production shop floor would immediately accept any solution packaged into a scheduling system (a 'black-box' sequence, see Cowling 2003), particularly if it does not follows his/her intuition. Even if this issue could be fixed (at least in part) by a better training of the schedulers (see Chap. 13), the understanding of the logic behind the solution procedures, even at a rough level, will surely increase trust in the system and will foster its use.

As a consequence, solution procedures based on well-defined manufacturing principles may provide a more transparent system. Some of these principles could be the following:

- Focusing on bottleneck resources. Focusing on the bottleneck is a well-understood principle that helps managers to overcome the complexity of the factory and concentrate of the part of the shop floor that greatly impacts the performance of the system. The preprocessor functionality described in Chap. 11 may help identifying whether there is a distinct bottleneck in the shop.

- Aggregation of non critical resources. Many heuristic procedures are based on the principle of aggregating machines and/or jobs in order to apply a solution procedure that works well—or it is even optimal—for the resulting (aggregated) problem. When these machines/jobs are not critical, this principle is easy to be understood. Some of the ideas stated in Sect. 9.3 may apply here.
- Employing simple dispatching rules. For some highly constrained shops, there is little space for improvement, and simple dispatching rules may provide good solutions. Even if this is not the case, dispatching rules are well-known (and employed) by schedulers, so they can be used as a benchmark to assess the improvement of more sophisticated algorithms and to help schedulers in deciding whether this increasing complexity (and thus a loss of transparency) is worth the effort.
- Implement heuristics extracted from schedulers' experience. In researching what schedulers did (see MacCarthy and Wilson 2001), a number of qualitative heuristics and 'rule of thumb' have been encountered, such as scheduling certain type of job as the next job to be processed in a machine that has just been repaired. While there are several studies to assess whether these 'heuristics' are an asset or a liability (see e.g. McKay et al. 2000), a good strategy may be to implement in the system some of these procedures. This would not only provide the scheduler with a link between his/her reasoning and the system, but also would eventually help him/her to evaluate the rest of the algorithms provided by the system.
- Restricting moves to non critical jobs. This is another classical strategy, as it may help in marginal improvements while keeping the feasibility (or desired properties) of the initial solution.

In addition, such procedures generally require less data and are more robust to shop floor variability than their more complex counterparts. The widespread success of dispatching rules despite their relatively poor performance speaks for the need of carefully balancing the sophistication in the design of solution procedures against their understanding and acceptance by decision-makers. In this sense, it is perhaps interesting to draw the attention onto research on general frameworks to develop heuristics, such as the one by Pinedo and Yen (1997).

14.7.5 Design of Algorithms

Many approximate algorithms (i.e. metaheuristics) employ a number of parameters that must be tuned in order to accomplish a good performance. As discussed in Sect. 7.6, the usual procedure for tuning the algorithms is to collect a set of instances and try different values of their parameters (using a Design of Experiments) and pick the one yielding the best performance. While this procedure may not represent a big issue in a laboratory, it is somehow problematic in a real setting. First, such set of instances may not be available, or it may be outdated with respect, e.g. to the processing times of the jobs to be scheduled in the future. Second, the logic of these parameters may not be clear to the decision-maker and thus may be of little use, or misuse them (see previous section). Therefore, we believe that either algorithms with many parameters should be discarded, or the details or their tuning should be made transparent to the decision-maker. Clearly, this does not include the running time of the algorithm, as we have already discussed its importance.

14.7.6 Manipulation of Solutions

If we accept that no mathematical model can pretend capturing all possible scenarios and exceptions, it follows that we should allow the decision-maker to manipulate the solutions obtained by the different procedures, as there may be extraneous constraints and/or objectives which are not considered by the system and must be satisfied by the resulting solution (Bensana et al. 1986). Several approaches could be adopted. As seen in Sect. 12.4.1, one option would be to allow drag and drop the solutions on the Gantt chart. Another option, more complex, would be to allow partly to freeze jobs and/or partial schedules. This latter option is also very suitable for considering jobs that are being manufactured, and for reducing the nervousness of changes in the schedule, and it brings the interesting problem of the advantages of rescheduling/partly scheduling the non frozen jobs. Unfortunately, this problem has not been—to the best of our knowledge—addressed in the scheduling literature, but lessons from the application of production planning systems, such as MRP and MRPII could be learnt.

As it was mentioned in Chap. 11, the excessive manipulation of the solutions proposed by the system may result into worsening the original objective function. As a means to control and to balance this issue, in Chap. 11 it was proposed that the tool provides reports indicating the deviation of the so-modified schedules with respect to the ones suggested by the algorithms.

14.7.7 Using Already Available Technology

As more and more complex scheduling settings result in slow algorithms, possible solutions are parallelisation and cooperative computing. By using GRIDS or in a much simpler level, the available multiple CPU cores of modern computers, results of very high quality can be obtained in a fraction of CPU time.

Similarly, very complex scheduling problems can be solved in fast CPU times. Research in this field is rather scarce as compared on the large body of research on 'sequential' scheduling. Although not as elegant as much of the scheduling research, this field can greatly help in improving the performance of existing algorithms.

14.8 Constraints

The determination and satisfaction of a large variety of constraints is considered to be the crux of scheduling (Fox and Smith 1984). The number of constraints that can be identified in the shop floor can be extremely high (see e.g. the survey on scheduling practices for 40 schedulers by McKay et al. (1988) where more than 200 constraints are identified). Obviously, not all of them are relevant for scheduling.

While scheduling literature is rich in dealing with certain types of constraints, there are constraints that are so common in real environments that it is surprising that most existing models do not deal with them. Most of them have been already discussed in Chap. 4, so what follows is a recap of these constraints grouped in different categories:

- Machine constraints. Apart from specific machine constraints, most shops are characterised by the non availability of machines for all purposes. On one hand, not all machines may be amenable to process all jobs (or even if this is the case from a technological viewpoint, there may be economic reasons preventing that). Therefore, there is a problem of machine eligibility. On the other hand, in most cases, eligible machines are not available during all planning period. This is not only motivated by breakdowns or planned maintenance, but also because machines are busy processing already scheduled jobs. However, existing research on this topic is merely anecdotal.
- Staff constraints. Staff is almost an ignored resource in many scheduling models (see in this regard the results of the survey by Laforge and Craighead 2000). However, it is of utmost importance for many real shops. Even in automatic or semi-automatic processes, staff plays an important role in set ups times. Existing literature on scheduling with additional resources is rather scarce, there are some studies for parallel machines like the one by Chen (2006) but not for complex scheduling problems.
- Routing constraints. Many shop floors require some type of job reentrance and/or precedence constraints. The reasons vary from pure process-related to quality (including scraps and the need for reworking). The complexity of this type of resources (usually affecting a small percentage of the jobs in the system) should be balanced against the effort of developing a system that explicitly takes them into account.
- Transportation constraints. Transportation times in between stages and the control of the AGVs is also necessary for many systems.
- Capacity constraints. Storage capacity (number of boxes) in between each stage is also a key concern as in some cases this is a limited resource.

14.9 Conclusions and Further Readings

In this chapter, we have focused on an area that has received relatively little attention in the scheduling field, at least as compared to the wealth of literature dealing with scheduling models and procedures. However, in order to put these models and procedures into practice, they must adequately capture the details of a company's scheduling process, which usually entails modelling scenarios, constraints and objectives not often treated in the literature. In addition, the resulting models and solution procedures should be embedded into a scheduling system that feeds them with reliable and timely data and provides an efficient way to interact with the users. This gives rise to the following interrelated questions: which are the components of a scheduling system? and how a scheduling system can be developed and deployed? An answer to the first question has been given in the previous chapters, and regarding the second question, we propose a number of guidelines to help driving the design and implementation of scheduling systems based on the previous analysis of the literature on case studies and on our own experience.

As mentioned in the beginning of the chapter, we do not claim that our guidelines are the only way (or even the best way) to address the aforementioned issues, although they have been validated both through literature analysis and through industrial practice. By presenting them, we hope to enrich the rather scarce literature on the topic and thus help bridging the gap between the development of scheduling models and procedures, and their implementation in a real-life industrial setting.

This chapter is heavily based on the work by Framinan and Ruiz (2012), who in turn review the—rather scarce—literature discussing implementation cases. Among these, a number of interesting ideas can be found in Freed et al. (2007) regarding modeling the layout. The study by Berglund and Karltun (2007) includes valuable insights on the quality of the data required by a manufacturing scheduling system. Cowling (2003) discusses some issues regarding the objectives to be considered in the system. Finally, contributions on solution procedures are present in several works (see e.g. MacCarthy and Wilson 2001; McKay et al. 2000).

References

Benavides, J. and Carlos Prado, J. (2002). Creating an expert system for detailed scheduling. *Int. J. Oper. Prod. Manage.*, 22(7–8):806–819.

Bensana, E., Correge, M., Bel, G., and Dubois, D. (1986). An expert-system approach to industrial job-shop scheduling. In *Robotics and Automation. Proceedings. 1986 IEEE International Conference on*, volume 3, pages 1645–1650.

Berglund, M. and Karltun, J. (2007). Human, technological and organizational aspects influencing the production scheduling process. *International Journal of Production Economics*, 110 (1–2):160–174.

Bhattacharyya, S. and Koehler, G. (1998). Learning by objectives for adaptive shop-floor scheduling. *Decis. Sci.*, 29(2):347–375.

Chen, J.-F. (2006). Unrelated parallel machine scheduling with secondary resource constraints. *International Journal of Advanced Manufacturing Technology*, 26(3):285–292.

Cowling, P. (2003). A flexible decision support system for steel hot rolling mill scheduling. *Computers and Industrial Engineering*, 45:307–321.

Fox, M. S. and Smith, S. (1984). Isis: A knowledge-based system for factory scheduling. *Expert Systems Journal*, 1(1).

Framinan, J. and Ruiz, R. (2012). Guidelines for the deployment and implementation of manufacturing scheduling systems. *International Journal of Production Research*, 50(7):1799–1812.

Freed, T., Doerr, K., and Chang, T. (2007). In-house development of scheduling decision support systems: Case study for scheduling semiconductor device test operations. *Int J Prod Res*, 45(21):5075–5093.

Gao, C. and Tang, L. (2008). A decision support system for color-coating line in steel industry. In *Proceedings of the IEEE International Conference on Automation and Logistics, ICAL 2008*, pages 1463–1468.

Graves, S. C. (1981). A review of production scheduling. *Operations Research*, 29(4):646–675.

Higgins, P. G. (1996). Interaction in hybrid intelligent scheduling. *International Journal of Human Factors in Manufacturing*, 6(3):185–203.

Kathawala, Y. and Allen, W. (1993). Expert systems and job shop scheduling. *International Journal of Operations & Production Management*, 13(2):23–35.

Kerr, R. M. (1992). Expert systems in production scheduling: Lessons from a failed implementation. *Journal of Systems and Software*, 19(2):123–130.

Knolmayer, G., Mertens, P., and Zeier, A. (2002). *Supply chain management based on SAP systems*. Springer.

Laforge, R. and Craighead, C. (2000). Computer-based scheduling in manufacturing firms: Some indicators of successful practice. *Production and Inventory Management Journal*, 41(1):29–34.

Leachman, R., R.F., B., Liu, C., and Raar, D. (1996). Impress: An automated production-planning and delivery-quotation system at harris corporation- semiconductor sector. *Interfaces*, 26:6–37.

MacCarthy, B. L. and Wilson, J. R., editors (2001). *Human performance in Planning and Scheduling*. Taylor & Francis.

McKay, K. N., Morton, T., Ramnath, P., and Wang, J. (2000). 'aversion dynamics' scheduling when the system changes. *J. Scheduling*, 3(2):71–88.

McKay, K. N., Safayeni, F., and Buzacott, J. (1988). Job-shop scheduling theory: What is relevant? *Interfaces*, 18(4):84–90.

McKay, K. N. and Black, G. W. (2007). The evolution of a production planning system: A 10-year case study. *Computers in Industry*, 58(8–9):756–771.

Missbauer, H., Hauber, W., and Stadler, W. (2009). A scheduling system for the steelmaking-continuous casting process. a case study from the steel-making industry. *International Journal of Production Research*, 47(15):4147–4172.

Pinedo, M. L. and Yen, B. P.-C. (1997). On the design and development of object-oriented scheduling systems. *Annals of Operations Research*, 70(1):359–378.

Storer, R., Wu, S., and Vaccari, R. (1992). New search spaces for sequencing problems with application to job shop scheduling. *Management Science*, 38:1495–1509.

Tang, L. and Wang, G. (2008). Decision support system for the batching problems of steelmaking and continuous-casting production. *Omega*, 36(6):976–991.

Wiers, V. (1997). *Human-Computer Interaction in Production Scheduling: Analysis and Design of Decision Support Systems for Production Scheduling Tasks*. Technische Universitiet Eindhoven.

Wiers, V. (2001). Design of knowledge-based scheduling system for a sheet material manufacturer. *Human Performance in Planning and Scheduling*, pages 201–215.

Wiers, V. (2002). A case study on the integration of aps and erp in a steel processing plant. *Production Planning and Control*, 13(6):552–560.

Chapter 15
A Case Study: Ceramic Tile Production

15.1 Introduction

In this chapter, a real case study dealing with the development and implementation of a manufacturing scheduling system is presented in detail. The goal is twofold. On one hand, we try to illustrate the application of some of the concepts described in earlier chapters, most notably those relating to the architecture and design of a scheduling tool and its deployment in an organisation. By doing so, we hope to 'close the circle' of the book and show how the ideas discussed in the previous chapters can be implemented in a real-life setting.

On the other hand, it has to be mentioned that despite the enormous size of the scheduling literature, papers dealing with real problems like the one depicted in this case study are much more scarce. Although the so-called 'gap' between scheduling theory recognised by many authors has been already commented in other parts of this book, it is interesting to recall it at this point. Graves (1981) already hinted at some solutions to bridge this gap. The famous studies about the usage of operations research techniques in production management by Ledbetter and Cox (1977) and by Ford et al. (1987) showed that there was very little application of scheduling theory in practice. Similar conclusions can be seen in McKay et al. (1988) and in Olhager and Rapp (1995). There are even papers where this gap is directly studied like Dudek et al. (1992), MacCarthy and Liu (1993) and McKay et al. (2002). Quantitative studies about scheduling research carried out by Reisman et al. (1997), showed that from a total of 184 reviewed papers, only 5 (less than a 3 %) dealt with realistic production settings. The conclusion is that when modeling realistic production scheduling problems, there is little hope to find the specific problem at hand being already studied in the scientific research. More recent reviews of hybrid flows hops, like those of Ribas et al. (2010) and Ruiz and Vázquez-Rodríguez (2010), also reach the same conclusion: realistic settings are seldom studied. Therefore, we feel that the book would be somehow uncomplete without adding at least one case study to the literature.

J. M. Framinan et al., *Manufacturing Scheduling Systems*,
DOI: 10.1007/978-1-4471-6272-8_15, © Springer-Verlag London 2014

This case study deals with a specific scheduling system development and deployment experience in the Spanish ceramic tile sector. Before entering into the details of the scheduling system developed, a brief description of the industrial problem and scheduling model is given. We first describe the ceramic tile production process. A model of the production process is also given. After this, we describe in detail the development of a scheduling system, which mimics most of the content that has been exposed throughout this book. For example, we provide emphasis on a crucial aspect of the development, which is the data gathering process. Finally, and after exposing the developed scheduling system, some open issues are discussed.

15.2 The Production Process of Ceramic Tiles

Ceramic tile production comprehends many different processes according to each different finished product. Basically, the process is composed of the following stages:

- Clay preparation, either by dry grinding or by wet milling and atomization.
- Forming or molding of the tile by either dry pressing or by extrusion.
- Glaze preparation.
- Drying, glazing and decoration of the tile.
- Kiln firing.
- Classification and packing.

A more detailed schematics of the process is pictured as a flowchart in Fig. 15.1.

The raw materials preparation, specially by the wet milling process (including spray drying) is usually a separated process, often simply called 'atomization' and is carried out by third companies, referred to as 'atomizers'. The resulting product comes in the form of a clay granulate of tiny hollow spheres, usually referred to as atomized clay. This clay behaves much like a fluid as it is formed from countless tiny spheres that confer a fluid behavior to the otherwise hard to work with raw clay.

There is another process that is often subcontracted to third companies and that is the glaze preparation. Glazes are made of grinded frits and pigments.

From all the processes depicted in Fig. 15.1, single kiln firing (also called fast single firing) is, by far, the most common ceramic tile production method. As a result, the previous production process of Fig. 15.1 can be simplified as shown in Fig. 15.2.

Here we concentrate only on the production of ceramic tiles, bypassing the production processes of the aforementioned raw materials like atomized clay, frits, pigments or glazes. In other words, we assume that all these raw materials are readily available. As a result, and as it is common in most ceramic tile production plants, the process starts with the molding or forming of the ceramic tile. From this point, the production of ceramic tiles is very much automated.

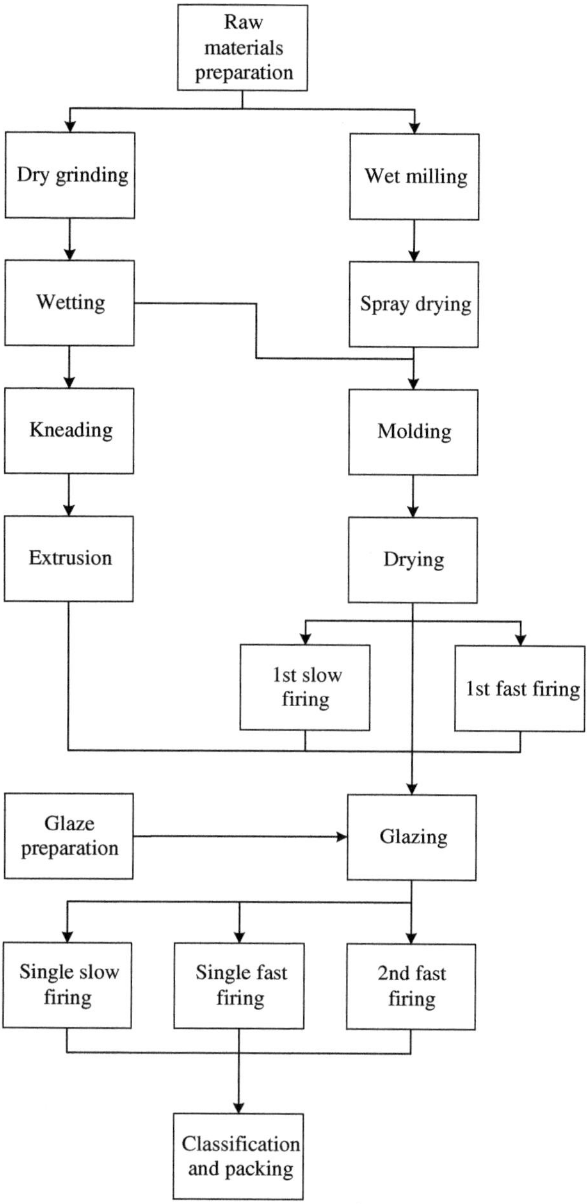

Fig. 15.1 Detailed flowchart of the production process of ceramic tiles with all variants

The fist step is the molding where the atomized clay powder is loaded into molds (here is where the fluidity of the atomized clay plays an important role) and it is pressed by large hydraulic presses to form what is called the bisque. This process is

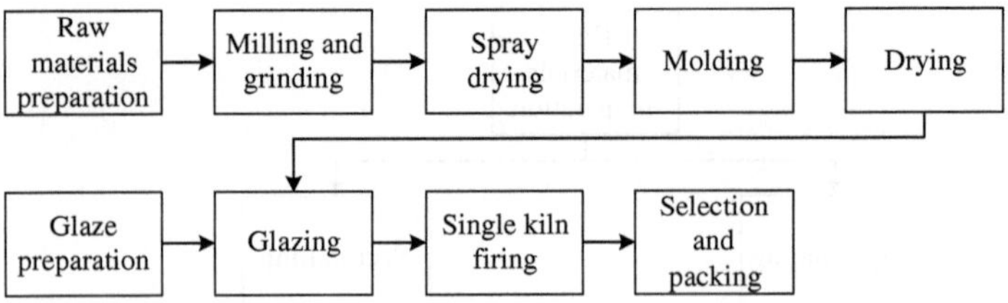

Fig. 15.2 Wet milling, atomization and single kiln firing flowchart: the most common ceramic tile production procedure

very important as all further operations depend on the correct formation of the tile. The physical dimensions of the tile, often referred to as the format of the tile, depend on the characteristics of the mold employed. Additionally, the visible side of the tile might not be entirely flat as it might have a 'stone-like' pattern. This is determined by the plate of the mold.

Right after molding, the bisques are quite fragile and should undergo further drying. This process is usually carried out in either vertical or horizontal dryers (small kilns). After drying, the bisque is hardened to what resembles a chocolate bar and therefore, can withhold further processing operations. The bisque or ceramic tile is then decorated in the glazing lines. Some specific products are not glazed, however, this is not very common. In a glazing line, tiles move over conveyor belts and different applications, decorations and other operations are performed over them. The most basic operation is the application of the base glaze by simple bell or waterfall machines to advanced random rotary decorations, referred to as rotocolors. Modern glazing equipment includes 'printers' which can basically print ceramic tiles with a large array of colours, patterns and decorations. Glazing lines can contain more than 50 different machines and can be up to 50 m long, sometimes even more. The different glaze applications determine the style, colour, surface and characteristics of the tile.

Decorated bisques are stored in big boxes or 'roller stillages' that contain many roller planes where bisques are fed into, stored and extracted from at later stages. The big boxes are transported from the glazing lines to an intermediate storage area by automated guided vehicles (AGVs).

The central stage of the production process is the single kiln firing. Ceramic kilns are large horizontal ovens with usually a single plane of rollers that move in one direction, advancing the glazed bisques through the different stages. Each type of clay and tile follows a different temperature curve inside the kiln. This temperature might be also affected by the physical size or format of the tile. A number of physical and chemical reactions undergo inside the kiln to form a hardened and finished tile. Most notably, the glaze fuses and hardens on top of the tile, conferring most known and desired characteristics of ceramic tiles.

After firing, AGVs transport again the fired tiles to an intermediate storage area. The final stage is the classification and packing of the fired tiles. Several classification and packing lines are usually available. Visual or automated quality control for defects is carried out. Depending on the number and importance of defects, tiles are usually classified into first or second quality and discards. Due to small differences in the raw materials and the complex production process, not all tiles result in the same tonality and physical dimensions (calibers) so the tiles are actually assigned to different batches, each one containing the same quality, caliber and tonality. As a final step, tiles are boxed and palletised for distribution.

Nowadays, there are many other optional production steps for ceramic tiles, for example, tiles might undergo one or more polishing steps to confer tiles with a stone or mirror finish. Other process include the rectification of the edges of the tile, or cutting processes to produce special pieces like edges or reliefs. Additional detailed decorations and further firing are also possible. As a result, the common production process depicted in Fig. 15.2 can only be regarded as a simplification. In any case, the bulk of the ceramic tile production undergoes this main production process.

In modern production plants, from start to finish, the product is rarely touched and very little human intervention is needed. This results in the mass production of ceramic tiles at affordable prices.

Given such high automatisation in the sector, companies have heavily invested in machinery to achieve maximum efficiency under this mass production context. The most expensive machinery in the ceramic tile production is the kiln. Small tile producers typically have 2–3 kilns, whereas large producers might have several production plants with several kilns each. Therefore, in order to minimise the unit costs due to machine depreciation, machinery usage should be maximised. Again, it is typical for tile producers to operate the kilns in a 24/7 schedule and stopping them only once per year for maintenance. This naturally leads to the minimisation of the makespan, which is the main objective chosen for the models.

15.3 Modeling the Ceramic Tile Production Process

15.3.1 Initial Model

The previous section detailing the ceramic tile production process, together with more detailed studies about ceramic tile production, as the one shown in Vallada et al. (2005) allow us to provide a precise modeling of the scheduling problem arising in ceramic tile manufacturing. The first observable feature of the ceramic tile production problem (CTPP from now on) is that all products follow a natural flow through the plant, which suggests a flow shop structure. Production plants are divided into sections, usually, molding, glazing, kiln and classification. Furthermore, even Small and Medium Enterprises (SMEs) have several glazing lines and typically, two or more kilns. As a result, we have a hybrid flow shop (HFS). A HFS, like already commented in Chap. 3, is a combination of two well known scheduling problems, namely the flow shop and the parallel machine problem. The HFS problem has been an active topic of research and many literature reviews have been published for it, like the ones of Linn and Zhang (1999) and Vignier et al. (1999). More recent related reviews include those of Wang (2005), Ruiz and Maroto (2006), Quadt and Kuhn (2007), Ribas et al. (2010) and Ruiz and Vázquez-Rodríguez (2010).

Let us recall that in a flow shop, there is a set N of n jobs that have to be processed on a set M of m machines. Each job visits each machine in the same order, which can be, without loss of generality, machines $1, \ldots, m$. As a result, each job is composed of m tasks. Common constraints apply, i.e. no machine might process more than one job at the same time and the tasks of the same job cannot overlap. The only input data needed to solve the problem is n, m and the processing times of each job on each machine, commonly referred to as p_{ij}, $i \in M, j \in N$ (more details can be consulted in Chap. 3, Sect. 3.2.3.3).

Conversely, in a parallel machine problem, the set of m machines are not arranged in series, but in parallel. Each job has to be processed on exactly one out of the m machines. The most general case is when the m parallel machines are unrelated, meaning that the processing time for a job depends on the machine it has been assigned to.

Given the two previous definitions, in a HFS we also have a set N of jobs, $N = \{1, \ldots, n\}$. However, each job has to visit not a set of machines but rather a set of M stages, $M = \{1, \ldots, m\}$. At every stage i, $i \in M$ there is a set $M_i = \{1, \ldots, m_i\}$ of parallel machines that can process the jobs where there is at least one stage with more than one machine, i.e. $\exists i \backslash |M_i| > 1, i \in M$. Every job has to pass through all stages and must be processed by exactly one machine at every stage. In the following, p_{ilj} indicates the processing time of job j, $j \in N$ at machine l, $l \in M_i$ inside stage i. Again, for more details on the HFS, the reader is referred to Chap. 3, Sect. 3.2.3.6.

With the above definitions, we can already model the different molding lines, glazing lines, kilns and classification lines in the CTPP. However, such modeling is extremely simplistic due to one important missing feature. In the CTPP, there are significant change over times in all stages, these times, frequently referred to in the scheduling literature as setup times are for example, very important in the molds.

When the production of a given ceramic tile with a format of 33×33 cm. is finished, a very large setup time is needed if the next batch of tiles to produce in that same mold has a different format, like for example 60×60 cm. Large portions of the molding machine have to be disassembled, several adjustments must be done and all this usually takes up to a full working shift. Depending on the batch size of tiles to produce, the setup time might be even larger than the processing time itself. Glazing lines are another clear example, even if two batches of tiles with the same format are being produced one after another, a slight difference in the decorations or colours for each batch might result in a full working shift for cleaning and adjusting the glazing line. As a matter of fact, mostly all stages in the CTPP have strong setup times. What is more, these setup times depend on the production sequence and on each machine since some machines (even within the same stage) are more easily reconfigurable than others. These types of setups are referred to as sequence dependent setup times as it was already indicated in Chap. 4. The scheduling literature considering setups is rich, albeit once again it lacks applications to real production problems. Reviews of scheduling research with setups are available from Allahverdi et al. (1999) and

Allahverdi et al. (2008). In the context of the CTPP, we have a machine- based sequence dependent setup time in every machine l at stage i when processing job k, $k \in N$, after having processed job j which is noted as S_{iljk}. Setup times were explained in Chap. 4, Sect. 4.2.2.

Another common situation found in the ceramic tile producers is that not every possible job—batch of a given quantity (in m^2) of a certain type of ceramic tile from the company's catalog—can be processed by every available parallel machine at a given stage. Once again, a clear example comes from the molding presses. Large tile formats like 60×120 cm. often require special molding presses with pressing forces in excess of 45,000 kN. As a result, not every possible job might be processed on every possible machine. This is known in the HFS problem as machine eligibility, also commented in Chap. 4. For every job j and stage i, we have a set E_{ij} of eligible machines that can process job j. Obviously, at least one machine must be eligible at each stage for all jobs, i.e. $1 \leq |E_{ij}| \leq m_i, \forall i \in M, j \in N$. Note also that p_{ilj} can be considered 0 if $l \notin E_{ij}$. Basically, the actual value of the processing time for a non-eligible machine is irrelevant. As with setups, machine eligibility was introduced in Chap. 4, Sect. 4.2.3.

The previous model, referred to as HFS with sequence dependent setup times and machine eligibility is obviously a very hard combinatorial problem. Employing the usual three field notation $\alpha|\beta|\gamma$ for scheduling problems commented in Chap. 3 and the extension for HFSs proposed by Vignier et al. (1999) (also commented in the same chapter, see Sect. 3.3.2), this initial formulation of the CTPP is noted as $FHm, ((RM^{(i)})_{i=1}^{(m)})|S_{sd}, M_j|C_{\max}$ if the optimisation objective is to minimise the maximum completion time of the jobs or makespan ($C_{\max}$). This problem was first studied by Ruiz and Maroto (2002) and later by Ruiz and Maroto (2006). We will refer to this first model as CTPP_Model_1. The details regarding the development and deployment of the scheduling system suitable for this first model will be given in the following sections.

15.3.2 A More Detailed Model

CTPP_Model_1, although probably containing more realistic features than mainstream literature, still does not capture many different situations and constraints arising in the CTPP. In order to deal with real problems, a much more comprehensive model is needed.

As commented before, there are some optional stages, like tile polishing, additional decorations and so on. This results in a production plant with many stages and not every job has to visit every stage. Therefore, stage skipping appears. In a general way, we define by F_j the set of stages that job j visits, $1 \leq |F_j| \leq m$. Stage skipping in the regular flow shop literature results in what is known as flexible flow lines. In the context of HFSs, this new problem with stage skipping is referred to as hybrid flexible flow line problem or HFFL. Note that if a given job j skips a stage i, ($i \notin F_j$) then $|E_{ij}| = 0$ and $p_{ilj} = 0, \forall l \in M_i$ (or again, these values are of no relevance).

Apart from the processing times, there is more information associated with a job, like the release date of job j, r_j. The first operation at the first visited stage of job j cannot start before r_j. Also we refer to d_j as the due date of job j. The last operation of job j must be preferably completed before d_j. A given job might also have an importance or weight, referred to as w_j.

Production floors are seldom found empty. When doing real scheduling, machines are at a given moment processing materials. This means that new schedules must start after existing jobs are finished. Therefore, rm_{il} represents the release date of machine l inside stage i. This date indicates when the machine is released from previous work and can start processing new jobs. No operation of any job of the new schedule might start on machine l before rm_{il}.

In the previous initial modeling attempt, we already defined sequence dependent setup times. For this enhanced model, we also consider the possibility of defining each setup as anticipatory or non-anticipatory. As the name implies, an anticipatory setup time, for example a cleaning, can be done in the corresponding machine as soon as the machine is free without actually waiting for the next job in the sequence to arrive. However, some setups on machines might require the jobs to be already present on the machine (i.e. fixing the product to be processed on the machine) and are therefore non-anticipatory. As a result, if $A_{iljk} = 0$ then the corresponding setup S_{iljk} is non-anticipatory and job k should be already on machine l for setting up. It should be noted that $S_{iljk} = 0$ if $j = k$ and that $S_{iljk} = 0, \forall j, k \in N, l \in M_i$ if one or more of the following conditions are satisfied: $i \notin F_j$, $i \notin F_k$, $l \notin E_{ij}$, $l \notin E_{ik}$ or $j \in P_k$. The value of A_{iljk} in these cases is irrelevant and will be assumed 0.

Another common assumption in scheduling theory is to assume that the n jobs are independent. While this might be acceptable in some situations, it is not for the CTPP. Many times some batches of ceramic tiles have to be finished prior to other batches since they are used as raw materials. Mosaics, stair steps, tile edges and many other special tiles are clear examples. In short, the CTPP contains precedence constraints. As such, P_j contains the set of predecessors of job j. Jobs may have zero or more predecessors indicating that the first operation of job j should not start until all the last operations of jobs in P_j have finished. All sets P_j define a precedence graph among jobs which is directed and acyclic.

A job, defined before as a batch of tiles does not need to be fully glazed before entering the kiln. A large production batch of tiles might need several days of glazing and several days of kiln firing. Batches are made out of thousands of tiles. This means that when a sufficient quantity of tiles have been glazed, these can already enter the kiln while the remaining lot is being glazed. This effectively represents an overlap between operations. In some other situations, we have a waiting time in between stages for a given batch. Some special glazes require an 'air drying' time before entering the kiln. As a result, we might have overlaps and/or waiting times. In order to model this situation, we denote by lag_{ilj} the time lag between the end of the processing of job j at machine l inside stage i and the beginning in the next stage. An overlap is represented by a negative lag ($lag_{ilj} < 0$) whereas a waiting time is given by a positive lag ($lag_{ilj} > 0$). Note that for any given job j, $lag_{ilj} = 0, \forall l \in M_i$, if $i \notin F_j$ or if for a given $l \in M_i$, $l \notin E_{ij}$.

Table 15.1 Eligibility (E_{ij}) of jobs on machines for the example

	i	1	2	3
j	1	{1, 2}	{2}	{1}
	2	{1, 2}	{1, 2}	–
	3	–	{1}	{1}
	4	{2}	–	{1}
	5	{1, 2}	{1, 2}	{1}

The joint consideration of lags with the previous formulation makes the problem specially tricky. For example, the input data has to satisfy some obvious relations, like for example, if i is the last stage for job j, then $lag_{ilj} = 0, \forall l \in M_i$. Overlaps can also be problematic since a given overlap has to be less than or equal to the processing time, i.e. $|lag_{ilj}| \leq p_{ilj}$ otherwise we could have successive operations starting before the starting time of preceding operations. Similarly, $|lag_{ilj}| \leq p_{i+1,l',j}, \forall l' \in E_{i+1,j}$, i.e. the overlap should be smaller than or equal to the processing time on any machine at the next stage, otherwise successive operations might finish before preceding operations. While these relations are straightforward, in huge input datasets, unless these relations are checked, strange racing conditions might arise in solution methods.

This last model is much more comprehensive and captures much of the existing constraints and situations arising in the CTPP. The full problem can be denoted as $HFFLm, ((RM^{(i)})_{i=1}^{(m)})/rm, lag, S_{ijk}, M_j, prec/C_{\max}$ and is obviously a very hard scheduling problem to solve. This problem was first studied from a mathematical programming and heuristic perspective by Ruiz et al. (2008). In this work it was shown that modern solvers are able to solve only small instances of up to a dozen jobs at most. Real problems have at least ten times as many jobs.

The following example taken from Ruiz et al. (2008) will help in illustrating this complex HFFL problem. We have an instance with five jobs and three stages with two machines in the first two stages and only one machine in the third. Therefore $n = 5$, $m = 3$, $m_1 = m_2 = 2$ and $m_3 = 1$. Jobs 1 and 5 visit all three stages whereas job 2 skips stage 3, job 3 skips stage 1 and job 4 skips stage 2. So we have that $F_1 = F_5 = \{1, 2, 3\}$, $F_2 = \{1, 2\}$, $F_3 = \{2, 3\}$ and $F_4 = \{1, 3\}$. Furthermore, jobs 4 and 5 are preceded by jobs 2 and 3, i.e. $P_4 = P_5 = \{2, 3\}$. Table 15.1 gives the eligibility for the jobs on the machines. Table 15.2 shows the processing times of jobs and the release dates for machines.

Tables 15.3 and 15.4 show the lag values and sequence dependent setup times, respectively. Table 15.4 shows the setup time and, between parentheses, the anticipatory value for every machine at each stage are given. Note that '–' means that both the setup time and the anticipatory value are 0.

Now, the example consists on a permutation of jobs $\pi = \{2, 1, 3, 4, 5\}$ for which we want to draw the sequence. The assignment of jobs to machines at each stage, is as follows:

Table 15.2 Processing times (p_{ilj}) and release dates for machines (rm_{il}) for the example

	i	1		2		3
	l	1	2	1	2	1
	rm_{il}	4	3	8	16	23
	p_{ilj}					
j	1	10	15	0	8	6
	2	6	9	11	4	0
	3	0	0	9	0	8
	4	0	10	0	0	6
	5	11	14	6	12	3

Table 15.3 Lag times (lag_{ilj}) of jobs for the example

	i	1		2	
	l	1	2	1	2
j	1	10	2	0	−4
	2	2	−2	0	0
	3	0	0	3	0
	4	0	1	0	0
	5	−5	−6	−3	8

$$\begin{pmatrix} 2 & 1 & 0 & 2 & 1 \\ 1 & 2 & 1 & 0 & 2 \\ 0 & 1 & 1 & 1 & 1 \end{pmatrix}$$

Each column in the previous table represents the machines assigned at each stage to the job at the corresponding position in the permutation. It should be clear that π is feasible from the perspective of the precedence constraints among jobs and that the assignment of jobs to machines at each stage is also feasible with respect to the eligibility constraints. The resulting Gantt chart is shown in Fig. 15.3.

It can be observed that jobs 4 and 5 must wait for their predecessors before starting. Also, sometimes the lag times allow for an overlap of the operations whereas in other situations these lags force a waiting period between successive operations of the same job. A closer look at job 4 reveals that, even with a missing stage 2, the lag still holds for the third stage. Some setups are anticipatory (see e.g. the setup between jobs 1 and 5 on machine 1 in the first stage), whereas other setups are non-anticipatory (see e.g. the setup between jobs 3 and 4 on machine 1 in the last stage). This means that the setup between jobs 1 and 5 in the first stage, for example, could be started anytime between 14 and 44, whereas the setup between jobs 3 and 4 in the last stage can start only after job 4 arrives at that stage (and, in this case, after lag_{124} is also considered).

Table 15.4 Sequence dependent setup times and anticipatory flags (S_{iljk} and A_{iljk}) for the example

$i=1$						$i=2$					$i=3$					
	k					k						k				
j	1	2	3	4	5	1	2	3	4	5	j	1	2	3	4	5
1	–	3(1), 6(1)	–	–, 8(1)	4(1), 2(1)	–	–, 6(1)	–	–	–, 6(1)	1	–	–	6(1)	3(1)	9(1)
2	4(0), 5(0)	–	–	–, 6(1)	1(1), 4(1)	–, 5(1)	–	6(0), –	–	4(1), 2(0)	2	–	–	–	–	–
3	–	–	–	–	–	–	8(0), –	–	–	5(1), –	3	4(0)	–	–	1(0)	8(1)
4	–, 8(0)	–	–	–	–, 2(1)	–	–	–	–	–	4	5(0)	–	–	–	2(1)
5	6(0), 10(0)	–	–	–, 4(0)	–	–, 4(1)	–	–	–	–	5	2(0)	–	–	6(0)	–

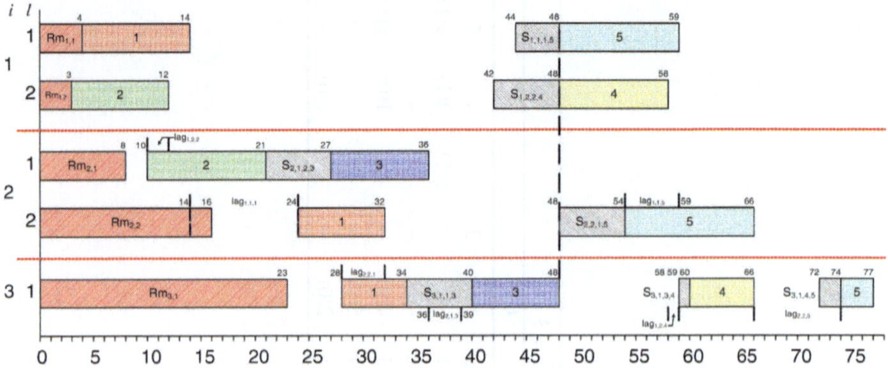

Fig. 15.3 Gantt chart for the example problem. $C_{\max} = 77$

Note that calculating the $C_{\max}$ value for this problem is rather costly. Let us define the expressions that are needed for such calculation. First, some additional notation is needed:

- G_i is the set of jobs that visit stage i, ($G_i \subseteq N$ and $G_i = \{j | i \in F_j\}$),
- $G_{il} \subseteq G_i$ is the set of jobs that can be processed on machine l inside stage i, i.e. $G_{il} = \{j | i \in F_j \wedge l \in E_{ij}\}$,
- S_k gives the complete and unchained set of successors of job k, i.e. $S_k = \{j | k \in P_j\}$
- FS_k (LS_k) is the first (last) stage that job k visits.

We also have that the machine assigned to job j at stage i is denoted by T_{ij} or by l in brief. The previous job that was processed by machine l is denoted by $k(l)$. Let stage $i-1$ be the previous stage visited by job j before stage i (note that is the previously visited stage, which does not have to be the previous stage in the shop, which might not be visited by job j). Similarly, stage $i+1$ is the next stage to be visited.

With all previous notations, we can calculate the completion times for job j at all visited stages with the following expressions:

$$\begin{aligned} C_{FS_j,j} = \max\{&rm_{FS_j,l}; \max_{p \in P_j} C_{LS_p,p}; C_{FS_j,k(l)} + A_{FS_j,l,k(l),j} \cdot S_{FS_j,l,k(l),j}\} \\ &+(1 - A_{FS_j,l,k(l),j}) \cdot S_{FS_j,l,k(l),j} + p_{FS_j,l,j}, \qquad j \in N \end{aligned} \tag{15.1}$$

$$\begin{aligned} C_{ij} = \max\{&rm_{il}; C_{i,k(l)} + A_{i,l,k(l),j} \cdot S_{i,l,k(l),j}; C_{i-1,j} + lag_{i-1,T_{i-1,j},j}\} \\ &+(1 - A_{i,l,k(l),j}) \cdot S_{i,l,k(l),j} + p_{ilj}, \qquad j \in N, i > FS_j \end{aligned} \tag{15.2}$$

We have assumed that $A_{i,l,k(l),j} = S_{i,l,k(l),j} = 0$ for $i \notin F_j$ or $i \in F_j$ but $l \notin E_{ij}$ and $A_{i,l,k(l),j} = S_{i,l,k(l),j} = C_{i,k(l)} = 0$ when no preceding job $k(l)$ exists.

The previous calculations should be made job-by-job to obtain the completion times of all tasks. For each job, the completion time for the first stage is calculated with Eq. (15.1), considering availability of the machine, completion times of the predecessors, setup and its own processing time. For the other stages Eq. (15.2) is applied, considering availability of the machine, availability of the job (including lag), setup and its processing time.

If job j is assigned to machine l inside stage i, the time at which machine l completes job j is denoted as L_{ilj}. Following our notation, $L_{ilj} = C_{ij}$ given $T_{ij} = l$. Furthermore, we refer to the job visiting stage i after job j as job q and to an eligible machine at the next stage for job j as $l' \in E_{i+1,j}$.

Suppose now that we are scheduling job j in stage i, $i \in F_j$. We have to consider all machines $l \in E_{ij}$ for assignment.

Obviously, now the makespan is easily calculated:

$$C_{\max} = \max\{C_{ij}\}, \qquad j \in N, i \in F_j \tag{15.3}$$

Note that as stated, for complex problems, the calculation of the objective is no longer straightforward.

For the sake of completeness, the mixed integer linear programming (MILP) model is described below (taken also from Ruiz et al. 2008). We first define the decision variables:

$$\begin{aligned} X_{iljk} &= \begin{cases} 1, & \text{if job } j \text{ precedes job } k \text{ on machine } l \text{ at stage } i \\ 0, & \text{otherwise} \end{cases} \\ C_{ij} &= \text{Completion time of job } j \text{ at stage } i \\ C_{\max} &= \text{Maximum completion time or makespan} \end{aligned}$$

The objective is the minimisation of the makespan:

$$\min C_{\max} \tag{15.4}$$

The constraints needed are the following:

$$\sum_{\substack{j \in \{G_i, 0\} \\ j \neq k, j \notin S_k}} \sum_{l \in E_{ij} \cap E_{ik}} X_{iljk} = 1, \qquad k \in N,\ i \in F_k \tag{15.5}$$

$$\sum_{\substack{j \in G_i \\ j \neq k, j \notin P_k}} \sum_{l \in E_{ij} \cap E_{ik}} X_{ilkj} \leq 1, \qquad k \in N,\ i \in F_k \tag{15.6}$$

$$\sum_{\substack{h \in \{G_{il}, 0\} \\ h \neq k, h \neq j \\ h \notin S_j}} X_{ilhj} \geq X_{iljk}, \quad j, k \in N,\ j \neq k,\ j \notin S_k,\ i \in F_j \cap F_k,\ l \in E_{ij} \cap E_{ik} \tag{15.7}$$

$$\sum_{l \in E_{ij} \cap E_{ik}} (X_{iljk} + X_{ilkj}) \leq 1, \; j \in N, k = j+1, \ldots, n, j \neq k, j \notin P_k, k \notin P_j,$$
$$i \in F_j \cap F_k \tag{15.8}$$

$$\sum_{k \in G_{il}} X_{il0k} \leq 1, \quad i \in M, \; l \in M_i \tag{15.9}$$

$$C_{i0} = 0, \qquad i \in M \tag{15.10}$$

$$\begin{aligned} C_{ik} + V(1 - X_{iljk}) \geq \max\Big\{\max_{p \in P_k} C_{LS_p,p}, rm_{il}, C_{ij} + A_{iljk} \cdot S_{iljk}\Big\} \\ +(1 - A_{iljk}) \cdot S_{iljk} + p_{ilk}, \\ k \in N, \; i = FS_k, \; l \in E_{ik}, \; j \in \{G_{il}, 0\}, \; j \neq k, \; j \notin S_k \end{aligned} \tag{15.11}$$

$$\begin{aligned} C_{ik} + V(1 - X_{iljk}) \geq \max\Big\{C_{i-1,k} + \sum_{\substack{h \in \{G_{i-1}, 0\} \\ h \neq k, h \notin S_k}} \sum_{l' \in E_{i-1,h} \cap E_{i-1,k}} \left(lag_{i-1,l',k} \cdot X_{i-1,l',h,k}\right), \\ rm_{il}, C_{ij} + A_{iljk} \cdot S_{iljk}\Big\} + (1 - A_{iljk}) \cdot S_{iljk} + p_{ilk}, \\ k \in N, i \in \{F_k \setminus FS_k\}, l \in E_{ik}, j \in \{G_{il}, 0\}, j \neq k, j \notin S_k \end{aligned} \tag{15.12}$$

$$C_{\max} \geq C_{LS_j,j}, \qquad j \in N \tag{15.13}$$

$$X_{iljk} \in \{0, 1\}, \;\; j \in \{N, 0\}, k \in N, j \neq k, k \notin P_j, i \in F_j \cap F_k, l \in E_{ij} \cap E_{ik} \tag{15.14}$$

$$C_{ij} \geq 0, \quad j \in N, \; i \in F_j \tag{15.15}$$

The first set of constraints (15.5) ensures that every job has exactly one predecessor on only one machine at each stage. Note that only the possible variables are considered as in this model, the binary variables that are zero are not defined (for example, if one job is a successor of another, it might never precede it, therefore, the corresponding variable is not even defined). Note that for every stage and machine we introduce a dummy job 0, which precedes the first job at each machine. This also allows for the consideration of initial setup times, should they be needed. Constraints in the set (15.6) are similar in the way that every job should have at most one successor. Constraint set (15.7) forces that if a job is processed on a given machine at a stage, then it should have a predecessor on the same machine. This is a way of forcing that assignments are consistent in the machines. Constraint set (15.8) avoids the occurrence of cross-precedences. Note again that only the possible alternatives are considered.

With constraint set (15.9) we enforce that dummy job 0 can only be predecessor of at most one job on each machine at each stage. Constraint set (15.10) simply ensures that dummy job 0 is completed at time 0 in all stages. Constraint set (15.11) controls the completion time of jobs at the first stage they start processing by considering all eligible machines. The value V represents a big number so to make the constraint redundant if the assignment variable is zero. Notice that precedence relationships are considered by accounting for the completion of all the predecessors of a given job. Note also that both types of sequence dependent setup times (anticipatory and non-anticipatory) are also taken into account. Constraint set (15.12) gives the completion time on subsequent stages. Here the completion time of the same job in the previous stage along with the lag time is considered. Constraint set (15.13) defines the maximum completion time. Finally, (15.14) and (15.15) define just the decision variables.

The above model was shown by Ruiz et al. (2008) to be mostly unsolvable for instances of 15 jobs. More precisely, among 1,536 instances with $n = 15$, only 5.21 % (80) could be solved to optimality in 15 min of CPU time with CPLEX version 9.1 on a Pentium IV running at 3.2 GHz. While this solver version and hardware is now dated, such percentages increase only slightly with modern solver versions and current solvers (including threaded solvers).

Due to this poor performance of exact methods, Ruiz et al. (2008) also proposed some effective heuristics. Later, Urlings and Ruiz (2007) presented some local search methods. Metaheuristic algorithms, from simple genetic algorithms to methods with complex solution representations were studied later by Urlings et al. (2010a) and by Urlings et al. (2010b). We refer to this extended formulation as CTPP_Model_2 in the following sections.

15.4 Development and Deployment of the Scheduling System

Initial research on the CTPP started as early as 2000 in Spanish ceramic tile producers like Cerysa Cerámicas S.A. and Halcón Cerámicas S.A. The first one is a SME while the second one is a large producer with two large production plants. Development and deployment work went through during 2000 until year 2005. For most of this time, CTPP_Model_1 sufficed a large part of the demands of the two previously mentioned companies. However, when other large companies entered collaborations like ColorKer Cerámicas S.A. in 2006, Porcelanosa S.A. in 2007 and Taulell S.A. in 2008, CTPP_Model_1 resulted insufficient and CTPP_Model_2 was needed. Later, other producers were also involved, like Keraben S.A. or Undefasa S.A.

In the following, we describe first how all necessary data was collected, how this data was stored in databases and the schematics of the scheduling system that was developed.

15.4.1 Gathering Initial Data for CTPP_Model_1

The set of straightforward initial data is the configuration of the shop, i.e. the number of stages m and the number and main characteristics of each machine per stage m_i.

This has to be carried out at a per-company basis. SMEs in the ceramic tile sector typically contain two or more kilns. The molds and glazing lines typically range from 3 to 6 and a somewhat smaller number of classification lines is also common (Vallada et al. 2005). The division of the different stages is not entirely clear and must be also studied case by case. For example, for most companies, the first three stages, namely molding, drying and glazing, are connected by conveyor belts. This means that one mold is directly connected to a dryer and this one is connected to the glazing line. In practice, if a ceramic tile is molded in that specific mold, it directly goes to that dryer and glazing line. Obviously, there is no sense in dividing these three machines in three stages. What we have is a large virtual stage (as far as CTPP_Model_1 goes) that includes molding, drying and glazing.

A second stage is the kiln firing. All interviewed companies from Vallada et al. (2005) had the bisques stored in big boxes in between glazing and firing. The third stage is the selection, classification and packing. However, for small SMEs, this is not always the case. For some companies, there is not a storage area between the kilns and the classification lines and the kiln is directly connected to the classification lines by conveyor belts. In this scenario, the initial CTTP problem consists in either two (molding + drying + glazing; firing + classification) or three stages (molding + drying + glazing; firing; classification). Once the stage divisions of the production floor are clear, one has to count and characterise the existing machines.

The next step is to gather all the possible products that the company produces, which later turn into 'jobs' for CTPP_Model_1. As simple as this might seem, obtaining a list of products for production purposes only is many times challenging. Depending on the brand and instantiation of the Enterprise Resource Planning (ERP) that each company uses, the problems arising might range from simple to solve to almost impossible to overcome. Inside a company, a product reduces usually to a code. Companies try to codify most important characteristics into the product code, like type of product, family, colour, quality and so on. in the CTPP problem, stored products in the inventory include information like the quality, tone and caliber, information that does not exist when the product starts production as these characteristics result from the production process itself. Other problems arise in the packaging. Companies usually produce products for bigger companies and therefore the packaging of the product changes from client to client and sometimes even from different export targets. As a result of all this, the same original product, i.e. a type of ceramic tile, might have dozens, sometimes even more than a hundred different codings in the databases.

The experienced reader might have realised that this situation poses no problem, since all the explosion of codes for the same products originates from the same *base code* and all that one needs to do is to differentiate the production products from the base codes. However, in practice this is many times challenging, as the data needed for characterising each product is scattered among many different references. To solve all these problems, a careful work has to be performed on each case, with intimate collaboration from the different companies.

At this stage, all information relating to the production floor, machine characteristics and products to produce is already available. The next step is to obtain the stages that each product follows in production. As with previous steps, this might be from very easy to somewhat challenging. In any case, CTPP_Model_1 is rather limited in this regard. We will discuss in more detail the stage visitation information for each product later when approaching CTPP_Model_2.

In order to sequence products, one piece of important information is the production times or p_{ilj}. One quickly realises the magnitude of the dataset that is needed here. Picture for example a three virtual stages medium company with six molding, drying and glazing lines, four kilns and four classification and packing lines. This makes a total of 14 machines. For a company of this size, it is not at all uncommon to have a total number of production references in excess of 15,000. Therefore, a total of $14 \times 15{,}000 = 210{,}000$ data is needed. Introducing each one of these figures in the scheduling system is an arduous and repetitive task, let alone maintaining this data at a latter stage. An smarter approach is needed in order to streamline this aspect of the scheduling system.

In a nutshell, the 'speed' that a machine can process a batch of ceramic tiles depends on a set of characteristics, many of which are common for large families of products. Therefore, the speed of the machines can be determined from these characteristics and it is not necessary to derive this information from the product code alone. A clear example is the molding press. The mold inside the molding press determines how many tiles are pressed on each pressing cycle. The base material of the tile (the type of clay) along with the thickness of the tile and the characteristics of the molding press, determine how fast each pressing cycle goes. As a result, together with the type of molding press, and type of mold, we can determine the speed of the molding operation. Notice that the type of material used and the thickness of the tile are already factored in the type of mold.

The speed of the conveyor belt is pretty much determined by the speed of the molding press. Ceramic tiles advance along the conveyor belt, with enough space in between for the remaining drying and glazing operations to be performed without problems. The dryer usually accepts this speed and no further calculations are needed in most cases. The glazing lines, as explained before, consist of large conveyor belts with the tiles advancing over them. Therefore the processing time here is just derived from the length of the glazing line.

From the above discussion, it is easy to calculate how many tiles are pressed per minute in a given molding press. Then, with the speed and length of the conveyor belt, we can calculate the *cadence* or the time it takes from a tile to reach the end of the glazing line after being pressed. With this data and the size of the ceramic tiles batch, we calculate the processing time.

A similar situation arises at the kilns. Here, the firing time depends mainly on the format and base material of the tile, with very small variations that depend on the glaze employed. Large format tiles and/or thick tiles need larger firing times whereas smaller and thin tiles can be fired in as little as 25 min.

Lastly, classification lines are even easier, as the classification speed varies (and not by a large margin) with the tile format.

At this point it is worth noticing that eligibility might help in reducing the number of data needed. Eligibility is in many cases determined by the base material and the format of the tile. Simply speaking, some molds do not have enough power to press large tiles. Special decorations are only available at some glazing lines and some kilns are specially devised for small or large tiles. Therefore, calculation of eligibility is carried out on each possible product and machine on the basis of matching characteristics. If these characteristics match, then we can calculate the processing time. All in all, and following the previous example, from the original and initial 210,000 processing times needed, the resulting *distinct* processing times amount to less than 2,500 for the molding press, less than 200 for the kilns and about 300 for the classification lines.

The most problematic data requirements of CTPP_Model_1 are the sequence dependent setup times or S_{iljk}. Following the previous example of 14 machines and 15,000 products, the possible setups go all the way up to $14 \times 15{,}000 \times 15{,}000 = 3{,}150{,}000{,}000$. It is easy to see that this amount of setup times is impossible to store in any existing database. Note however, that not all setup combinations are possible, for example, the setup time value S_{iljk} is irrelevant—and can be assumed to be zero—in the following cases: $j = k, \forall j, k \in N$ and $l \notin E_{ij}$ or $l \notin E_{ik}$.

As a result, the real number of needed setups is much smaller and can only be calculated once eligibility data is known. Furthermore, some ceramic tile production sequences require such large setups that make them impracticable in real life. For example, producing a very small white tile after a large black tile with many other additional differences needs probably half a week of setups in the mold, glazing line and kiln. As a result of this, such setups can be approximated to a very large value, without really caring about the real value.

Furthermore, the setup times can also be calculated following the approach we employed with the processing times. However, calculating the amount of setup times is much more daunting. For each pair of different products that are eligible for a given machine we can calculate the differences in their characteristics. For example, consecutive products sharing the same mold in the molding press require no setup time on the molding press and dryers. However, if they differ in the colour or in the number of glazing applications, significant cleaning and adjustment is needed in the glazing lines. As a result, the setup operations can be accounted for from the differences in the product's characteristics. For example, changing a large mold needed for a large format tile requires operators to use jib or even overhead traveling cranes due to the weight. Smaller molds can be changed directly by the operators. Hence, large differences in setup times occur. Similarly, some decoration machines do have self cleaning routines (like the tile 'printers') whereas other machines require manual and comprehensive cleaning.

The internal organisation of the production shop differs from company to company. Some companies have many setup squads that can re-configure the molding press, dryer and glazing lines simultaneously. In other cases, these setups are done sequentially by a single squad. In summary, we 'only' need to know the time it takes to unmount and to mount every usable mold at each press. The time it takes to change the visible plate from the mold (given the same format), time to change the drying curve from format to format, and all the changes in applications from a product to another in the ceramic glazing line to quantify the setup. This still results in a very large set of data but that is many orders of magnitude smaller than the original number of possible setups. For example, a typical medium size company might have 80 different formats. Each molding press is usually able to press a 60 % of these formats. As a result, we approximately need a minimum of 48 unmounting and 48 mounting times to come up with a fairly accurate sequence dependent setup time matrix at the molding press.

15.4.2 Gathering Additional Data for CTPP_Model_2

CTPP_Model_2 significantly extends the previous model in many ways. However, from the data point of view, most of the necessary data was already gathered in CTPP_Model_1.

In CTPP_Model_2, we have now much more accurate description of the stages, where each job might skip stages. Therefore, additional and optional stages like cutting and polishing might be included in the stages information. Of course, processing and setup times of all the machines at these new stages are needed, which are approximated as explained in the previous section.

The accurate description of the stages visited by each product of F_j is now needed. This information is usually available in most companies. Release times, due dates and importance or weight for each job or batch of ceramic tiles is just additional data that is commonly available at the planning phase, along with the size of the batch. In any case, none of these are mandatory, i.e. if no priorities are given, we can assume the same priority for all jobs.

Along with the stage visitation sequence of all jobs and the information provided at the planning phase, it is possible to obtain the information regarding the precedence relationships for each job or P_j. This can be simply done at the planning stage by linking batches of jobs. An interesting result of the addition of the F_j and P_j datasets is that much less processing and setup times are needed. For example, setups are not needed when the following conditions are met: $i \notin F_j$ or $i \notin F_k$ and $j \in P_k$ or $k \in P_j$. Other data, like the release dates for the machines (rm_{il}), can be easily obtained from the release times of the schedule that is currently being followed at the production plants.

A different story is that of the lag times (lag_{ilj}). Additional information for each product and stage is needed. Namely, information about cooling or drying times in between stages or the capability of overlapping production. Usually, when a given portion of a batch of ceramic tiles is already glazed, kiln firing can occur. Kiln firing does not usually happen as soon as the first tiles reach the end of the glazing line in order to avoid possible problems. If an early production glitch is spotted at the mold, production has to stop at the glazing line and if the kiln had started production, a significant problem arises at the plant since stopping kilns is a costly procedure. Therefore, lags are mainly approximated from some additional product characteristics, sizes of the batches, the already known p_{ilj} processing times and some additional characteristics of the machines.

Although we have reduced in a significant way the otherwise almost impossible to gather amount of data needed for both models, in some situations gathering this reduced set of data might be still highly problematic.

As noted in Vallada et al. (2005), many companies do not carry out detailed scheduling at all and mainly follow a production plan that is more or less manually performed and roughly followed by plant shop personnel. This results in a large myriad of additional problems. Production data is seldom kept in the ERP system and is scattered (if available at all) in a number of spreadsheets or small databases. Usually, the information in these data repositories is sketchy, incomplete and rarely matches the remaining information in the ERP system.

Other situations are even more complicated. It is not unlikely to find several ERP systems working inside a company, resulting from failed or partial implantation of new systems or new systems that had to coexist with old software. On top of that, we often find ad-hoc software developed by the company or third party software that is interconnected at many different points with the existing ERP systems. Under such circumstances, deploying a scheduling system becomes a medium or even long term project.

As proposed in Sect. 14.5.4, SCADA systems along with the accompanying software can be extremely useful tools when deploying scheduling systems. An SCADA system can automatically measure the processing time for a given product or the time lapse between the production of consecutive products (raw data for the setup times). However, these systems often require terminals at both ends of each machine and production line along with sensors and other hardware that is both hard and expensive to maintain. We are not aware of any existing SCADA system capable of providing the production data that we need for both proposed models.

15.4.3 Storage, Retrieval and Maintenance of the Data

With such enormous data sets, special care should be put on the storage and retrieval. Carefully crafted relational databases, where performance of the queries is considered at the design stage, were developed. The details of such databases fail outside the scope of this book. Suffice to say that important implementation specific details like distributed tables and multiple indexes where employed. Resulting databases exceeded 20 GB of information in some cases.

Information retrieval is another key issue. Depending on the size of the company, and on the span of the scheduling setting, typically from as little as 50 to as more than 500 batches of ceramics tiles are to the scheduled. This means that all these 500 products, along with their predecessors structure, their stages visitation, processing times, setup times, and so on must be loaded from the database. In order to carry this operation in an acceptable time frame, we launch several SQL queries in parallel, using thread programming and the multiple processing cores available in modern desktop CPUs. The relational databases are stored in powerful servers that are capable of serving the SQL requests in parallel. By doing so, all the data required for a given scheduling scenario can be retrieved in less than a minute. Comparatively, a more common approach with a single SQL query could take as long as 40 min.

Data maintenance represents another major hurdle. The addition to new products, shop reconfigurations (new and/or replaced machines) can significantly alter the acquired data. In order to cope with these situations, a carefully crafted user interface is developed. It is not enough with giving the user the possibility of updating and removing old information. Advanced data wizards are developed which are capable of introducing large amounts of information in the database with little user intervention. For example, a new product might be essentially similar to an existing one except for some bits of information. In such a case, when entering this new product the user only has to specify the similar existing product and the new characteristics. This saves the user the introduction of lots of redundant data and, at the same time, improves the quality of the data in the system, which was highlighted as a concern in Sect. 14.5.2.

15.4.4 Final Implementation and Results

The implementation of the scheduling tool which contained the CTPP_Model_1 presented in Sect. 15.3.1 was called ProdPlanner; Fig. 15.4 shows a capture of one of the Gantt diagrams that this software was capable of producing as a result of the scheduling optimisation process. In a nutshell, Prodplanner incorporates the genetic algorithms detailed in Ruiz and Maroto (2006), together with some simple heuristics based on the NEH for initialisation. Simple local search schemes are also employed. The reader is referred to Ruiz and Maroto (2006) for more details on the algorithms employed.

ProdPlanner was shown to work in practice and viable to several scheduling settings. Nevertheless, this software lacked all the data acquisition and management abilities discussed in Chap. 14, and these proved to be absolutely necessary to be of actual use to companies. As a result, a complete rewrite of the tool was necessary. Additionally, the model CTPP_Model_1 did not quite capture every possible situation found in practice and CTPP_Model_2 was developed. With all previous experience, a new much more powerful and capable software—named SeKuen—was developed along the lines of the modular architecture of a manufacturing scheduling tool described in Chap. 11. Indeed, this software was used as a real test of the architecture proposed, so many of the final contributions of the architecture presented came after the implementation of the software in some companies. Figure. 15.5 shows a screenshot of the tool.

SeKuen is now in use in several of the aforementioned ceramic tile companies and is being deployed in several other additional firms. To do so, additional modeling effort was required and extensions are being actively incorporated.

From our experience, only a small fraction of the time was devoted to the actual research of the scheduling algorithms needed to schedule the production in the CTTP. This may be a striking result of this case study, but it is absolutely in line with the

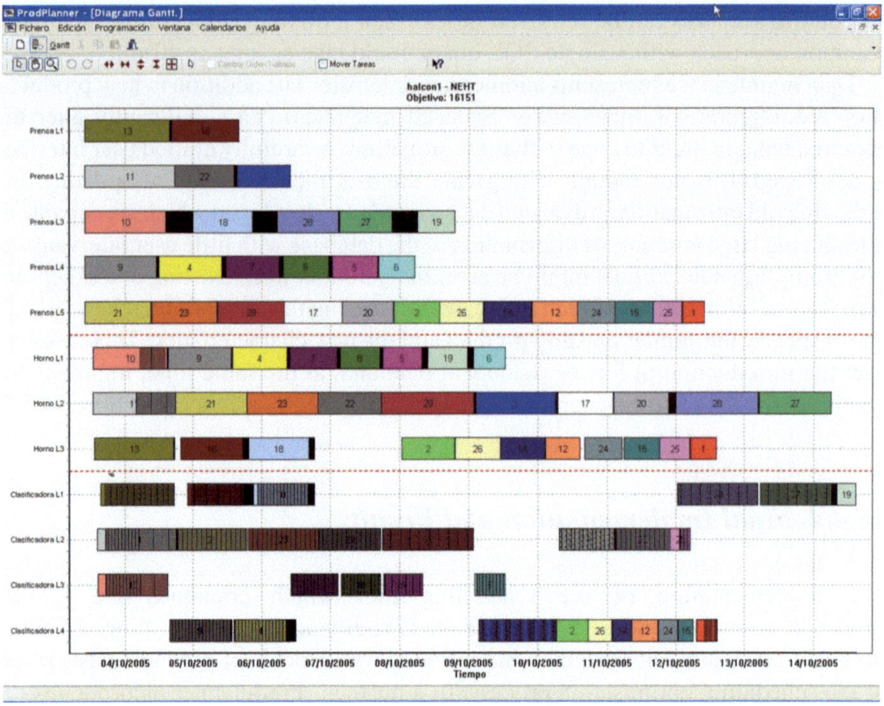

Fig. 15.4 Screenshot of ProdPlanner software, capable of producing solutions for CTPP_Model_1

guidelines described in Sect. 14.7. Most of the work was needed in the scheduling system development and data gathering.

The outcome was really encouraging at the time since up to 16 % makespan reductions for the CTTP could be observed. Development of the user interface and the data management modules were equally challenging as providing the required flexibility results in complex software applications. Overall, and considering all the modules in the proposed extended architecture, only about 10 % of the time was devoted to the development—not research—of the optimisation algorithms. More than 50 % of the time was devoted to the User Interface, whereas the Database Management and DAM required about 40 % of the time. Similarly, when a new company is interested in SeKuen, most of the implantation efforts go to the User Interface and Database Management modules, specially in the Import/Export/Synchronisation submodule. More information about SeKuen is available from http://www.sekuen.com.

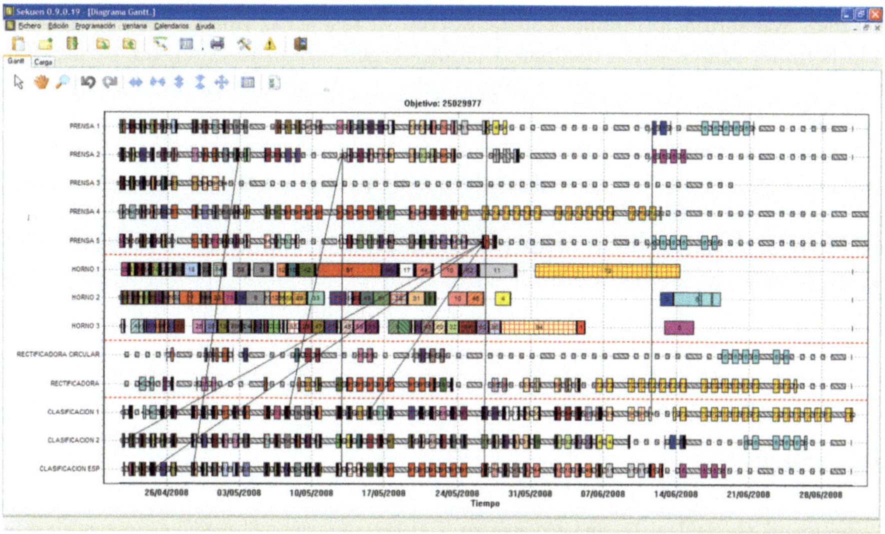

Fig. 15.5 Screenshot of SeKuen software, based on ProdPlanner, suitable for CTPP_Model_2

15.4.4.1 Open Issues

Even though CTPP_Model_2 model has been just recently developed, there are still some open issues in the modeling. First of all, the machines are not the only resource that must be considered in the production floor. Personnel, tools and other resources are also necessary. These resources are limited and must be also considered in the scheduling. For example, even though we might be considering sequence dependent setup times at the molding presses, these setups are actually carried out by setup squads, which might be limited in number. A setup needed a squad not currently available will need to be postponed and therefore might be likely to result in a bad schedule. Transportation times in between stages and the control and of the AGVs is also necessary. Storage capacity (number of boxes) in between each stage is also a key concern as in some cases this is a limited resource. Additional work is needed in order to significantly increase the number of realistic situations that SeKuen is able to handle. As of 2012, we are working on a third installment of SeKuen that is able to sequence production considering additional resources. This new version is also being rewritten from Delphi to C# and the .NET platform 4.0.

More and more complex scheduling settings result in slow algorithms. Calculating the production sequences in the presence of so many constraints is really slow. However, the metaheuristic techniques employed need to evaluate thousands of schedules in order to produce high quality results. The solution is parallelisation and cooperative computing, as already pointed out in Sect. 14.7.7. By using GRIDS or in a much simpler level, the available multiple CPU cores of modern computers, results of very high quality are obtainable in a fraction of CPU time. Similarly, very complex scheduling problems can be solved in fast CPU times. Research in this field is rather scarce when compared on the large body of research on 'sequential' scheduling. However, some promising results are already appearing, like the work of Vallada and Ruiz (2009).

Single objective optimisation is controversial in practice. A sequence obtained after makespan minimisation will result in a high violation of due dates and usually large number of setups. Minimising tardiness results again in large setups and large makespan. In practice, single objective optimisation is of little use. We are not aware of any attempts at solving multi-objective complex scheduling problems. However, promising results already exist (Minella et al. 2008), and more work is needed in this direction.

References

Allahverdi, A., Gupta, J. N. D., and Aldowaisan, T. (1999). A review of scheduling research involving setup considerations. *OMEGA, The International Journal of Management Science*, 27(2): 219–239.

Allahverdi, A., Ng, C. T., Cheng, T. C. E., and Kovalyov, M. Y. (2008). A survey of scheduling problems with setup times or costs. *European Journal of Operational Research*, 187(3): 985–1032.

Dudek, R. A., Panwalkar, S. S., and Smith, M. L. (1992). The lessons of flowshop scheduling research. *Operations Research*, 40(1):7–13.

Ford, F. N., Bradbard, D. A., Ledbetter, W. N., and Cox, J. F. (1987). Use of operations research in production management. *Production and Inventory Management*, 28(3):59–62.

Graves, S. C. (1981). A review of production scheduling. *Operations Research*, 29(4):646–675.

Ledbetter, W. N. and Cox, J. F. (1977). Operations research in production management: An investigation of past and present utilisation. *Production and Inventory Management*, 18(3):84–91.

Linn, R. and Zhang, W. (1999). Hybrid flow shop scheduling: A survey. *Computers & Industrial Engineering*, 37(1-2):57–61.

MacCarthy, B. L. and Liu, J. (1993). Addressing the gap in scheduling research: A review of optimization and heuristic methods in production scheduling. *International Journal of Production Research*, 31(1):59–79.

McKay, K., Safayeni, F., and Buzacott, J. (1988). Job-shop scheduling theory: What is relevant? *Interfaces*, 18(4):84–90.

McKay, K. N., Pinedo, M. L., and Webster, S. (2002). Practice-focused research issues for scheduling systems. *Production and Operations Management*, 11(2):249–258.

Minella, G., Ruiz, R., and Ciavotta, M. (2008). A review and evaluation of multi-objective algorithms for the flowshop scheduling problem. *INFORMS Journal on Computing*, 20(3):451–471.

Olhager, J. and Rapp, B. (1995). Operations research techniques in manufacturing planning and control systems. *International Transactions in Operational Research*, 2(1):29–43.

Quadt, D. and Kuhn, D. (2007). A taxonomy of flexible flow line scheduling procedures. *European Journal of Operational Research*, 178(3):686–698.

Reisman, A., Kumar, A., and Motwani, J. (1997). Flowshop scheduling/sequencing research: A statistical review of the literature, 1952-1994. *IEEE Transactions on Engineering Management*, 44(3):316–329.

Ribas, I., Leisten, R., and Framinan, J. M. (2010). Review and classification of hybrid flow shop scheduling problems from a production system and a solutions procedure perspective. *Computers & Operations Research*, 37(8):1439–1454.

Ruiz, R. and Maroto, C. (2002). Flexible manufacturing in the ceramic tile industry. In *Eight International Workshop on Project Management and Scheduling, PMS 2002. Abstracts*, pages 301–304, Valencia. Universitat de València.

Ruiz, R. and Maroto, C. (2006). A genetic algorithm for hybrid flowshops with sequence dependent setup times and machine eligibility. *European Journal of Operational Research*, 169(3):781–800.

Ruiz, R., Sivrikaya-Şerifoğlu, F., and Urlings, T. (2008). Modelling realistic hybrid flexible flowshop scheduling problems. *Computers and Operations Research*, 35(4):1151–1175.

Ruiz, R. and Vázquez-Rodríguez, J. A. (2010). The hybrid flowshop scheduling problem. *European Journal of Operational Research*, 205(1):1–18.

Urlings, T. and Ruiz, R. (2007). Local search in complex scheduling problems. In Stützle, T., Birattari, M., and Hoos, H. H., editors, *First Workshop on Stochastic Local Search (SLS 2007). Engineering Stochastis Local Search Algorithms. Designing, Implementing andAnalyzing Effective Heuristics*, volume 4628, pages 202–206, Berlin. Springer-Verlag.

Urlings, T., Ruiz, R., and Sivrikaya-Şerifoğlu, F. (2010a). Genetic algorithms with different representation schemes for complex hybrid flexible flow line problems. *International Journal of Metaheuristics*, 1(1):30–54.

Urlings, T., Ruiz, R., and Stützle, T. (2010b). Shifting representation search for hybrid flexible flowline problems. *European Journal of Operational Research*. In review.

Vallada, E., Maroto, C., Ruiz, R., and Segura, B. (2005). Analysis of production scheduling in spanish tile industry. *Boletín de la Sociedad Espanola de Cerámica y Vidrio*, 44(1):39–44. In Spanish.

Vallada, E. and Ruiz, R. (2009). Cooperative metaheuristics for the permutation flowshopn scheduling problem. *European Journal of Operational Research*, 193(2):365–376.

Vignier, A., Billaut, J.-C., and Proust, C. (1999). Les problèmes d'ordonnancement de type flowshop hybride: État de l'art. *RAIRO Recherche opérationnelle*, 33(2):117–183. In French.

Wang, H. (2005). Flexible flow shop scheduling: optimum, heuristics and artificial intelligence solutions. *Expert Systems*, 22(2):78–85.

Index

J. M. Framinan et al., *Manufacturing Scheduling Systems*,
DOI: 10.1007/978-1-4471-6272-8, © Springer-Verlag London 2014

MIX
Papier aus verantwortungsvollen Quellen
Paper from responsible sources
FSC® C105338

If you have any concerns about our products,
you can contact us on
ProductSafety@springernature.com

In case Publisher is established outside the EU,
the EU authorized representative is:
Springer Nature Customer Service Center GmbH
Europaplatz 3, 69115 Heidelberg, Germany

Printed by Libri Plureos GmbH
in Hamburg, Germany